Children

of

Ezekiel

Aliens,

UFOs,

the Crisis

of Race,

CHILDREN

and the

of
EZEKIEL

Advent

of

Michael Lieb

End Time

DUKE UNIVERSITY PRESS · DURHAM AND LONDON 1998

© 1998 Duke University Press

All rights reserved

Printed in the United States of America on acid-free paper ∞

Typeset in Minion by Keystone Typesetting, Inc.

Library of Congress Cataloging-in-Publication Data appear

on the last printed page of this book.

For

Kathryn

Grace

and

Nicholas

Samuel

Contents

Acknowledgments   viii

Introduction   1

I

Cultural

Transactions and

the Poetics of

Aggression

1. Technology of the
Ineffable   21

2. The Psychopathology
of the Bizarre   42

3. Prophecy Belief and the
Politics of End Time   74

4. Arming the Heavens   100

II

Ideology,

Eschatology,

and Racial

Difference

5. Heralding the Messen-
ger   129

6. The Eschatology of the
Mother Plane   155

7. Visionary Minister   178

8. Armageddon and the
Final Call   198

Conclusion   230

Notes   235

Index   299

Acknowledgments

I am deeply indebted to those whose guidance has aided me in the production of this book. These include the readers for Duke University Press: Albert C. Labriola, Michael E. Zimmerman, and Joseph Anthony Wittreich Jr. For their wise evaluations, judicious comments, and careful reading, I extend my heartfelt thanks. I also take this opportunity to express my gratitude to Stanley E. Fish for his confidence in my project and for his continuing encouragement and support. Finally, I am grateful to Reynolds Smith for his guidance in helping me see my book through the press.

Several of my colleagues assisted me in the research and writing of this book, and I thank them here. Noel W. Barker first alerted me to the presence of Ezekiel's vision in the writings of the Nation of Islam. Martin Riesebrodt proved helpful in sharing the benefit of his deep knowledge. David Jackson proved himself most responsive to my inquiries and offered suggestions for further research. James Hall, Ned Lukacher, and Mae Henderson were generous with their time and advice. William A. Covino provided essential insights into the world of cyberspace. Virginia Wexman and John Huntington offered generous suggestions and support. Susan Wadsworth-Booth's advice proved invaluable. A good portion of the manuscript received the careful and gracious attention of Mary Beth Rose, and Regina Schwartz offered advice and understanding.

Much of my research for the second part of the book involved personal contacts and discussions with members of the Nation of Islam. In addition to undertaking library research, I made a point of becoming acquainted with members of the Nation willing to share with me the benefits of their insights and experiences. Among those with whom I talked at length, Munir Muhammad, cofounder of the Coalition for the Remembrance of Elijah (CROE), was extremely helpful. I shall long remember my several visits to the headquarters of CROE, where Munir Muhammad not only greeted me warmly but took time from his busy schedule to discuss with me my project and its implications. I also acknowledge here the openness and genuine good humor of Sidney Muhammad, whose insights proved invaluable to my undertaking. Finally, I note with thanks the efforts of Claudette Marie Muhammad on my behalf.

This is a project that required extensive research in a wide range of libraries. I acknowledge the staff, as well as the collections, of the University of Illinois at Chicago and its sister campus in Urbana; the Regenstein Library of the University of Chicago; the Northwestern University Library; the libraries of the various theological seminaries that populate Chicago; the Newberry Library; and the Henry E. Huntington Library. For my work in the area of evangelicalism and fundamentalism, I have had the benefit of the Moody Bible Institute with its fine collections of end-time literature and its helpful staff, and I am grateful to the librarians at Aurora University, Aurora, Illinois, for allowing me access to rare material bearing on the Jehovah's Witnesses. For research on the Nation of Islam, I acknowledge here the Vivian G. Harsh Research Collection of Afro-American History and Literature, Chicago Public Library. The staff of the Harsh Collection proved very astute in hunting down important source material.

This book had its inception in 1983, when I was appointed as a fellow to the Institute for the Humanities at the University of Illinois at Chicago, my home institution. During a period of almost three decades at this institution, I have benefited from the ideal environment it has provided for both teaching and research. In its later stages, my project was also supported by a fellowship from the National Endowment for the Humanities, and I am grateful indeed for this support.

Substantially revised, two of my previously published articles have been incorporated into the first chapter. The first is "Milton's 'Chariot of Paternal

Deitie' as a Reformation Conceit," *Journal of Religion* 65 (1985): 359–77. © 1985 by The University of Chicago. All rights reserved. The second is "Children of Ezekiel: Biblical Prophecy, Madness, and the Cult of the Modern," *Cithara* 26 (1986): 12–22. Permission to incorporate these articles as part of my study is gratefully acknowledged. Extracts from the Authorized Version of the Bible (the King James Bible), the rights in which are vested in the Crown, are reproduced by permission of the Crown's Patentee, Cambridge University Press.

I reserve the last word for my family. My wife, Roslyn, has been my mainstay. Her affection, support, and assurances over the years have meant everything to me. She has been ever my best friend. Reading large portions of my book in its various stages of composition, she offered timely and helpful advice. My sons, Mark and Larry, aided me with their encouragement and proved to be excellent sources of information concerning various aspects of my project. This book is dedicated to Kathryn Grace and Nicholas Samuel, my grandchildren. Although they are still too young to read what is written here, perhaps one day they will come to appreciate what it means to be "children of Ezekiel."

Aliens, UFOs, the Crisis of Race, and the Advent of End Time: what unites this disparate set of terms? And who are the so-called children of Ezekiel? Answers to these questions are found in the vision of God that inaugurates the biblical prophecy of Ezekiel. My discussion of this vision and its aftermath will take us on a journey into the history of the culture of alien encounters, racial crisis, and the fear of apocalyptic annihilation. This is a world in which unidentified flying objects and "mother planes" descend from the heavens. It is a world distinguished by fiery chariots and such technological wonders as tanks, laser weapons, and space-based "defense" systems. In this world, the emergence of technology confirms the ability of humankind to gain control of its environment and to overwhelm its enemies. This ability is grounded in the "will to power" that derives its impetus (as well as its putative legitimacy) from the ultimate source of power, God himself.[1]

As the very embodiment of God's power, the vision of Ezekiel is one of the most remarkable events in all Hebrew Scripture.[2] Appearing in the first chapter of the prophecy (and therefore known as the "inaugural vision"), Ezekiel's *visio Dei* is a revelation of momentous import. At the time that the prophet experiences his vision, he is in exile on the shores of the river Chebar in Babylon.[3] As the text of the vision unfolds, Ezekiel declares:

> And I looked, and, behold, a whirlwind came out of the north, a great cloud, and a fire infolding itself, and a brightness *was* about it, and out of the midst thereof as the colour of amber, out of the midst of the fire. Also out of the midst thereof *came* the likeness of four living creatures. And this *was* their appearance; they had the likeness of a man. And every one had four faces, and every one had four wings. And their feet *were* straight feet; and the sole of their feet *was* like the sole of a calf's foot: and they sparkled like the colour of burnished brass. And *they had* the hands of a man under their wings on their four sides; and they four had their faces and their wings. Their wings *were* joined one to another; they turned not when they went; they went every one straight forward. As for the likeness of their faces, they four had the face of a man, and the face of a lion, on the right side; and they four had the face of an ox on the left side: they four also had the face of an eagle. Thus *were* their faces: and their wings *were* stretched upward; two *wings* of every one *were* joined one to another, and two covered their bodies. And they went every one straight forward: whither the spirit was to go, they went; *and* they turned not when they went. As for the likeness of the living creatures, their appearance *was* like burning coals of fire, *and* like the appearance of lamps: it went up and down among the living creatures; and the fire was bright, and out of the fire went forth lightning. And the living creatures ran and returned as the appearance of a flash of lightning. Now as I beheld the living creatures, behold one wheel upon the earth by the living creatures, with his four faces. The appearance of the wheels and their work *was* like unto the colour of a beryl: and they four had one likeness: and their appearance and their work *was* as it were a wheel in the middle of a wheel. When they went, they went upon their four sides: *and* they turned not when they went. As for their rings, they were so high that they were dreadful; and their rings *were* full of eyes round about them four. And when the living creatures went, the wheels went by them: and when the living creatures were lifted up from the earth, the wheels were lifted up. Whithersoever the spirit was to go, they went, thither *was their* spirit to go; and the wheels were lifted up over against them: for the spirit of the living creature *was* in the wheels. When those went, *these* went; and when those stood, *these* stood; and when those were

lifted up from the earth, the wheels were lifted up over against them; for the spirit of the living creatures *was* in the wheels. And the likeness of the firmament upon the heads of the living creature *was* as the colour of the terrible crystal, stretched over their heads above. And under the firmament *were* their wings straight, the one toward the other: every one had two, which covered on this side, and every one had two, which covered on that side, their bodies. And when they went, I heard the noise of their wings, like the noise of great waters, as the voice of the Almighty, the voice of speech, as the noise of an host: when they stood, they let down their wings. And there was a voice from the firmament that *was* over their heads, when they stood, *and* had let down their wings. And above the firmament that was over their heads *was* the likeness of a throne, as the appearance of a sapphire stone: and upon the likeness of the throne *was* the likeness as the appearance of a man above upon it. And I saw as the colour of amber, as the appearance of fire round about within it, from the appearance of his loins even upward, and from the appearance of his loins even downward, I saw as it were the appearance of fire, and it had brightness round about. As the appearance of the bow that is in the cloud in the day of rain, so *was* the appearance of the brightness round about. This *was* the appearance of the likeness of the glory of the Lord. And when I saw *it,* I fell upon my face. (Ezek. 1:4–28)

Within this text, Ezekiel's vision of God represents an impulse through which the will to power finds expression in the wonders of technology that define the modern world. In 1970 Herbert J. Muller called those responsible for technological advancement the *children of Frankenstein,* an allusion that implicitly centers the emergence of technology in a "secular" text.[4] With the matrix of my investigation in a "sacred" text, the Book of Ezekiel, I call those who seek to harness the power that gives rise to technology the *children of Ezekiel.* Inventors, scientists, technologists, evangelicals, and poets, they are visionaries all. For them Ezekiel's *visio Dei* represents the wellspring of the impulse to fashion a technology out of the ineffable, the inexpressible, the unknowable. Drawing on the forces within the vision, they reinvent it, re-create it, "technologize" it in their own terms. It is the history of the impulse to technologize the ineffable centered in Ezekiel's vision of God that this study seeks to record. To provide a context for the study as a whole, I rehearse here the main lines of that history, followed by a discussion of Ezekiel's vision and the theoretical implications on which the study is founded.

I begin with one of the most notable children of Ezekiel, the poet John

Milton. Through his epic *Paradise Lost,* the act of fashioning a poetic "vehicle" from a biblical source becomes tantamount to envisioning the original vision anew. To that end, Milton "invents" the "Chariot of Paternal Deitie," which, as a pivotal representation, emerges from the midpoint of his epic. The Chariot of Paternal Deitie is the vehicle on which the Son of God as sublime charioteer embarks to overwhelm the rebellious angels in the celestial battle that Milton portrays. Reconceptualizing Ezekiel's vision, the Miltonic chariot embodies the technological impulse on a grand scale. As such, it is correspondingly imbued with a sensibility that only one engaged in a lifetime of combat with those opposed to the fulfillment of the Reformation ideology could envision. Inspired by the all-pervasive zeal that permeates this ideology, the Chariot of Paternal Deitie represents the desire to draw on the immense powers of the divine in order to oppose all who would stand in its path.

If Milton's epic provides a poetic context for the reconception of Ezekiel's vision, the children of Ezekiel that succeed Milton reinvent or technologize the vision in their own terms. The emergence of these technologies is delineated through narratives that encompass events such as the invention of the flying machine, the founding of the railway system, and the deployment of the military tank in modern warfare. Products of the spirit of enterprise that emerged during the Industrial Revolution, technologies of this sort attest to the dissemination and transformation of Ezekiel's vision in the modern world. In this capacity, they function as the occasion through which Ezekiel's vision becomes a source of discovery, a means of knowing, of perceiving the true nature of the ineffable and how the ineffable operates in the world as we know it, that is, how the ineffable is transformed into actual machines.

In the narratives that these transformations or "technomorphoses" encode, one such machine is the "unidentified flying object" or UFO, a phenomenon that represents a crucial aspect of the desire to appropriate the ineffable, to master it, and to conceive it in technological terms. With Ezekiel's *visio Dei* as its source, the "UFO phenomenon" has become an all-pervasive feature of modern culture. As I shall discuss, the transformation of the vision of Ezekiel into a UFO is discernible not only in "popular" forms of communication (tabloid, motion picture, and television) but in "scholarly" or officially sanctioned forms of investigation (government reports, scientific studies, and university-sponsored symposia) as well. Here, I shall examine figures ranging from Erich von Däniken, the famous UFO popularizer, to Josef F. Blumrich, a former NASA engineer. As a reflection of New Age fervor, the enthusiasm that infuses the visionary experiences of such children of Ezekiel is correspondingly reflected in the cyberspatial universe of emergent information technologies. This, too, becomes the site of confluence between popular or mass culture and high or elite culture. Exploring the ramifications of that confluence, I take the opportunity to bring to the fore a host of witnesses who eminently fulfill our desire to become acquainted with the modern purveyors of Ezekiel's vision as a wellspring of sublime enactment.

My discussion will demonstrate that the history of these purveyors is made evident on other fronts as well. Exploring the traditions of evangelicalism that emerged in the nineteenth century, I shall argue that Ezekiel's vision is crucial to the fervor that has since become a distinguishing factor in contemporary notions of apocalypse or end time. One of the most outstanding cases is the movement that has come to be known as the Jehovah's Witnesses. Under the auspices of Charles Taze Russell and later Joseph Franklin Rutherford, this movement conceived itself as a "chariotlike organization" inspired at once by a belief in the efficacy of progress and by an apocalyptic sense of the inevitability of Armageddon. These two strains of thinking bestowed on Ezekiel's vision a renewed impetus in the evangelical traditions.

Consistent with this impetus, the evangelical devotion to end time located a new site of interpretation on which to found the apocalyptic outlook. Complementing the vision of God that emerges from the first chapter of Ezekiel is the oracle against Gog, articulated in the thirty-eighth chapter.[5] On the basis of this chapter, there arose the idea of "prophecy belief." This idea transformed the biblical text into a blueprint for the invention of nuclear armaments that threaten to annihilate the world. With the Gog oracle, one discovers a mode of interpretation known as the "nuclearization of the Bible." This is the inclination to view the warfare delineated in Ezekiel's oracles as a prophecy of the nuclear armaments that have become the dread of the modern world. Made evident in the religious life surrounding government facilities that produce weapons of mass destruction, the act of "nuclearizing" the Bible assumes paramount importance to end-timers of every stripe. Among the most influential of these is Hal Lindsey, whose works have attracted an immense reading public in the past twenty-five years. Incorporating end-time thinking into his wide-ranging publications, this purveyor of eschatology is one among a host of prophecy believers for whom Ezekiel is germane. Of corresponding importance are such television evangelists as Jerry Falwell and Pat Robertson, both of whom enjoy immense followings in the end-time industry.

With the advent of apocalyptic thinking grounded in the oracles of Ezekiel, a concomitant phenomenon is brought to the fore. Complementing the act of "nuclearizing" the Bible is that of "racializing" the Bible. Such an act is the product of the anxieties associated with Ezekiel's oracle against Gog as an enemy prepared to swoop down from the northern regions and attack Israel. As a result of the fears to which this impending siege gives rise, apocalypticists transform the conflict between aggressor and victim into a final racial confrontation of cosmic significance. Initiating this confrontation are the alien hordes, under whom are marshaled the red forces of Russia and its allies, the yellow peoples of China and its neighbors, and the black African armies of the "Pan-Arab alliance." Under certain circumstances, even the "misled" white Europeans are conceived as aggressors. As powers that have always hated Israel and its inhabitants, these nations prepare for an all-out brutal attack of race against race, but, we are assured, the armies of the Anti-Christ will ultimately be destroyed through divine intervention.

The politics implicit in such an outlook make their presence known not only among influential preachers who espouse the end-time cause but also among those who have held high public office. As one for whom the oracles of Ezekiel assumed consummate importance throughout his career, Ronald Reagan became a true believer in the efficacy of the prophetic word. Incorporating the ideology of prophecy belief into his own notions of end time, both as governor and as president Reagan became the means by which the specter of end-time thinking found its way most compellingly in the corridors of power. Because of the extent to which Reagan's life was reimagined through his own acts of self-fashioning within his films, the relation between the cinematic personae and the public personae that he fostered as actor and as politician must be taken into account in any assessment of the role of end time in his thought. The fantasies of power that underlie such conceptions of end time influence the enactment of public policy. Indebted to cinematic renderings, these fantasies are of seminal importance to the establishment of the Strategic Defense Initiative (commonly known as "Star Wars"), which in its own way finds its counterpart in the technologizing of the milieu associated with Ezekiel's vision.

The second part of this book focuses on the Nation of Islam, a movement driven from the outset by powerful, controversial, and charismatic personalities that must be addressed in some depth if a true understanding of Ezekiel's vision in its modern cultural setting is to be realized. At issue are the Honorable Elijah Muhammad and the Honorable Louis Farrakhan. During the period of their respective dispensations, each has experienced his own calling and has had his own encounter with the vision that inaugurates the prophecy of Ezekiel. In fact, their individual callings are deeply implicated in their visionary encounters. To appreciate the nature of those encounters, one must explore the forces that shaped the lives and personalities of these two figures from the very foundations of the Nation to the present time. Having done that, we may then come to terms more fully with what became the racial dimensions of Ezekiel's vision, which the Nation appropriated as a foundational event in its history and culture.

In its conception of that event, the Nation transforms the vision into a wonder of technology, one in which it is figured forth as that machine of machines, the Mother Plane or Mother Wheel. For the Nation, this machine becomes an apocalyptic phenomenon of momentous import, indeed, the UFO par excellence. In the technological transformation that such a configuration entails, the vision as Mother Plane becomes the embodiment of racial difference as the crucial determinant of being and identity. A dynamic and ever-changing phenomenon, the Mother Plane comes to reflect the growth of the Nation as a potent religious and political force in the history of the struggle for recognition and independence among black nationalists. In this capacity, the Mother Plane is infused with a resonance that is by its very nature both combative and conflicted.

At issue is how the Mother Plane is conceived as the vehicle through which the Nation struggles against and ultimately overwhelms the so-called powers of the world. Whether in the form of the white race, the Christian community, or the Jewish community, those perceived by the members of the Nation of Islam as the "enemy" must face the consequences of their actions. Held responsible for the long period of affliction that the Nation seeks to counter, the powers of the world face ultimate obliteration through the workings of the Mother Plane as a vehicle both of destruction and of salvation. In its annihilation of the enemy and its resuscitation of the faithful at the time of Armageddon, the Mother Plane will have fulfilled its true calling. Such is the bearing that Ezekiel's vision assumes in the Nation of Islam, a movement that fully assimilates and reconfigures all that the vision implies as both an apocalyptic and a racial entity of monumental proportions.

This study accordingly encompasses a broad spectrum of interests and concerns. Whether in the form of poetic renderings, blueprints for the construction of flying machines, the deployment of armaments in modern warfare, or speculations on the visitations of alien beings from other worlds, these interests and concerns collapse distinctions between elite and mass culture to suggest the centrality of visionary enactment to all areas of life. With the technologizing of the ineffable, the prophecy of Ezekiel comes to assume a crucial place in the history of culture through which the idea of the "modern" is defined and redefined from the Miltonic era to the present. As the process develops momentum during this period, the anxieties produced by the manifestation of deity in Ezekiel's prophecy give rise to such alarming events as the "nuclearization" and "racialization" of the Bible. It is in such a setting that the apocalyptic impulse underlying visionary enactment moves disturbingly to the fore. At the core of Ezekiel's *visio Dei* reconceived in technological terms is the specter of end time. Emerging in all its dreadfulness from the anxieties that define the present, such a specter haunts us as our century moves inexorably toward the millennium.

Having provided this overview, I need to say a few words about Ezekiel's vision itself and the theoretical implications that underscore my argument. Confronting the prophet with its "thunderous otherness," the vision of Ezekiel records the moment in which the profound distance between the realm of the divine and the realm of the human is "portrayed in all its awesome enormity."[6] Imbued with its own symbolism, the vision underscores its otherness with representations that confound at the very point that they reveal. Rather than clarifying what the prophet sees, a plethora of details prevents the vision from ever coming into sharp focus. Although pictorial renderings of the vision abound (see figs. 1–2), the precise *form* of what the prophet sees finally eludes our grasp.[7] The storm, the cloud, the fire, the creatures, the faces, the wings, the wheels, the firmament, the throne: what in fact does the prophet behold? At the very point of perception, the vision reveals only to hide. If the creatures stretch two of their four wings upward in a posture of disclosure, they also cover their bodies with their two other wings in a posture of concealment (Ezek. 1:11). The paradox implicit in this posture establishes a metaphysics of vision: disclosure is concealment, concealment disclosure. The more we behold, the less we see. We are never permitted to penetrate the vision to its core. This paradox is further compounded by another element that underscores the complexity. For this is a vision that is not only seen: this is a vision that likewise *sees.* Not only the wheels but (in a subsequent delineation) the living creatures themselves are replete with eyes (Ezek. 1:18, 10:12). That which is beheld also beholds. The result is one that undermines all attempts to "fix" the vision, to stabilize it, to master it. This is a vision that confounds both by defying full perception and by seeing back. What results is an immensity that not only separates seer and seen but also causes the vision itself to remain forever elusive.[8]

Despite the profound elusiveness of the vision, there is indeed a quality about it that prompts one to view it in distinctly technological terms. In fact, one might suggest that the technological dimensions of Ezekiel's vision are all pervasive.[9] Implicit in the vision is a self-propelled object that moves irresistibly toward "thingness." What Ezekiel himself admires so much about it is what the original Hebrew calls its *maʿaseh,* "workmanship" or "construction" (Ezek. 1:16). The brilliance of the "substances" (bronze, chrysolite, crystal, sapphire)

Figure 1. Ezekiel's vision of God.
Reproduced from a rendering of Ezekiel 1 in *Biblia sacra* (1566),
by permission of the Newberry Library, Chicago.

Figure 2. "The Likeness of Four Living Creatures."
Reproduced from an engraving of Ezek. 1:5 by B. Picart (seventeenth century)
in the Kitto Bible, 26:4938, by permission of the Huntington Library,
San Marino, California.

that appear to compose it, the intricacy of its "mechanical parts" (e.g., the "rims" and "spokes" of the wheels), the complexity of its movements: submerged within the "mysterium" of the unknowable vision is that which cries out for *object*ification, for individuation, for the bestowal of a name.[10]

It is no doubt for this reason that the inaugural vision has traditionally come to be known in Hebrew as the *merkabah* or throne-chariot.[11] Despite the imposition of a name on the vision, Ezekiel himself never designates the "theophany" or revelation of deity he beholds as that of the *merkabah*. For the prophet, the vision remains unnamed, just as the deity whom the vision portrays is unnameable (cf. Exod. 3:13–15). It is others who impose the name *chariot* or *throne-chariot*.[12] So Ben-Sira, in the apocryphal book of Ecclesiasticus, refers to it as "the chariot of the cherubim" (49:8), a designation that the Chronicler applies to the ark of the covenant in the holy of holies (1 Chron. 28:18).[13] As historians of religion have documented, the inaugural vision might well derive some of its inspiration from temple paraphernalia of this sort. A priest of the temple, Ezekiel would have been intimately familiar with objects such as the ark with its covering cherubim.[14] At the same time, the inaugural vision finds its counterpart in the various throne-chariots that appear everywhere in the culture of the ancient Near East, whether among the Persians, the Egyptians, the Phoenicians, the Hittites, or the Babylonians.[15] The point is that the impulse to associate the theophany that Ezekiel beholds with some identifiable object is understandable, if not inevitable, considering the nature of the vision itself and the way it invites (indeed, compels) concretization. In this respect, the vision represents the very wellspring of the impulse to technologize the ineffable.

An obvious example of such an impulse is discernible in the afterlife of the language through which the vision is delineated. The language employed to describe its workmanship or construction (such as the rims and spokes of its wheels) is in itself already sufficient to justify the interpretive act of technologizing the vision. More is at stake in the act of technologizing the ineffable, however, than discovering there the originary mechanism of a nascent vehicle. It is when the language of mystery assumes an afterlife that is totally imbued with the characteristics of a later culture that the transformation of "vision" into "thing" becomes particularly remarkable. This transformation reconceives the quality of "thingness" in a new form, as the manifestation of a new order, an entirely "other" dimension of knowing. This act at once concretizes and harnesses the mysterium of the language in which the vision is cast in order to provide a nomenclature for a phenomenon in the modern world that is entirely absent from the world of its origins.

In the vision of Ezekiel, such an event centers in that most awesome and mysterious of terms, *hashmal*. This is the word used to describe that "something" that emanates from the middle of the fire surrounding the vision as a whole: "and out of the midst thereof as the colour of amber [*hashmal*], out of the midst of the fire" (Ezek. 1:4).[16] The word *amber* (*hashmal*) is attested only twice thereafter in Ezekiel, first in 1:27 ("and I saw as the colour of amber [*hashmal*]"), then in 8:2 ("as the appearance of brightness, as the colour of amber [*hashmal*]"). It is attested in no other passages either in Ezekiel or elsewhere in Hebrew Scripture, for that matter. That is why the actual meaning of the term is dubious at best. Such is the conclusion reached by the Brown, Driver, and Briggs edition of the lexicon of Gesenius, the standard source: "etym. and exact mng. dub.; evidently some *shining* substance," perhaps "a brilliant amalgam of gold and silver."[17] Whereas the Authorized Version and the New Revised Standard Version render it as "amber," the New International Version renders it as "glowing metal" and the Revised Standard Version as "gleaming bronze." In the Septuagint, it appears as *elēktron* and in the Vulgate as *electrum*.[18]

According to the rabbinic traditions, an unauthorized encounter with Ezekiel's vision can be devastating because of the unpredictable nature of *hashmal*.[19] The ancient rabbis taught that there was once a child who happened on the Book of Ezekiel in the house of his teacher. There, the child decided to read the opening chapter of the prophecy. Arriving at that point in which the text mentions *hashmal*, and apprehending its meaning, the child was suddenly consumed by the fires that surged forth from the mysterious phenomenon. In response to this devastating experience, the rabbis sought to suppress the Book of Ezekiel (Ḥagigah 13a, BT, 8:77–78).[20] The power of the ineffable that resides within sacred texts can have horrendous consequences for those who are not prepared. The very sacredness and therefore the potential danger surrounding not just the inaugural vision but the entire prophecy are, according to the rabbinic traditions, centered in the mysterium of *hashmal*. If one is not destroyed by *hashmal*, one risks at the very least being driven insane. One thinks of Rabbi ben Zoma, who, in the mysterious realm known as *pardes*, looked and became demented as the result of his encounter with the *merkabah* (Ḥagigah 14b, BT, 8:90–91). In the talmudic narratives (*'aggadot*) that are a primary source of lore concerning Ezekiel's vision, such events underscore that most holy of phenomena, the *ma'aseh merkabah* (work of the chariot), on which it is dangerous to speculate unless those who are speculating are truly sages and fully in possession of their own faculties. To anyone who would violate the interdictions surrounding the *merkabah*, "it were a mercy," the rabbis declare, "if he had not come into the world" (Ḥagigah 11b, *BT,* 8:59–60). Such is particularly true for those who are not "old" enough to protect themselves from the power of "the holy."[21] *Ḥashmal* is not for "children" of any age. Although there have been speculations on precisely how old one must be in order to survive the ordeal of speculating on the *maʿaseh merkabah,* the point is that one who is a child and lacking in those faculties that distinguish the true sage risks immolation or, at the very least, insanity.[22] In the rabbinic traditions, such is the aura surrounding the *maʿaseh merkabah* and such the power of *ḥashmal.*

The technological underpinnings of Ezekiel's vision are particularly apparent in the meanings that are ascribed to the original language of the vision in contemporary discourse. *Ḥashmal,* for example, assumes the form of the modern Hebrew word for *electricity,* a meaning that recalls the Septuagint (*elēktron*) and Vulgate (*electrum*) renderings. Whether as *ḥashmal zirmi* (current electricity), *ḥashmal galvani* (galvanic electricity), *ḥashmal magneti* (magnetic electricity), or *ḥashmal setati* (static electricity), in its various forms and combinations *ḥashmal* derives its impetus directly from the biblical usage as a pivotal moment in the unfolding vision that inaugurates the prophecy of Ezekiel.[23] What in the original prophecy is a term that encodes the mysterium of brilliance and majesty that emanates overwhelmingly from the *visio Dei* beheld by the prophet becomes in contemporary discourse a term that encodes the power through which modern culture implements and activates its world of technomastery: its generators, its dynamos, its automobiles, its aircraft, its factories, its telecommunications systems, its weaponry, its very mechanistic sources of empowerment and domination. A most compelling instance of the impulse to technologize the ineffable, this is a transformation of the profoundest sort.

From a philosophical point of view, the nature of the impulse to technologize is best summed up by Martin Heidegger in his discourses on technology. As he argues in "The Question Concerning Technology," for example, the concept *technology* (from *technē,* skill, art) is realized in the act of bringing forth (*Hervorbringen*). As such, technology is not the result of that which is simply "produced" through the act of causing something to be formed or constructed; rather, it is that through which the *being* of a thing is "manifested" or "disclosed." In the process of bringing forth, technology causes that which is concealed to be unconcealed or disclosed. The experience is tantamount to a revelation. For Heidegger, the act of bringing forth, then, is conceived as "revealing" or *alētheia.* As a "mode of revealing," *technē,* says Heidegger, "comes to presence in the realm where revealing and unconcealment take place, where *alētheia,* truth, happens."[24] As Michael E. Zimmerman observes, the being of an entity for Heidegger implies its "presencing" in the sense of its "appearing" or its "self-manifesting." Zimmerman terms this the "*alētheia*-logical, or truth-like, conception of being." In keeping with that idea, *alētheia* as truth is specifically "unconcealment."[25] In its ability to bring forth into unconcealment, *technē*, for Heidegger, finds its correspondence in *poiēsis*, the highest form of *technē*. Accordingly, *technē* is something "poetic." So Heidegger maintains in "The Origin of the Work of Art" that poetry, too, is an "illuminating projection toward unconcealment." Specifically, "poetry is the saga of the unconcealment of what *is*."[26] It is, Zimmerman observes, "an ontologically disclosive mode," one that has affinities with the idea of seeing, of revealing, of revelation itself. In the perception of that self-disclosure, the philosopher becomes seer, one gifted with the "ontological vision" required for insight into "the various modes of presencing that constitute the history of being."[27] With his ontological vision, the philosopher as seer is attuned to "the saga of the unconcealment." If *poiēsis* is the highest form of *technē*, then that which is truly technological in the Heideggerian sense is that which is truly visionary. Although the *merkabah* does not figure into Heidegger's discourses on *technē* and *poiēsis*, one might nonetheless observe that, as visionary poem par excellence, the text of Ezekiel's inaugural vision embodies "the saga of the unconcealment of what *is*."

Exploring various events in the history of that saga, I shall argue that the act of technomorphosis reveals as much about those who attempt to interpret the vision as it does about the vision itself. At the same time, I propose to explore the psychogenetic foundations of the vision, initially, at least, from the point of view of an experience that has affinities with the aberrant, the disturbed. Approaching the vision from this point of view confirms Jacques Ellul's contention not only that technology in its broadest sense "makes a fundamental appeal to the unconscious" but also that, conceived psychologically, the technological expresses itself as a "derangement," one that finds its correspondence in an "aesthetic of madness."[28]

If such is true of technology in general, it is especially true of the psychological foundations of Ezekiel's vision. Both the prophet and the vision have provoked no end of controversy in this regard. First, there is the curious behavior of the prophet himself. Scholars marvel at Ezekiel's experience of bodily paralysis and periods of trances (Ezek. 3:15, 4:4–6); his accounts of levitation (Ezek. 3:12–14, 8:3, 11:1); his cutting, weighing, dividing, burning, binding, and scattering his hair (Ezek. 5:1–4); his sudden clapping of the hands and stamping of the feet (Ezek. 6:11); and his belief in his power to destroy with speech (Ezek. 11:13). His behavior perplexes in the extreme, and his place of work is the sickroom. He is told to shut himself within his house. He is bound with cords, and his tongue cleaves to the roof of his mouth so that he is dumb (Ezek. 3:24–26). He is given to prepare his food with dung (Ezek. 4:15) and to accuse his enemies of worshiping dung balls (the term *dung ball* is found more often in Ezekiel's prophecy than anywhere else in the Hebrew Bible).[29]

Such circumstances have prompted some scholars to see in Ezekiel evidence of an especially pronounced pathology.[30] Edwin C. Broome "diagnoses" the prophet as one plagued by paranoid schizophrenia, characterized by catatonic symptoms and an inclination toward masochism, exhibitionism, narcissism, and delusions of grandeur. Within this context, the inaugural vision becomes an "anxiety dream" in which the dreamer envisions a kind of "influencing machine" that Broome finds typical of recorded cases of paranoia. In one such case cited by Broome, the patient "drew a scheme in which the top 'level' was represented by a square, containing three types of rays; below this was a 'search-light extension' represented by a triangle, and below this, a trapezoidal figure called the 'machine.' This, the patient believed, had been used by his 'enemies' since his birth, to read and control his thoughts, and to govern his actions through 'electronic' waves." As a prime instance of a case in which such an influencing machine assumes theological proportions, Broome singles out Sigmund Freud's classic case of paranoia, Daniel Paul Schreber, whose conception of God's omnipotence, Broome argues, is not dissimilar to Ezekiel's own. God appears to Schreber as a bipartite figure whose overwhelming presence emits both rays and voices that spin out toward Schreber's head like telephone wires. In his analysis of Ezekiel's so-called psychosis, Broome notes other points of comparison between the prophet and Schreber as well.[31]

If much of what Broome maintains has come under attack from several quarters, Ezekiel's personality remains a cause célèbre in biblical scholarship. Most recently, David J. Halperin has devoted a major, full-scale study to the subject. Although critical of several of Broome's assumptions, Halperin has demonstrated the appropriateness of the psychoanalytic approach to an understanding of Ezekiel's personality. If the *merkabah* remains outside the immediate bounds of Halperin's study, his analysis nevertheless makes it clear that this phenomenon invites the kinds of speculations evinced by Broome's attempt to understand the putatively "disturbed" nature of the psyche that might produce so remarkable a vision as that which inaugurates the prophecy of Ezekiel.[32]

Broome's speculations at the very least confirm the correspondence between the psychological and the technological that not only coalesces in the visionary but also expresses itself in what Ellul calls a "derangement." *Technē*, to use Heidegger's formulation, moves inevitably toward (and, in fact, reemerges as) *psychē*. Expressed as a "derangement," the visionary produces an "aesthetic of madness," a technology of the bizarre. Because Ezekiel's own personality has been characterized as "nonnormative," perhaps even "deranged," it should not come as a surprise that the inaugural vision lends itself to reformulations that might be considered correspondingly "deviant." If such is the case, these reformulations are eminently worthy of our consideration for they will be seen to shed as much light on the vision as any that the opening chapters of Ezekiel might otherwise be thought to engender.

It is through the "popular" culture of the inaugural vision that we are finally able to gain insight into what is really implied by the term *alētheia*. In our consideration of that culture, we shall encounter those inspired by an enthusiasm to translate visionary material into "scientific" schemes. As technological wonder, the inaugural vision will become a machine whose forces can be harnessed and channeled in whatever direction the person who has conquered these forces sees fit. Under such circumstances, the Schreberian psychotic of Edwin C. Broome will have found a way to overcome his "anxiety dream." Gaining control of his "influencing machine," he will be in a position to place that machine at his own disposal or at the disposal of those to whom he would bequeath its tremendous powers. Having harnessed those powers, he will assume the exalted role of "scientist," one totally in control of all that the visionary experience has to offer.

In this respect, he will become the new *merkabah* mystic, determined to transform the old mysticism into a new technology. Through him and his kind, the *ma'aseh merkabah* will take on renewed meaning, consistent with all those "mystics" who attempted to ascend to the realm of the *merkabah* in later Judaic lore. Conceived variously as Enoch, Abraham, Rabbi Akiba, or Rabbi Ishmael, these mystics have been designated the *yordei merkabah* (putatively, riders in the chariot).[33] They are heroic figures indeed. Their adventures have been mapped in detail. Extending from the apocryphal and pseudepigraphic literature through a whole series of texts that recount the journeys of the *yordei merkabah* through the *hekhalot* or celestial halls to encounter the throne of God, the mystical experiences of the riders in the chariot grow out of the rich lore surrounding the *ma'aseh merkabah*.[34]

The means of their ascent are variable. Borne upward on the wings of birds or on the wings of the wind, the mystics are transported to their destination in wagons of light and carriages of fire. They are even known to be dragged on their knees as if abducted in their journeys to the celestial realms.[35] Whatever the means of transport, the purpose of their journeys is ultimately that of spiritual transcendence: the *yordei merkabah* desire to transcend the realm of the body in order to see and worship God "face to face." This is a dangerous business, to be sure, one that risks a range of disasters resulting in the possibility of madness or immolation: "The mystic is liable to be stoned or thrown into molten lava. His flesh is transformed into fiery streams that threaten to consume him. His eyelashes become flashes of lightning. He is in danger of being deluged by waves of water that represent the ethereal luminescence of the marble plates that adorn the individual palaces." To ascend through the heavens to the highest realm, he must placate angelic guards that stand watch at the entrance gates to the palaces. For that purpose, he fashions magic seals made of secret names. Reciting these names, he seeks entrance into ever-higher levels of gnosis. He who succeeds enjoys remarkable rewards. Transformed into a powerful being stationed beside the divine throne, he becomes a guardian of the secrets and treasures of the other world. Imbued with wondrous theurgic powers, he is able to behold the entire scope of history in a moment. The cosmos is at his disposal. He becomes a figure of utmost power. His is the will to power that transforms the desire to "see" God into the desire to "be" God, or at least to be as godlike in his possession of power as one can possibly imagine.[36]

The old *yordei merkabah* anticipate the new. If the old *merkabah* mystics appropriated Ezekiel's vision in order to achieve a spiritual transcendence, the new avatars of *technē* appropriate the vision in order to achieve a material transcendence.[37] For the old *merkabah* mystics, Ezekiel's vision became a source of the triumph of spirit over matter; for the new avatars of *technē*, Ezekiel's vision becomes a source of the triumph of matter over spirit. What for the old *merkabah* mystics is trope for the technologists is letter; what for the mystics is word for their counterparts is thing. What for the mystics is a revelation of that divine "otherness" that distinguishes itself from the merely human for the new interpreters of the vision is a revelation of that by which the divine otherness is to be appropriated, transformed, and placed in the service of those who are no longer human but tantamount to gods.

Through these new interpreters, *poiēsis* assumes consummate form as "machine." Theirs is a "technopoetics" by means of which the *deus ex machina* becomes a *machina ex deo*.[38] What they embody is the impulse to mechanize, to control. In them, one finds the sensibility of the "modern" in all its forms for they are the moderns, the prophets of the New Age. Theirs is the religion of the New Age, the religion of the modern through which earlier forms of devotion, archaic modes of worship, discover a new outlet. These are the children of Ezekiel. If they (like the child caught reading the prophecy of Ezekiel in the house of his teacher) risk the prospect of madness or even immolation, they remain confident in their abilities to confront and overcome the dangers of the vision. Whatever the respective "ages" of the children, they must be given their due. Visionaries of all persuasions, they augur a brave new world of locomotion, of flight, of mechanized conquest, of interplanetary space travel. They herald the era of the machine. What unites them is that most profound of biblical events, the *visio Dei* that inaugurates the prophecy of Ezekiel.

I

Cultural

Transactions

and the

Poetics

of

Aggression

Technology

of the

Ineffable

1

Any consideration of the appropriation of Ezekiel's *visio Dei* in the modern world might well begin with the most crucial of poetic renderings that this vision assumes. As an expression of the visionary imagination, the rendering in question is that of "the Chariot of Paternal Deitie," the sublime vehicle that represents the centerpiece of John Milton's *Paradise Lost*.[1] Emerging at the very midpoint of this great epic, the Chariot of Paternal Deitie accrues to itself both narrative and cultural implications of seminal importance to the work as a whole.[2] As it rushes forth "with whirlwind sound" on "the third sacred Morn" of the War in Heaven, the chariot is described as

> Flashing thick flames, Wheel within Wheel undrawn,
> It self instinct with Spirit, but convoyd
> By four Cherubic shapes, four Faces each
> Had wondrous, as with Starrs thir bodies all
> And Wings were set with Eyes, with Eyes the wheels

Of Beril, and careering Fires between;
Over thir heads a chrystal Firmament,
Whereon a Saphir Throne, inlaid with pure
Amber, and colours of the showrie Arch.

Within this vehicle, the Son of God, armed "in Celestial Panoplie," sets out to overwhelm the rebel angels: "At his right hand Victorie / Sate eagle-wing'd, beside him hung his Bow / And quiver with three bolted Thunder stor'd, / And from about him fierce Effusion rowld / Of smoak and bickering flame, and sparkles dire" (6.748–66).

More than any other poetic construct that distinguishes the visionary imagination, this chariot represents a defining moment in the history of representation through which the vision of Ezekiel is manifested from the seventeenth century to the present time. To gain an understanding of the meanings that accrue to the Chariot of Paternal Deitie in the history of representation, one must at the outset address the Reformation underpinnings of the Miltonic chariot.[3] Through a reconstruction of those underpinnings, it is possible to come to terms with the import of the chariot not only to the narrative from which it springs but also to the afterlife that it engenders. This afterlife includes all those phenomena that become the staple of the children of Ezekiel: locomotives, automobiles, tanks, aircraft, unidentified flying objects, and the wonders of cyberspace. Manifested in religious movements ranging from the Jehovah's Witnesses to the Nation of Islam, it is an afterlife distinguished as much by the anxieties that beset a civilization on the verge of annihilation as by the marvels that distinguish the technology of the modern world.

The Reformation underpinnings of the Chariot of Paternal Deitie are already discernible in Milton's antiprelatical tracts, such as the *Apology* for Smectymnuus (1642). There, Milton conceives of "Zeale whose substance is ethereal, arming in compleat diamond" and "ascend[ing] his fiery Chariot drawn with two blazing Meteors figur'd like beasts, but of a higher breed than any the Zodiack yeilds, resembling two of those four which *Ezechiel* and S. *John* saw, the one visag'd like a Lion to express power, high autority and indignation, the other of count'nance like a man to cast derision and scorne upon perverse and fraudulent seducers." Armed with these weapons, "the invincible warriour Zeale shaking loosely the slack reins drives over the heads of Scarlet Prelats, and such as are insolent to maintain traditions, brusing their stiffe necks under his flaming wheels" (*YP,* 1:900). In his depiction of this chariot, Milton considers himself a poet who soars to transcendent heights.[4]

Through his reformulation of the throne-chariot as a polemical device,

Milton demonstrates not only the centrality of the biblical antecedent to his outlook but also the extent to which that antecedent permeated his own Reformation sensibility. At the same time, his reformulation becomes a pivotal moment in the anticipation of its epic counterpart. Because it is conceived specifically as a poetic rendering within a polemical context, the chariot of the antiprelatical tract suggests how the chariot of the epic is to be read. The Chariot of Zeale, in effect, provides a hermeneutic for the Chariot of Paternal Deitie. That hermeneutic is one in which the Chariot of Paternal Deitie will be viewed most profitably (if not inevitably) as a Reformation conceit. To understand the full significance of this idea, one must explore the Chariot of Zeale in greater depth.

Of immediate significance is Milton's citing not only Ezekiel but also Saint John the Divine as the primary source for his conception of the Chariot of Zeale. Milton has in mind here Revelation 4, which is commonly looked on as a New Testament redaction of Ezekiel 1.[5] That it was so viewed among Reformation theologians may be seen in Henry Bullinger's *A Hundred Sermons upon the Apocalypse* (1573). Responding to Revelation 4, Bullinger observes: "The goodliest beastes do drawe the triumphant chariotes of Princes. Therefore by a lyke kinde of speache as is used among men, beastes are set to the throne of God. For God in hys Prophetes is caryed upon Cherubin, that is, in hys heavenly chariot. And Ezechiell . . . nameth openly Cherubin, beastes: and the whole texte proveth, that the place must be understoode of Gods chariot, drawen by beastes."[6] Milton's own association of Ezekiel's throne-chariot with the apocalyptic setting envisioned by Saint John the Divine is wholly consistent with what Katharine Firth sees as a major trend in Reformation thought. Among other crucial texts, Ezekiel and Revelation serve as the basis for the outlook espoused by the Reformers.[7]

In the *Apology* for Smectymnuus, the specific occasion that prompts Milton to invoke the Chariot of Zeale is his desire to defend his stance as Reformation polemicist. Having been castigated for using language unbecoming an orator, Milton maintains that his "vehement vein [of] throwing out indignation, or scorn upon an object that merits it" is supported not only "by the rules of the best rhetoricians" but also by the "true Prophets of old," on the one hand, and "Christ himselfe," on the other. Indeed, this "fountaine of meeknesse found acrimony anough to be still galling and vexing the Prelaticall Pharisees." Christ's "sanctifi'd bitternesse against the enemies of truth," in turn, inspired the father of the Reformation, Martin Luther, "whom God made choice of before others to be of highest eminence and power in reforming the Church." Luther, comments Milton, "writ so vehemently against the chiefe defenders of old untruths in the Romish Church, that his own friends and favourers were many time offended with the fierceness of his spirit." As antiprelatical polemicist, Milton becomes the heir to the Reformation fervor that inspired the "sanctifi'd bitternesse" of Luther. Infused with this bitterness, Milton transforms Ezekiel's throne-chariot into a polemical weapon that drives over the heads not only of the prelates but also of all who are "insolent to maintaine traditions" (*YP*, 1:899–901). In this respect, the Chariot of Zeale is used as a means to explode tradition. Under the flaming wheels of the chariot, the stiff neck of conformity (cf. Exod. 33:5; Acts 7:51) gives way to the new spirit of a reforming zeal.

For Milton, that zeal assumes meteoric proportions in the blazing figures that emerge from the zodiac of his imagination. Reducing from four to two the number of creatures envisioned in his sources, Milton singles out the lion and the man as best suited to his polemical purposes. Whereas the lion expresses "power, high autority and indignation," the man signifies "derision and scorne" (*YP*, 1:900). These qualities embody the essence of the Reformation fervor that Milton attributes to the chariot as a polemical device. Upholding the cause of the Reformation, the polemicist assumes at once the role of lion, the regal symbol of military prowess, and the role of man, the rational symbol of intellectual superiority. In this manner, the polemicist embarks on an enterprise of the most exalted sort, one in which the Chariot of Zeale is his ultimate weapon. It is his engine of destruction, and he wields it with a vehemence befitting one whose calling is to do battle with idolaters and corrupters of the church.

In *Paradise Lost,* the spirit of idolatry, corruption, and blasphemy is, of course, dramatically manifested in the enemy of God, Satan, whose "argument blasphemous" Abdiel excoriates as a prelude to the War in Heaven (5.809). This blasphemous argument, in turn, assumes concrete form in the conduct of the war itself. Indeed, at the very outset of the war, Satan as consummate Blasphemer emerges in a vehicle that represents both an anticipatory parody of and an affront to the Chariot of Paternal Deitie. So the Blasphemer is beheld approaching in idolatrous splendor:

> High in the midst exalted as a God
> Th'Apostat in his Sun-bright Chariot sate
> Idol of Majestie Divine, enclos'd
> With flaming Cherubim, and golden Shields.
> (6.99–102)

By means of this "Sun-bright Chariot" Satan exalts himself in his idolatrous quest to unseat the true God, the true Lord of the Chariot. Enthroned in his own vehicle of blasphemy, the Blasphemer enters the fray in what might be called the Chariot of Satanic Deitie.[8] From the perspective of Reformation polemic, the scene recalls Milton's castigation of the prelates in *The Reason of Church-Government*. As an example of their "profane insolence," the prelates, says Milton, have mounted the church as a false idol on a "Prelaticall Cart," "drawn with rude oxen their officials, and their owne brute inventions" (*YP*, 1:754–55).[9] At the sight of Satan so mounted, Abdiel, newly returned from his first encounter with the Blasphemer, reasserts his zeal by delivering a blow to "the proud Crest of *Satan*," who recoils backward in amazement (6.188–94). As the first official act of physical retribution, Abdiel's blow anticipates that of the Son, just as Abdiel's zeal foreshadows the full manifestation of that attribute in the coming forth of the Son within the Chariot of Paternal Deitie. The conflict as a whole, in turn, is framed by the parodic chariot at the beginning of the battle and by the divine chariot at the end of the battle. Within the parodic chariot, Satan is enthroned as the very symbol of idolatry; within the divine chariot, the Son is enthroned as the very symbol of true worship. So framed, the conflict is waged by the adherents of idolatry, on the one hand, and the adherents of true worship, on the other. Ascending the Chariot of Paternal Deitie, only the Son himself can settle the conflict. In the portrayal of that event, Milton's Chariot of Zeale is transformed into a poetic construct.

The language that portrays this remarkable event is replete with nuances that are endemic to the Reformation frame of mind. In fact, in the very wording of the argument to book 6, one hears the polemicist of the prose tracts speaking: "God on the third day sends Messiah his Son, for whom he had reserv'd the glory of that Victory: *hee in the Power of his Father* coming to the place . . . with his Chariot and Thunder *driving into the midst of his Enemies,* pursues them unable to resist" (italics mine). It is the rhetoric of the *potentia Dei* that infuses both the language and the syntax here. The fiery zeal of the Reformation enthusiast inflames even the prose description that precedes the poetic account.

What is anticipated in the argument is amply fulfilled in the poem. There, the rhetoric of power and might so crucial to the Reformation point of view is already present in God's commissioning of the Son: "Go then thou Mightiest in thy Fathers might, / Ascend my Chariot, guide the rapid Wheels . . . / Pursue these sons of Darkness, drive them out / From all Heav'ns bounds into the utter Deep" (6.710–16). Obeying his Father's command, the Son responds in a language infused with zealous indignation: "Scepter and Power, thy giving, I assume" for "Whom thou hat'st, I hate, and can put on / Thy terrors, as I put thy mildness on / Image of thee in all things" (6.730–36). As I have established elsewhere, the Son's response is consistent with the transformation of the *ira Dei* (the Son as an embodiment of his Father's anger) into what might be called the *odium Dei* (the Son as the embodiment of his Father's hatred).[10] Perfectly in keeping with Reformation theology, such a transformation is viewed as the ultimate manifestation of a divine zeal to overcome all who would blaspheme God and his ways.[11] The *odium Dei* is the transcendent expression of God's anger: it is anger divinized as hate. The servants of God who possess it are empowered to exercise it to its fullest: "Do not I hate them, O Lord, that hate thee? . . . I hate them with a perfect hatred" (Ps. 139:21–22). As Milton avers in *De doctrina christiana*, the exercise of such hatred, as occasion warrants, is "a religious duty; as when we hate the enemies of God or the church [*odium etiam pium est; ut cum hostes Dei aut ecclesiae odio habemus*]" (*YP*, 6:741–43; *CM*, 17:258–61).[12] Rendered indomitable as an expression of the zeal to rid God of his enemies, such hatred assumes the form of utmost power by one who is commissioned to go "Mightiest in [his] Fathers might" as he ascends the Chariot of Paternal Deitie and drives out the "sons of Darkness" from "all Heav'ns bounds into the utter Deep." Assuming God's scepter and power, the Son as embodiment of both the *ira Dei* and the *odium Dei* accordingly puts on his Father's terrors and, armed with God's might, sets forth to "rid heav'n of these rebell'd" (6.730–37).

The vehicle in which he rides is the dynamic embodiment of God's omnipotence. "Instinct with Spirit," the chariot rushes forth "with whirlwind sound." "Careering Fires" envelop the vehicle, which emits a "fierce Effusion" of "smoak and bickering flame and sparkles dire." Under the "burning Wheels" of the chariot, "the stedfast Empyrean" trembles. The propelling force is the Spirit of God, which infuses the fourfold creatures and gives them life. Their eyes glare lightning and shoot forth pernicious fire (6.749–850). This is an engine to be reckoned with indeed! As much as Milton spiritualizes the Chariot of Paternal Deitie, the language that is used to describe it ironically causes the vehicle to find its profane and debased counterparts in the very weapons of destruction invented by Satan and his crew. With its emphasis on "careering fires" and the "fierce Effusion" produced by "smoak and bickering flame and sparkles dire," the divine chariot becomes in effect God's answer to Satan's cannons. God might be said to counter demonic technology with a divine technology all his own.

Once again, the idea is already present in the argument to book 6 of Milton's epic: Satan "calls a Councel, invents devilish Engines, which in the second dayes Fight put *Michael* and his Angels to some disorder." The faithful angels respond by pulling up mountains to overwhelm "both the force and Machins of *Satan*," but, of course, the tumult is not resolved until "God on the third day sends *Messiah* his Son" to conquer Satan through the power of the divine chariot.

Clearly, the pulling up of mountains in a kind of Hesiodic *titanomachia* is not sufficient to destroy the forces of modern warfare in the form of those cannons. The clash of "ancient" and "modern" methods of warfare results only in chaos. To resolve the conflict, divine intervention is called for. In the very conception of that intervention resides the awareness of an engineering feat that finds its debased correspondence in the kind of perverse machinations that consume Satan and his crew as they set about to invent the implements of modern warfare, the technology of the new order.[13]

Probing deep beneath heaven's soil to discover the "materials dark and crude" with which to fashion "hollow Engins long and round / Thick-rammd," they forge combustible weapons capable of sending forth "from far with thundring noise" such "implements of mischief" that they "dash to pieces, and orewhelm whatever stands / Adverse." Satan hopes that the effect will be one of causing his enemies to fear that he and his crew have "disarmd / The Thunderer of his only dreaded bolt" (6.469–91). When completed, Satan's chariots of destruction are revealed disarmingly to the faithful angels, who suddenly behold "A triple-mounted row of Pillars laid / On Wheels." As these pillars are discharged, "immediate in a flame / But soon obscur'd with smoak, all Heav'n appeerd, / From those deep-throated Engins belcht" as they foully disgorge their "devilish glut, chained Thunderbolts and Hail / Of Iron Globes" (6.572–89).[14] In this manner, the wheeled conveyance is brought forth to overwhelm the enemy.[15]

Consistent with this impulse is the polemical justification afforded by all that the Chariot of Zeale represents. Milton invokes that chariot, we recall, to justify the "sanctifi'd bitternesse" that distinguishes his stance as polemicist. Through a zealous exercise of "power, high autority and indignation," on the one hand, and "derision and scorne," on the other, the polemicist counters the blasphemy, idolatry, and duplicity of the adversary. It is just this polemical dimension that underscores the Chariot of Paternal Deitie. If Milton does not conceive of that chariot as being propelled specifically by creatures visaged like a lion and a man, the polemical qualities that these creatures embody are implicit in the Chariot of Paternal Deitie nonetheless. We have already encountered at least two characteristics of the lion (authority and power) in God's commissioning of the Son to assume his Father's scepter and power. The third is discernible in the Son's command to the faithful angels before he attacks the Satanic crew: "stand onely and behold Gods indignation on these Godless pourd / By mee" (6.810–11). That indignation is aroused as a result of satanic despite and envy. "Not you but mee they have despis'd, / Yet envied" (6.812–13), the Son declares to the assembled faithful. The response of the Son is justified outrage at Satan's behavior. By means of the Chariot of Paternal Deitie, the Son answers despite with despite in an outpouring of divine indignation. Such a response is in keeping with the characteristics that distinguish the second creature that propels the Chariot of Zeale: derision and scorn. If power, authority, and indignation are the hallmarks of the creature visaged like a lion, derision and scorn are the hallmarks of the creature visaged like a man. The counterpart of the regal is the supremely rational, here cast in its sublimest form. As we are well aware, derision and scorn underlie God's response to the foolishness of his enemies. "Mightie Father," says the Son, "thou thy foes / Justly hast in derision, and secure / Laugh'st at thir vain designes and tumults vain" (5.735–37). If God's enemies engage in their own form of derision and scorn (e.g., 6.628–33), it is God who has the last laugh. Accordingly, the Son not only overthrows the rebel host but effectively humiliates them as well.

Overthrown, they are compared to "a Herd / Of Goats or timerous flock" (6.856–57), a simile that suggests just how pathetic Satan and his crew have become as a result of their encounter with the Chariot of Paternal Deitie. Whether the simile alludes to Jesus' exorcism of the devils who subsequently take up residence in a herd of swine that perish in the sea (Matt. 8:28–32) or to the separation of the sheep from the goats at the Last Judgment (Matt. 25:31–33), as Arnold Stein long ago recognized, this moment represents "the grand finale of physical ridicule" that has been heaped on the rebel angels throughout.[16] One need only recall the crucial lines from Milton's translation of Psalm 2 to recognize the tone in which the event as a whole is to be understood: "He who in Heav'n doth dwell / Shall laugh, the Lord shall scoff them, then severe / Speak to them in his wrath, and in his fell / And fierce ire trouble them" (lines 8–11). When in the preface to the *Animadversions* Milton defends anger and laughter as the two most noble weapons of the polemicist (*YP*, 1:663–64), he no doubt has in mind that, from the divine point of view, they are characteristics of God's response to his enemies as well. In their most extreme form, they become an expression of the *odium Dei*. The transition from the political world of the polemicist to the poetic world of the epic poet could not be more compelling. In both respects, the Reformation underpinnings of the chariot are fully in evidence.

If such is the case, then nowhere are those underpinnings more apparent than in the relation between Zeale as charioteer and the Son as charioteer. Ascending their respective chariots, each performs a function of fundamental importance to the Reformation frame of mind. As theological attribute, zeal for Milton is characterized by an ardent desire to sanctify the name of God, combined with an indignation against whatever might violate or betray a contempt of religion.[17] Its purpose is to cleanse religion of blasphemy. With its ethereal substance and diamond armor, Zeale finds its theological justification in the gospel. For those receptive to the dazzling power of the gospel, the Reformation experience will be one that results in a true *visio Dei*. For those corrupted by idolatry, the Reformation experience will be one that results in annihilation. In either case, zeal as theological attribute is exalted to a position of highest eminence as it "ascends" the "fiery Chariot" that goes forth to overwhelm the "Scarlet Prelats."

When the Son "ascends" the Chariot of Paternal Deitie in *Paradise Lost* (6.762), he, in turn, is reenacting Zeale's "ascent" into the fiery chariot of Milton's prose tract. Attired in "Celestial Panoplie" of "radiant *Urim*" and (implicitly) Thummim (cf. Exod. 25:16, 28:30), the Son represents a fitting counterpart to his polemical forebear attired in "compleat diamond." If the armor of Zeale is such to "dazle, and pierce" the "misty ey balls" with its radiance, the "divinely wrought" stones that grace the Son's armor do no less. Signifying, respectively, "light" and "perfection," Urim and Thummim appear on the "breastplate of judgment" worn by the high priest as emissary of God. As such, the function of these gems is oracular: through them, God manifests his will. In accord with the Son's ascent into the Chariot of Paternal Deitie in *Paradise Lost*, their purpose is to provide a *visio Dei* for the faithful and to enact a *privatio Dei* for the enemies of God, who are "cast out" from "blessed vision" into "utter darkness, deep ingulft," their place "ordained without redemption, without end" (5.613–15). Like the diamond armor of Zeale, then, the oracular armor of the Son reinforces his task of becoming the means of divine vision, on the one hand, and the source of purification, on the other.

Given these circumstances, the implications of the Son's "ascent" in *Paradise Lost* are very much in keeping with the apocalyptic bearing of Zeale's earlier "ascent" in the *Apology*. So the Son ascends the Chariot of Paternal Deitie, "conquering, and to conquer":

> Attended with ten thousand thousand Saints,
> He onward came, farr off his coming shon,
> . . . Hee on the wings of Cherub rode sublime
> On the Chrystallin Skie, in Saphir Thron'd,
> Illustrious farr and wide . . .
> When the great Ensign of *Messiah* blaz'd
> Aloft by Angels born, his Sign in Heav'n.
> (6.767–76)

The eschatological basis of the Son's approach is reinforced by its biblical underpinnings: "And then shall appear the sign of the Son of man in heaven: and then shall all the tribes of the earth mourn, and they shall see the Son of man coming in the clouds of heaven with power and great glory" (Matt. 24:30). The Son's ascent into the Chariot of Paternal Deitie presages the coming forth of the risen Christ to enact a final judgment at the end of time (cf. *PL*, 6.842–43; Rev. 6:16).[18] The event is foretold by God to the Son in book 3 of *Paradise Lost*: "All knees to thee shall bow, of them that bide / In Heav'n, on Earth, or under Earth in Hell; / When thou attended gloriously from Heav'n / Shalt in the Sky appeer" to "judge / Bad men and Angels," who "shall sink / Beneath thy Sentence" (lines 321–32). Those who originally refused to bow their knees in response to the Son's begetting (5.600–616) will surely bow them in response to his renewed exaltation at the Last Judgment.

In the salvation history that the Son's ascent into the Chariot of Paternal Deitie encapsulates, the eschatology that was so much a part of the fervor of Milton's prose tracts makes itself felt yet once more as epic event. In its transmuted form, that event, however, is no longer bound by the political aspirations that distinguished Milton's prose. Transcending the eschatology that looked forward to the establishment of a *civitas Dei* in this world, the eschatology of the epic embraces a Reformation fervor far more profound. It is this fervor that the Chariot of Paternal Deitie embodies. Tracing its lineage to Reformation theology in general and to Milton's polemical formulations in particular, the Chariot of Paternal Deitie assumes renewed significance within the context that the epic poet provides for it. If its forebear is the Chariot of Zeale of the prose tracts, its new impetus is the result of a Reformation fervor chastened by the collapse of an entire political system. From the perspective of Milton's early aspirations for political reform, its milieu, then, is marked by the darkness and dangers that encompass the epic poet in his latter days. From the perspective of Milton's renewed faith in a polity that cannot be compromised, however, its milieu is marked by an illumination that represents a source of lasting hope. Despite what Milton in *Paradise Lost* calls the "savage clamor" wrought by "the barbarous dissonance / Of *Bacchus* and his revellers" (7.34–36), the Chariot of Paternal Deitie rushes forth with a momentum undreamed of by the polemicist who originally conceived a fiery Chariot of Zeale that would crush the corrupt and idolatrous enemies of the Commonwealth.

If the act of *poiēsis* transforms the ineffable into trope, the zeal that drives this act finds its counterpart in the belief that trope can be realized as "thing," that the "machine" dwelling within the mysterium can be brought forth into the realm of being. What is required is the disposition to "invent." For Milton, this is a Satanic disposition, to be sure. It results in things like gunpowder and cannons. As much as the Chariot of Paternal Deitie might appear to eschew such an impulse, there remains a residue of gunpowder in God's smoke. Invention is a *poiēsis* in which *technē* is realized not just as trope but as thing. Ezekiel's vision exists as much by virtue of its "thingness" as it does by virtue of its tropology. Satanic or not, the impulse to invent actual things has its place as well.

This fact is given full expression in the centuries following the production of the Chariot of Paternal Deitie. Following hard on that chariot are other chariots equally as remarkable and equally as indebted to Ezekiel's *visio Dei* for their inspiration. An account of those chariots will take us on a journey of exploration into the realms of invention. Those realms will disclose the wonders of *technē* that are the products of a quest for mastery through which those who seek to harness the powers latent in the visionary view themselves as gods. With their grand designs, their grand schemes, their grand machines, the purveyors of *technē* are empowered to subdue, if not to annihilate, all who stand in their way. In their hands, the throne-chariot of Ezekiel as a fully mechanized vehicle knows no bounds, no limits. It is able to accomplish whatever its inventor sets forth to perform.

A case in point is Melchior Bauer, an eighteenth-century German inventor of flying machines.[19] Born in 1733 at the village of Lehnitzsch near Altenburg, Bauer journeyed to London in 1763 to seek the patronage of the recently crowned King George III. Hoping to finance construction of his design for a man-powered aircraft, he was, according to his own account, not able to get beyond the official scribe, who refused to copy out the submission because of its apparent folly. Undaunted by this rebuff, Bauer returned to his native Germany, where he attempted to gain the ear of Frederick the Great in Potsdam. It was not, however, the best of times to appeal for royal favor, Frederick being consumed with the responsibilities arising out of the aftermath of the Seven Years' War and with the troubles of a battered economy. Although it was customary for Frederick's officials to forward the plans of any promising new invention to the Academy of Science for investigation and assessment, Bauer was not even granted this sign of recognition and accordingly fared worse on his own native soil than he did on foreign soil. Failing to gain royal favor, Bauer desperately attempted one last time to acquire venture capital for his fledgling enterprise. He appealed to Count Heinrich XI of Reuss, the ruler of his local district of Thuringia. In a fresh submission, he recounted his earlier disappointments, included his unsuccessful letter to Frederick the Great, and provided diagrams for his invention. Along with the tribulations that he encountered in gaining support for his design, he also detailed at great length the scope and purpose of the invention itself. This is the substance of his manuscript "Die

Flugzeughandschrift des Melchior Bauer," now located in the Staatsarchiv, Weimar.[20]

According to the testimony of the manuscript, Bauer sought to construct with his own hands a flying machine or *Cherubwagen* "just in the shape in which the holy prophet Ezekiel saw it with his godly eyes." The various creatures that distinguish the carriage (the eagle, the lion, the ox, and the human) bring to mind for Bauer the universality of what he would replicate. In this sense, the carriage finds its counterpart in the ark of Noah: "Just as all mankind now travels over the water on the instrument of Noah," comments Bauer, "so it will be with this carriage." Moving from its universality to the materials of its construction, Bauer, the prototypical engineer and avatar of *technē*, proceeds to literalize the spirit of vision into the matter of substance: "The things from which I can make this carriage," he declares, "are fir wood, woven silk, and brass wire." All these materials are readily available to the artisan of the visionary. As a true technocrat faced with the prospect of deadlines, he boasts that he can complete the work within three or four months' time, but to do so he would require the appropriate construction facilities (factory, tools, and the like) afforded by the venture capital he seeks.[21] One assumes that, under these circumstances, mass production would not be out of the question.

With the material he specifies for construction, he plans to construct a canopy that resembles two outspread wings of a bird, with each wing "7 ells long and 5 ells wide." By means of such a canopy, the whole carriage would be able to soar into the air with a man in it. With his hands steering it wherever he wishes, the pilot propels it through the air with two other movable wings. In keeping with the tumultuous sounds of Ezekiel's vision, Bauer's carriage would likewise make "a great rushing and thundering" noise. This noise would be produced, observes Bauer, "because everything is light and drawn tight with wire, and is made so taut that the wires ring." When the pilot who stands in the carriage flaps the wings strongly, the carriage moves on its wheels so quickly that, with its outstretched canopy, it cuts its way up into the air and soars like a bird that holds its wings steadily spread. Because the carriage stands on four large wheels, it should be launched from a smooth hill or from a broad, smooth expanse of ice, whereupon it will fly swiftly into the air like an ascending swan, at sharp angles, as it carries the rest of the structure along with it. When the work is completed, promises Bauer, "it will in all things most clearly resemble what the holy prophet Ezekiel saw and described." Accompanying Bauer's proposals are detailed illustrations of the fuselage, the canopy and stay wires, the hinges, and the wings. Cross sections, side elevations, and front elevations provide a sense of dimensionality as well as "scientific" verisimilitude.[22] Had the flying machine ever been constructed, the result would have been "a parasol-winged monoplane of low aspect ratio." Standing in a short, four-wheeled fuselage fixed below, the pilot would have been the sole source of power, as he flapped an additional pair of compound wings, the span of which would have been about half that of the main plane. Unlike most other ornithopter wings, Bauer's flappers would have been firmly secured from tip to tip on rigid spars, with the result that one wing would have risen while the other descended.[23]

As we might well expect, the enthusiasm with which Melchior Bauer described and illustrated his plans was intensified by a religious fervor of the most strident sort. A zealous evangelical Protestant with an obsessive hatred of Catholicism, Bauer drew inspiration from the biblical original not simply to construct an aerodynamically sound mechanism. He likewise sought inspiration from Ezekiel as a biblical prophet in order to channel his own fierce hatred of the popish enemy, indeed, his hatred of all those he looked on as adversaries of the church he embraced. Clive and Helen Hart observe that Bauer "did not want to fly merely for the joy of it. He expressly and emphatically believed that he was inventing a new and virtually invincible war machine which would enable any Protestant prince to eradicate the curse of the ungodly Catholics." His own testimony represents ample evidence of his feelings on this score: his invention "is nothing other than a victory or triumphal carriage over the antichristian Pope who is still ruling with his accomplices, and over the unbelieving, idolatrous heathen."[24]

Through the use of this carriage, Bauer declares, "the word of God and the pure gospel will be kindled and spread throughout the whole world, as God himself promised, among the Jews, the Turks, and the heathen. For as all commentators explain, it is a carriage of victory." The carriage of God will disperse a thousand enemies, two of them will make ten thousand fly, and before five such carriages all the enemies of the Lord will flee. Accordingly, the construction of such weapons of mass destruction "may with justice be called the greatest of all the arts, since by its means God will destroy and overthrow the kingdom of Antichrist and will help mankind to reach complete salvation and righteousness."[25] Melchior Bauer would have been no stranger to the military-industrial complex. In fact, he might be looked on as its founding father. Drawing on my discussion of Milton's own Chariot of Paternal Deitie, I might suggest that in Melchior Bauer one discovers the material embodiment of the *odium Dei*. For Bauer, all the destructive forces of that divine *odium* find release in his invention. Deriving inspiration from its visionary wellspring in the prophecy of Ezekiel, Bauer's *merkabah* as flying machine is thereby the consummate heir of that technological mastery that Jacques Ellul views as a derangement.

When Melchior Bauer sought support from Frederick the Great in Potsdam, Frederick's war counselor rudely dismissed the inventor in the following terms: "The fiery fever has turned your head. . . . My dear man, are you not now fearful for your sanity? But I pity you from the heart for having fixed such a mad scheme into your head, for to all appearances you are a sensible fellow. If you had not given me your text I would not have believed you to be so great a fool."[26] Despite what Frederick's war counselor might have thought of Bauer and the enthusiasm he embodied, however, the German inventor remains important as a symbol of the impulse to technologize the ineffable and thereby to harness its forces. As presumptuous (if not sacrilegious) as his designs might appear to be to those who would cast aspersions on his putative invention, there is something noble in what Bauer would attempt, something eminently consistent with that sense of "workmanship" or "construction" implicit in the vision that Ezekiel himself beheld.

As *poiētēs* or "maker," Bauer resorts to the original visionary material in order to allow *alētheia* to occur. Doing so, he becomes a technologist of the visionary. Through him, we experience in all its glory one event in "the saga of the unconcealment of what *is.*" In terms of the traditions of *merkabah* mysticism, Bauer affiliates himself with those riders in the chariot that underscore the quest for transcendence so characteristic of their kind. As a modern, however, Bauer is not simply content to ride in the chariot. True to his calling, he feels compelled to invent and construct it as well. Presuming to discover in Ezekiel's vision the means of harnessing its astounding forces, he distinguishes himself as the "technologist" par excellence. In their study of Melchior Bauer, Clive and Helen Hart accordingly view his designs as prophetic of later aeronautical advances. For their part, Bauer's designs, in fact, are "an invention of true genius," and, had his manuscript been more widely read, "its many insights might well have accelerated the design of heavier-than-air machines in the late eighteenth and nineteenth centuries."[27]

The point is that inherent in the inaugural vision as originally conceived and executed is an impulse that inclines us to suspend our judgment regarding what can be considered normative and what not. Although we are disposed to offer sympathy for the point of view expressed by Frederick the Great's war counselor, we must also take account of the fact that Bauer's designs actually do have technological substance. As much as that which can be considered technological is indebted to the visionary, there is obviously a fine line between sanity and madness, normative and nonnormative. Assuming with Martin Heidegger that it is the function of the technological to bring forth what is concealed to "the realm where revealing and unconcealment take place, where *alētheia*, truth, happens," in his own unique way Melchior Bauer certainly confirms the Heideggerian proposition. Correspondingly, if an account of the transformations that Ezekiel's inaugural vision has undergone in the modern era is a chapter in the history of madness, then the children of Ezekiel confirm the idea that fools and prophets really do share the same characteristics. When Erasmus wrote his *Moriae Encomium,* he knew very well that he was dealing with a double-edged sword: the Apostles themselves, Folly reminds us, "seem'd to be drunk with new wine, and . . . Paul appeared to Festus to be mad."[28] Should the present study likewise be considered a *moriae encomium,* let us remember that the fools it canonizes might well be prophets, too.[29]

What they prophesy is the advent of all those technological wonders that distinguish the "modern," particularly that aspect characterized by the onset of the Industrial Revolution. One manifestation of this idea is discernible as the result of the advent of travel by railroad in nineteenth-century England, an event that is characteristically viewed as a crowning achievement of the Industrial Revolution. T. S. Ashton observes, "The locomotive railway was the culminating triumph of the technical revolution: its effects on the economic life of Britain and, indeed, on the world have been profound."[30] This observation has been endorsed by a host of studies that have explored the railway system and its impact on future generations.[31]

For our purposes, no study is more nearly germane than Leo Marx's *The Machine in the Garden.* Although Marx's focus is primarily on America, his concern with the effect of the Industrial Revolution on nineteenth-century England is correspondingly compelling. A crucial symbol of that effect is the locomotive as the supreme manifestation of the impulse to technologize. Associated with fire, smoke, speed, iron, and noise, the locomotive is the "leading symbol" of industrial power. If the invention of the steamboat was exciting, it paled by comparison to the railroad. What was true of the 1830s in America was no less true of the same period in England: the locomotive, "an iron horse or fire-Titan," was becoming a national obsession. The Industrial Revolution incarnate, the locomotive gave rise to its own mythology. "Stories about railroad projects, railroad accidents, railroad profits, railroad speed fill[ed] the press." As a subject of discourse, the locomotive became the focus of songs, political speeches, and magazine articles, both factual and fictional. It symbolized the "unprecedented release of human energy in science, politics, and everyday life." Equipped with this new power, humans were able "to realize the dream of abundance." In fact, "the entire corpus of intoxicated prose" that extolled the virtues of the locomotive made it clear that all hopes for happiness and well-being were dependent on technology. This discourse is what Marx aptly calls "the rhetoric of the technological sublime."[32]

It is this rhetoric that unabashedly celebrated the forces unleashed by the Industrial Revolution and its faith in the efficacy of progress. Such faith even went so far as to look on the invention of the locomotive as an act tantamount to the creation of a poem. Indeed, "one recurrent assertion is that all the other arts rest upon the mechanic arts." As such, inventions are seen to be "the poetry of physical science, and inventors are the poets." So it was maintained by one contemporary account in 1831 that very much the same sort of genius and intellect was required "to invent a new machine, as was necessary for the inspiration of a poem, and whether a man be a poet or inventor of machinery, is more the result of circumstances, or the age in which he chances to live, than in a difference of mental organization." That zeal assumes a mythic, indeed, oracular bearing in the rhetoric of the time. It is as if "machine power" fulfills what Marx calls "ancient mythic prophecy." Such is especially true for the railroad, which is seen as the true liberator of the New World.[33]

Among the locomotive systems in nineteenth-century England able to lay claim to the distinction of fostering the spirit of enterprise and discovery that infused the Industrial Revolution is the Liverpool and Manchester railway. It was on this newly constructed railway in 1829, says Ashton, "that the potentialities of steam transport were fully realized."[34] In a vivid contemporary account of the opening of the Liverpool and Manchester railway, John Francis describes the high drama of the event. "The 15th of September, 1830, will be memorable in the history of railways," he declares, for on that day the Liverpool and Manchester railway was officially opened. People flocked before sunrise to the parts where the best view could be obtained. Nobility was present; members of the senate were present. "The engines with waving flags and bright colours, added to the scene, and curiosity was at its height when the carriages started." As they gathered in groups that lined the tracks and mixed sociably together, all were innocently "unsuspicious of the extraordinary power which they were witnessing" and thus totally ignorant of "the danger which menaced them" with the approaching train, appropriately called the *Rocket*.[35] Despite the ensuing accidental deaths of those who were not careful, the enterprise was eventually a resounding commercial success. With this remarkable triumph of technology, happiness and prosperity ensued!

In an encomium that celebrates the many successes of the Liverpool and Manchester railway, an anonymous tract fittingly places the advent of this phenomenon within the context of the visionary mode. Titled *An Illustration of Ezekiel's Vision of the "Chariot,"* . . . *Its Literal Meaning, Utility, and Fulfilment, in the Nineteenth Century. Respectfully Dedicated to all the Tribes of Israel, Wherever Dispersed; and to the Public in General* (1843), this tract bears witness to the concept of *technē* as a visionary construct in the nineteenth century. According to the author, the chariot of Ezekiel prefigures the invention of "railroads and railway conveyance by locomotive carriages." In accord with this idea, Ezekiel "shows clearly" that the "component parts" of the chariot "were of *iron* and *burnished brass,* containing inwardly *fire,* without consuming itself— 'fire of coals!' sufficiently large and active to send upwards a lengthened wreath upon wreath, a crystal-coloured cloud, and their centre to be of burnished brass, sparkling, as with lightning speed they [the living creatures] winged their way, emitting sparks as from forged *iron,* instinct with a vital spirit, *unknown till steam* and its powerful effects were disclosed to man, by the manifold wisdom of God." In our own day, *"we see* the 'living creatures' . . . rushing straight forward, with lightning speed, panting as with life and voice of speech; doubts and conjectures flee before incontestible proof, and in the *living creature* before us, our astonished senses behold the perfect vision" of the locomotive. The author thereafter concludes by observing that the prophet Ezekiel has with his vision of the chariot inspired "an almost supernatural zeal and enthusiasm" in "all classes of scientific men" and has thereby excited an "extraordinary confidence" in the "minds of capitalists." The Liverpool and Manchester railway cost an enormous sum "advanced with a willingness, such indeed, as Omnipotence only could instill, command, or extend at a *given period,* not confining this period of enterprise and adventure to place or nation, but carrying out *His* vast designs into all lands, wheresoever the scattered tribes of his people, Israel are sojourning, until their warfare is accomplished and they return in triumph to the land of their forefathers."[36]

Remarkable in its own right as a testimony to the powerful, if not visionary, coincidence of technological innovation and the capitalist spirit, this treatment of Ezekiel's inaugural vision is even more remarkable as an act of "techno-poetics." Not content simply to rely on Ezekiel's vision as a prefiguration of the wonders it extols, the tract invokes a poetic document to support its contentions. Specifically, it calls on *Paradise Lost,* the epic of that prophet of prophets, John Milton, who is asked to take his place beside Ezekiel as the oracle of the locomotive. What the author of the tract has in mind, of course, is Milton's own Chariot of Paternal Deitie. Responding to this vehicle, the author declares: "Milton's inspired mind had conceived" an engine comparable to the locomotive "long before, in all its bearings and utility, as a conveyance for tens of thousands." This declaration is then followed by a lengthy quote from *Paradise Lost* depicting the Chariot of Paternal Deitie itself.[37]

Milton would no doubt have been amused. Although he might well have had second thoughts about endorsing such a reading of his poem, he would have been sympathetic to the notion of seeing in his own conception the sublime "machine" that he himself derives from the original visionary material underlying Ezekiel's vision. As I have already suggested, what is so noteworthy about the Miltonic vehicle is its "machineness." Although sublimated, spiritualized, and deified, it is a machine nonetheless: self-propelled, it harnesses incalculable energy and force that give off a "fierce Effusion" of "smoak and bickering flame, and sparkles dire" (6.766–72). This is an astounding mechanism indeed! As a war machine, Milton's chariot is simply not to be withstood. Every eye within it glares lightning and shoots forth "pernicious fire / Among th'accurst," who are likewise overcome by the thunder that the charioteer sends before him (6.836–38).

As we have seen, that which renders the Chariot of Paternal Deitie so effective as an example of sublime technology is its assimilation of the paraphernalia of modern warfare, which it sublimates and spiritualizes to achieve its own ends. Specifically, the terrible thunder and the "fierce Effusion" of smoke and flame and sparkles suggest precisely the sort of armaments with which Milton was familiar in the battles that raged during the Civil Wars. Incorporating and transforming these armaments to accord with its conception of the divine, the Chariot of Paternal Deitie establishes a context in which the original prophetic substratum of the inaugural vision is to be understood anew. This renewed understanding reflects (and implicitly comments on) what might be called the technological innovations of the modern age.

It is no accident that, when G. Wilson Knight published the *Chariot of Wrath* in the significant year 1942, he saw in the Chariot of Paternal Deitie the embodiment of *The Message of John Milton to Democracy at War,* as he subtitled his book. In this respect, Knight's enthusiastic response to Milton's chariot is particularly appropriate: "Messiah's God-empowered chariot is a transcendental conception deriving from Old Testament prophecy, but also incorporating and driving to the limit Milton's habitual fascination with the military and the mechanical. It is at once a super-tank and a super-bomber, forecasting contemporary inventions just as the Greek myth of Pegasus, or of Daedalus and Icarus, forecasts air-mastery in general." The Chariot of Paternal Deitie, Knight concludes, thereby "remains a superb conception and one to which our own experience of mechanised war, on land and in the air, may serve as an approach."[38]

Knight is not alone in responding in this manner to the technological implications of visionary material. In more recent times, the Israelis engaged in a technopoetics of their own. Although they obviously did not have Milton's chariot in mind, their thinking was certainly in accord with Milton's when they elected to resort to the biblical text in order to provide a nomenclature for their own artillery. Doing so, they applied the name *merkabah* to the powerful tanks that swept into Lebanon and routed the Palestinians in 1982. Equipped with observation periscopes, gunners' optics, white and infrared driving lights, a laser range fighter, and a special tank driver audio command system, the Israeli *merkabah* tanks are a force to be reckoned with.[39] As such, they represent compelling evidence that the impulse to technologize the ineffable is as powerful in our own times as it was in Milton's age and beyond.

In order better to understand the rationale underlying the decision to adopt the biblical nomenclature for these particular tanks, I wrote directly to Major General Israel Tal, the assistant defense minister of the Israeli Defense Force (IDF). In my letter of 5 June 1985, I informed General Tal that I had called the Office of the Consulate General of Israel in Chicago, which informed me that, as the designer of the tanks, he was the person with whom I should consult. In response to my inquiry, I received a letter dated 15 July 1985 from Ze'ev Klein, assistant to General Tal. Klein informed me that my inquiry was referred to him because of his "extensive knowledge of the development and manufacture of the Merkava tank."[40]

According to Klein, this tank dates back as early as 1968, a period in which he "had the honor to serve in General Tal's armoured division." Under the leadership of General Tal, Israel determined, in August 1970, to develop and manufacture its own tank, modified to the "special and unique needs" of the nation. The development of such a tank would help establish Israel's tank industry as well. Because of a desire to underscore the significance of that decision, General Tal, together with Major General Goren, then the IDF's chief rabbi, chose a biblical name for the tank. This name would symbolize not only the strength embodied in the tank but also its salvific role. These qualities, Klein observes, are embodied in the name *Merkava*, as the tank was designated, "with the hope that the tank would provide the I.D.F. the especially needed strength and salvation in whatever future war [might be] forced upon Israel by its enemies." Evincing a sense of religious and prophetic fervor, Klein concludes with this flourish: "Israel now has a defensive weapon, a tank, that gives expression to her physical power and spiritual determination to defend herself until peace with all her neighbors will prevail."

To provide insight into the interpretive processes through which the *mer-*

kabah tanks received their names, Klein accompanies his observations with an act of biblical exegesis. He cites those instances in the Hebrew Bible in which the actual term *merkabah* appears specifically as a manifestation of the qualities of strength and salvation with which the IDF sought to imbue its tanks. The first instance arises from 2 Sam. 15:1 (". . . Absalom prepared him chariots and horses . . ."). According to Klein, this text underlies the *merkabah* as a chariot of war, that is, a war vehicle that is a man-made product. The second instance arises from Isa. 66:15 (". . . and his chariots like the stormwind") and Hab. 3:8 (". . . upon thy chariot of victory—salvation in Hebrew . . ."). These texts, according to Klein, underlie the *merkabah* as a chariot of salvation, "a chariot of fire, a chariot of victory, a concept which expresses the creation of the Almighty, much like the creation of the Universe." Conceived in this way, salvation, Klein comments, "will come with the LORD, and the chariot is merely a symbol of God's strength and salvation."

Noticeably absent from Klein's references is the first chapter of Ezekiel. Considering the contexts that have been developed in this study, one might immediately be inclined to question the omission of so crucial a text. The obvious reason for such an omission is, nonetheless, the most compelling one. Klein does not mention Ezekiel's vision as a biblical source for the naming of the IDF's *merkabah* tanks because, in the most literal sense, there is no *merkabah* in Ezekiel 1, as there is in the biblical sources he mentions. In his citation of texts as sources of the *merkabah*, Klein is being eminently scrupulous. His is an omission that attests to the fact that, from a biblical perspective, the *merkabah*, as an artifact both human and divine, exists by name elsewhere in Hebrew Scriptures but is nowhere mentioned by name in the Hebrew text of Ezekiel.

This absence of denomination returns us to the observations ventured earlier. Despite the striking technological dimensions of Ezekiel's vision as a phenomenon that moves irresistibly toward concretization and cries out for objectification, for individuation, for the bestowal of a name, the term *merkabah* is an appellation that is the product of later generations. The *merkabah* that resides within Ezekiel's vision is the chariot that subsumes all other chariots, biblical and postbiblical alike. In technological terms, it is the profoundest expression of *technē* to have emerged from the substratum of the biblical text. It is present even more compellingly by virtue of its trace, that is, by virtue of its very absence from the exegetical discourse of Ze'ev Klein, who may neglect to acknowledge it as an immediate biblical source for the naming of the Israeli tanks but who cannot deny it as in fact the ultimate source of all sources.

With the emphasis that Klein places on the qualities of strength and salvation in the conceptualization of the IDF's *merkabah* tanks, moreover, one finds what appears to be a remarkable instance of the throne-chariot as a modern wonder of technology. In keeping with Milton's chariot, the *merkabah* tanks bring to the fore all the zeal that distinguishes the Commonwealth polemicist as the most ardent defender of the Good Old Cause. Both the Israeli spokesman and the poet of *Paradise Lost* view the biblical text as a means of extolling the regenerative and salvific virtues of what G. Wilson Knight calls the chariot of wrath. Lurking beneath that conception, however, are all the anxieties, if not the excesses, that distinguish the likes of Melchior Bauer with his plans for a flying machine, invented for the purpose of ridding the world of all those unbelieving, idolatrous heathens. In this respect, one is again reminded of the psychopathology of such figures as Edwin Broome's Schreberian psychotic, plagued by visions of influencing machines, with their rays, searchlights, and other paraphernalia of mental disintegration. Responding to the disturbing implications of these influencing machines, one might question the extent to which the *merkabah* as a modern instrument of warfare represents a concrete realization of the anxiety dream experienced by the exiled prophet on the shores of the Chebar. Whatever such instruments of warfare do represent, they are one more compelling instance of the way in which *technē* finds full expression in today's world as a profound manifestation of the visionary experience.

The

Psycho-

pathology

of the

Bizarre

2

If the Israeli *merkabah* tanks represent one extreme to which the impulse to technologize the ineffable is liable to extend in modern times, there are other extremes that must be taken into account, extremes that reflect the immense range of responses to which a reading of Ezekiel's inaugural vision as *technē* gives rise. Among those responses, none is more significant than the unidentified flying object (or UFO), conceived as an expression of the inaugural vision of Ezekiel.[1] To undertake a discussion of UFOs in general and Ezekiel's vision as a UFO in particular is not without its dangers as a scholarly enterprise.[2] Despite the vast literature to which the subject of UFOs has given rise, the very idea of venturing into this area threatens to render any serious discussion of such material immediately suspect.[3] In part, this is because of the extent to which the whole domain of UFOs has been appropriated by legions of mountebanks, an appropriation that has succeeded in causing the subject and those inclined to explore it to become guilty by association. The consensus is generally that the so-called UFO phenomenon is most properly the domain of the *National En-*

quirer and tabloids of that nature. Even if this were the case, the culture distinguished by such a venue is itself entirely worthy of exploration on its own terms. Mountebanks and tabloids have their place, too, as a crucial dimension of what defines popular culture. As we shall see, this dimension becomes particularly important in any assessment of Ezekiel's vision as an expression of the technological impulse in the modern world.

To be sure, the UFO phenomenon is hardly the exclusive domain of mountebanks and tabloids. It is also the domain of those legitimately interested in the sociological, psychological, and religious implications of this phenomenon throughout history and in more recent times. In this respect, the work of Leon Festinger, Henry W. Riecken, and Stanley Schachter is a case in point. Their book *When Prophecy Fails* is a landmark in the sociology of the UFO phenomenon and those who have embraced it.[4] It is crucial as an actual case study of a group of contactees active during and after 1949. The prophet of the group is a suburban housewife named Marian Keech, and its main proponents include Dr. and Mrs. Thomas Armstrong, adepts in the area of mysticism, the occult, and ufology. Establishing direct contact with this group, the authors provide a detailed description of the dynamics through which the group operates, attracts proselytes, spreads the word of its prophecy of impending doom, and sustains the burden of demise when the prophecy does not come to pass.

At the center of Mrs. Keech's experience of extraterrestrial life stands the impending visitation of the "tola" or spaceship. To prepare Mrs. Keech for this visitation, the extraterrestrials cause her hand to write the following message: "The cast of light you see in the southern sky is of our direction and is pulsating with a turning, spinning motion of the craft of the 'tola.'" The message makes clear that this vehicle is to land at a precise time and place: "It will be as if the world was coming to an end at the field when the landing occurs." Those who behold the craft "will not believe their senses when they see the craft of outer space in the midst of the field."[5] The arrival of the craft with its turning, spinning motion will be tantamount to a *visio Dei*.

As the contactee who first makes the impending presence of the spacecraft known and who disseminates the knowledge of its arrival, Marian Keech the suburban housewife becomes the true prophet of the new order. In the terms that have become familiar to us in this study, she assumes the role of the prophet on the banks of the Chebar who articulates her own vision of that vehicle thereafter known as the *merkabah*. That Mrs. Keech's prophecy fails is of little consequence to the present undertaking. That she experiences a *visio Dei* whose lineage can be traced to her biblical forebear is of the first moment. Glancing at her ancestry, Festinger, Riecken, and Schachter, in fact, sketch a kind of messianic history of millennial movements stemming from groups such as the Montanists up to the Shabbetaianists.[6]

In one form or another, the experiences of these millennialists populate history. It has been the particular concern of the computer scientist and astro-physicist Jacques Vallée, among others, to plot their course.[7] Vallée's *Passport to Magonia* and *Dimensions,* for example, provide fascinating narratives of such UFO experiences, extending from the Middle Ages to more recent times. The originary milieu of these experiences is the vision of Ezekiel. In fact, Vallée characteristically initiates his historical discourse on the whole subject by citing Ezekiel's vision and then proceeding to delineate various transformations of it in the recorded experiences of contactees from the Middle Ages onward.[8]

Among the examples of "ancient encounters" that Vallée cites is Agobard, archbishop of Lyons, one of the most celebrated and learned clerics of the ninth century. Agobard wrote of coming on a mob in the process of stoning three men and a woman accused of landing in a "cloudship" from the celestial region known as Magonia.[9] Is this cloudship in some sense related to the chariot Ezekiel beheld? Vallée asks. Are the mysterious creatures who fly through the sky and land in their cloudships of the same kind as the angelic creatures seen by Ezekiel? Questions of this sort engage Vallée throughout his narrative of the so-called ancient encounters. Vallée concludes with the following questions, to which he provides the only answer he finds suitable: "Is the mechanism of UFO apparitions, then, an invariant in all cultures? Are we faced here with another reality that transcends our limited notions of space and time? I see no better hypothesis at this point of our knowledge of UFO phenomena." What results is "a pattern of manifestations, opening the gates to a spiritual level, pointing a way to a different consciousness, and producing irrational, absurd events in their wake."

For Vallée, all the recorded incidents of encounters throughout history "seem to imply a technology capable of both physical manifestation and psy-chic effects, a technology that strikes deep at the collective consciousness, con-fusing us, molding us—as perhaps it confused and molded human civilizations in antiquity."[10] It is in the literalizing of the visionary experience, that is, in the conviction that Ezekiel's vision actually *was* the result of a visitation by astro-nauts from outer space, that the impulse to technologize the ineffable reveals what might be called its most aberrant form. Here, the inaugural vision as-sumes paramount importance as a technological phenomenon of truly remark-able proportions, particularly among those inclined to foster their own myth of the machine based on the unimpeachable authority of biblical precedence.

This idea becomes ironically apparent in the arena of popular culture, that is, of the tabloid and the motion picture. As an expression of Ezekiel's vision, the UFO phenomenon has indeed become such a staple of these venues that it makes its appearance in the most remarkable and, indeed, unexpected of forms. One need only recall the appearance of Ezekiel's vision in the film version of *The Best Little Whorehouse in Texas* (1982). Lying out under the stars on a beautiful Texas night, Mona Strangely (acted by Dolly Parton), captivating proprietor of the Chicken Ranch, and her lover, Sheriff Ed Earl Dodd (acted by Burt Reynolds), admire the evening sky. Suddenly, they witness a shooting star, as Mona Strangely excitedly points heavenward. Recalling her "strange" girlhood fantasies as an abductee of sorts transported to heaven by alien creatures, Mona is then moved to recite with the consummate accuracy of a religious acolyte the opening verses of Ezekiel: "And I looked, and, behold, a whirlwind came out of the north, a great cloud, and a fire infolding itself, and a brightness *was* about it, and out of the midst thereof as the colour of amber, out of the midst of the fire. . . ." Utterly perplexed by this performance, Ed Earl exclaims, "What the hell you talking about!" To this, Mona righteously responds: "That's from the Bible. That's what the Bible says about the spaceships in Ezekiel. Don't you know nothin' 'bout the Bible?" Of course, he knows about the Bible, Ed Earl protests. It's just that he hasn't come into contact with this Ezekiel fellow![11]

On a more exalted plane, the world of popular culture has been made familiar with UFOs through the work of Steven Spielberg, whose fascination with the phenomenon of extraterrestrial visitations has a long history. Most notably recorded in his films *Close Encounters of the Third Kind* (1977) and *E.T.: The Extra-Terrestrial* (1982), this fascination has brought to life a sense of childlike wonder in the otherworldly realms that are suddenly within reach of the mundane. *Close Encounters* is a case in point. Here, Spielberg was able to project onto the experience of contact with childlike but mysterious creatures from outer space a remarkable feeling of dreamlike serenity. As Frank Rich observes, what animates this film is the "breathless sense of wonder" that Spielberg brings to each frame. The film "is a celebration not only of children's dreams but also of the movies that help fuel those dreams."[12]

Much of the fascination of the film derives from the extent to which the quotidian circumstances that define the lives and experiences of the characters (particularly those of the protagonist, Roy Neary, played by Richard Dreyfuss) are transformed through the otherworldly encounter with the visionary. Culminating in that transformative experience, the film moves to a point of apotheosis in which all dreams of ascent to the supernal realms are actually realized. Roy Neary walks forward "deep into the fiery heart of the mystery," as the "brilliant opening" of the mother ship slides shut and the vehicle ascends "through layer after layer of clouds" until "this great city in the sky" becomes "the brightest of the brightest stars."[13]

The vehicle that transports the protagonist, among others, to the realm beyond the stars is conceived as a mysterious phenomenon that emits a splendid array of lights swirling in circular motions around a cloudlike center, as wheels within wheels. Exuding light from within the cloud, and surrounded by swirling lights, the phenomenon becomes "deep amber." This, the narrator of the novelized form of *Close Encounters* observes, "was an extraordinary sight, *a vision* that seemed to flash and swirl with meaning."[14] A phenomenon specifically designated "a vision," that which is beheld by Spielberg's fugitive characters ensconced within the foothills of Devil's Tower, Wyoming, replicates in its own way that awesome sight encountered by the exiled prophet on the shores of the Chebar: "a great cloud with brightness round about it and fire flashing forth continually." In the middle of this fire is that which appears like "gleaming amber" (Ezek. 1:4). Surrounded by its own fires, the "deep amber" mysterium that Spielberg envisions emanates a *hashmal* all its own (figs. 3–4).

In keeping with its visionary antecedents, moreover, that mysterium derives its impetus as much from the impulse to concretize and, indeed, technologize the ineffable as it does from the impulse to imbue it with ethereal qualities. In giving vent to this impulse, Spielberg bestows almost a comic quality on the experience of the vehicle's coming into full view. When that occurs, the vehicle ironically has the appearance of a gigantic old machine, the top of which looks like "an oil refinery, with huge tanks and pipes and working lights everywhere." As it slides across the canyon, the phantom mass seems "somehow old and dirty." It even appears to be "junky," rather like "an old city or an immense old ship that had been sailing the skies for thousands of years." In response to the attempt on the part of the assembled scientists and technicians to communicate with it through the use of musical notes, the vehicle makes a sound that at first resembles that of a pig grunting.[15]

On Spielberg's part, one might suggest, this "demystification" becomes in effect a self-reflexive act of ironic commentary that paradoxically serves to reinforce what is finally the ineffable and otherworldly quality of its bearing. Lurking beneath the ineffable is the "machine" that in its very mundaneness transforms the technological into a science of the divine. In that act of transformation, Spielberg brings into play his delight in the "scientific" dimensions of ineffability. It is here that the character of Claude Lacombe (played in the film version by François Truffaut) becomes so important. As a scientist cum ufologist of impeccable credentials, this figure adds a technological "credibility" to the whole affair. Lacombe, of course, is Spielberg's version of Jacques Vallée,

Figure 3. Vision of the mother ship, Devil's Tower, Wyoming.
Reproduced from *Close Encounters of the Third Kind*, Columbia Pictures, 1977.

Figure 4. The mother ship.
Reproduced from *Close Encounters of the Third Kind*, Columbia Pictures, 1977.

whose theories, writings, and presence resonate throughout the dramatization that underlies *Close Encounters of the Third Kind*.

In the epilogue to the novelized version of *Close Encounters*, J. Allen Hynek, then professor of physics and astronomy at Northwestern University and director of the Center for UFO Studies, Evanston, Illinois, lends further "scientific" credence to the experiences recounted in Spielberg's work by reiterating the distinctions he draws in his various writings among the three established kinds of UFO encounters. These include close encounters of the first kind, in which a UFO is seen very close up but in which there are no occupants or any discernible interaction with the immediate environment; close encounters of the second kind, in which UFOs leave some definite trace of their presence, such as scorched patches on the ground, signs of matted vegetation, broken tree branches, and the like; and, finally, close encounters of the third kind, in which the presence of "occupants" in or about the UFO is reported. This category involves actual, specific reports of allegedly genuine firsthand experiences of encounters with the occupants of the UFOs.[16] There are many reports of such "abductions" that Hynek and others have collected and recorded and on which Spielberg presumably based the experiences recorded in *Close Encounters*.

Such experiences have occasioned a great deal of debate recently. In *Abduction*, John E. Mack, professor of psychiatry at Harvard University and director of the Center for Psychology and Social Change, provides a wide-ranging analysis of a group of thirteen patients who disclose in graphic detail the experiences they underwent as "abductees" aboard the spacecrafts of extraterrestrials.[17] Mack's purpose is not to endorse the notion that his patients were actually abducted. Rather, he wishes to explore the states of mind of these "experiencers" and what those mental states imply. Drawing on the work of Thomas Kuhn, Mack seeks to establish a "paradigm shift" in the way abduction experiences are understood. For Mack, this represents a shift in consciousness on the part not only of abductees but also of those who analyze them.[18] "It would appear," he observes, "that what is required is a kind of cultural ego death, more profoundly shattering . . . than the Copernican revolution." The phenomenon of the UFO abduction suggests that "humans are not the preeminent intelligent beings in a universe more or less empty of conscious life." Rather, we are part of a universe with beings far more advanced than we are and in possession of the power to exert their influence over us. Attesting to this fact, each abductee becomes "a pioneer on a hero's journey." Each undergoes his own "ego-destroying terror," and, in sharing his experience, he reveals the existence of "unknown dimensions of the cosmos and the human psyche."[19] This is a pattern that we have already seen operative in the complex experiences of the riders in the chariot. According to some researchers, such a pattern is correspondingly evident in the experiences of shamans, the prototypical "otherworld travelers."[20] For Mack, experiences of this sort are, in effect, replicated in those of his abductees. In the context of the abductees, the experience of transcendence (whether as *merkabah* mysticism or as shamanic mysticism) assumes a new form, one consistent with the technocentrism and psychocentrism of the modern world.

Describing the methodology that underlies his clinical approach, Mack categorizes the information that can be gleaned from his analysis of the "abduction trauma." Such information includes an assessment of the facts of the narrative and an interpretation of the phenomena. Mack is particularly taken with the extent to which abductees attest to the higher powers possessed by beings capable of engaging in "thought travel," telepathy, and even shape-shifting. The vehicles through which the abductees are transported vary in appearance and characteristics, but they are usually accompanied by smaller craft of one sort or another that descend from the mother ship and reascend to it with the abductee. Once within the mother ship, the abductee becomes the subject of experiments by the occupants of the ship. These experiments are both physical and psychological. Physically, they may include the introduction of foreign instruments into the body or the possibility of sexual assault that results in rapelike trauma. Psychologically, they may produce in the abductee a visionary consciousness given to prophecies of apocalyptic events that recall the import of biblical narratives of end time. Such pathological aftereffects are commonplace. Of corresponding importance, however, is the transformational nature of the experience. Here, one finds the ability of the patient to "push through" to higher levels of consciousness, an "unveiling" of appearances that mask an ultimate reality. The abductee may experience the ability to return to an earlier existence, to a past life, or to some final destination or source called "Home," which is "an inexpressibly beautiful realm" beyond "space/time" as we understand it. For the abductee, the cosmos is reimbued with a sacredness that he seeks at all costs to maintain after he returns to the confines of his present existence. In this respect, the abduction is tantamount to a religious experience.[21] The work of John Mack is significant in its ability to elevate the debate about UFOs and to place it in the context of the experience of those who have undergone the so-called process of abduction. Although Mack says nothing per se about the relation between this process and the vision of Ezekiel, he is aware that the relation has been developed by others.[22]

Of corresponding importance to the investigations of John Mack are those of C. D. B. Bryan, a novelist and journalist whose recent book *Close Encounters of the Fourth Kind* presents an evenhanded account of the kinds of abduction experiences that engage Mack. The immediate occasion of Bryan's book was the Abduction Study Conference held 13–17 June 1992 at what has been designated the "high church of technology," the Massachusetts Institute of Technology.[23] Cochaired by John E. Mack and David E. Pritchard, an MIT physicist and researcher in atomic and molecular physics, the conference was able to boast not only the credibility of a distinguished setting but a roster of well-regarded researchers on the phenomenon of abduction as well.

As a means of defining the nature of the conference, Bryan draws on and extends Hynek's distinctions to arrive at what he calls "close encounters of the fourth kind." These are encounters that involve instances in which "occupants" of a spacecraft initiate actual contact between themselves and those individuals they abduct. This form of contact, Bryan observes, involves the "transportation of the individual from his or her terrestrial surroundings into the spacecraft, where the individual is communicated with and/or subjected to an examination before being returned." Although it might be argued that Bryan's "fourth kind" category is in fact an extension of Hynek's "third kind," such discriminations are not germane here. What is germane is that, after having attended the conference, Bryan arrived at a point of endorsing Jacques Vallée's view that one must move beyond the narrow and finally obsessional concern with the question of whether UFOs are actually extraterrestrial in origin and determine instead the extent and nature of the phenomenon's impact on culture and our "collective psyches." Approaching the subject from the anthropological, religious, and mythic points of view, Bryan is led immediately to Ezekiel's *visio Dei* as a foundational event for the generation of these perspectives. Citing the biblical account at some length, Bryan then enlists the visionary enactment that underlies Spielberg's *Close Encounters* to provide a means of interpreting Ezekiel's own *visio Dei*. An "admirer" of Spielberg's work, he declares, might encounter Ezekiel's vision and, as a kind of "interpretive abductee," offer the following report: "I looked up at the churning lightning-filled clouds to the north and saw a huge, bright, fiery-orange, rotating object emerge; it was the *mother ship!* Four smaller scout craft emerged from the mother ship. They were star-shaped: like a man with his arms and legs outstretched. Each craft had four windows and four wings. Their landing gear pointed straight down, then flattened out into pads. The cloud had a polished metallic appearance." With a little imagination, Bryan continues, one might interpret the vision of Ezekiel to mean that, in addition to the huge mother ship and small starships, the prophet saw "some large, cupolated discs with windows spaced above their rotating rims." When the craft moved, these discs emitted deep resonant sounds, and, when hovering, they were silent. During the time the discs and the scout craft were darting about, the immense, towering mother ship remained high above them. The bright orange light shining from the interior of the ship was so brilliant, Ezekiel informs us, that he collapsed in awe. The creatures that occupied the mother ship next communicated with Ezekiel telepathically, telling him to stand up and not to be afraid. Finally, these beings abducted the prophet.[24]

Informed by the elaborate technological underpinnings of Spielberg's vision, such a reading represents an act of technopoetics as it approaches Ezekiel's *visio Dei* through the eyes of the filmmaker. What results is a transformation of the originary substratum of the biblical event into a new phenomenon, one in which the inexpressible features of Ezekiel's vision can be discerned as concrete forms: the fire "infolding itself" and the *hashmal*-like appearance of the mysterium become a huge, bright, fiery-orange object that is really the mother ship or receptacle of four smaller starlike scout craft that are sent out on reconnaissance missions. The faces of the creatures or *hayyot* in Ezekiel's vision become windows and the extensions of the *hayyot* landing gear with flattened-out pads. Other features are interpreted in the same manner. Communicating with the prophet telepathically, alien beings exert control over him, then abduct him.

Such is the language of modern technology, such the language of ufology. In Bryan, that language becomes part of an interpretive stance that attempts to read abduction accounts from a larger cultural perspective receptive to the pull of trope, to the dynamics of figurative representation. Bryan is in no way claiming that Ezekiel's vision is an actual abductee account (although he would not rule this out). Rather, in an act of playfulness, he poses the possibility of reading Ezekiel's vision in Spielberg's terms. Bryan's purpose is to reach beyond the concretization of the present in order to embrace far-reaching theories of consciousness as analogues to the abduction experience. This he does in his discourse on "various theories" concerning the meaning of abduction experiences.[25] For our purposes, this act becomes important because it places Ezekiel's vision at the dead center of such inquiries. That, to be sure, is precisely where the vision belongs in any attempt to assess the theoretical bearings of the UFO phenomenon.

If, in the spirit of Jacques Vallée, the studies of John Mack and C. D. B. Bryan elevate the milieu of ufology to a higher level, one must not forget that the efforts of such investigators are complemented by those of the opportunists and mountebanks. They, too, have something to teach us because of the extent of their influence on modern culture. Admirably representing them is that strange character Mona Strangely; with her girlhood fantasies of being abducted and transported to heaven by alien creatures, she is the ultimate consumer of the tabloids through whom the spaceships of Ezekiel are given eloquent voice. To suggest the extent to which the interpretation of Ezekiel's vision in the spirit of Mona Strangely has assumed the status of a locus classicus in the contemporary world, one need only attend to the figure of Erich von Däniken. He, more than any other, has fostered an environment in which the likes of Mona Strangely are able to thrive.

Born in Zofingen, Switzerland, on 14 April 1935, von Däniken had a strict Catholic upbringing, one that involved an education at the international Catholic school Saint-Michel, in Fribourg, Switzerland.[26] When he was nineteen, he experienced an intense extraterrestrial vision of sorts, the details of which he refuses to discuss. As the result of a "communications breakdown" with his father and the church, he was removed from formal education and apprenticed instead to a Swiss hotelier. Pursuing his philosophical interests during that period, he studied archaeology and astronomy and their relation to religion. While he managed the hotel Rosenhügel in Davos, von Däniken wrote his first book, *Erinnerungen an die Zukunft* (Memories of the Future), which was published in 1968 and subsequently serialized by the Swiss newspaper *Weltwoche*.[27] The von Däniken movement spread from Switzerland to Germany, and, by December 1968, *Erinnerungen an die Zukunft* became one of the most popular books in West Germany. Translated into English as *Chariots of the Gods?* the book became so popular that it was even serialized in the *National Enquirer*. Meanwhile, film and television productions were undertaken in Germany, Canada, and the United States. In 1973, the film *In Search of Ancient Astronauts* was aired on NBC-TV, and a movie version premiered in Buffalo, New York. Follow-up television specials and movies such as *The Outer Space Connection* reinforced the public exposure to von Däniken's ideas.[28] Despite this notoriety, von Däniken found himself seriously in debt. In 1968, he was arrested for fraud and two years later convicted of embezzlement, fraud, and forgery. As a result of his conviction, he received a prison sentence of three and a half years (but did not serve the full sentence).[29] Such is the context in which the theories of Erich von Däniken are to be considered.

What makes von Däniken's *Chariots of the Gods?*—along with his other books—so important for our purposes is its espousal of the so-called ancient astronaut hypothesis.[30] Specifically, this hypothesis maintains that what in earlier times was looked on as a visitation from God, in one form or another, was in fact an interplanetary visitation from ancient astronauts who were mistakenly worshiped as God (or as gods). These astronauts had no choice but patiently to accept the adoration that was bestowed on them—an adoration, von

Däniken says, for which our own astronauts must be prepared when they visit unknown planets. Implicit in von Däniken's hypothesis is a form of euhemerism, the idea that the deities of Hellenistic mythology were originally men and women deified by those who came to worship them in later periods. Euhemerists practice a method of mythological interpretation that regards myths as accounts of real incidents in human history.[31] In his analysis of the vision of Ezekiel, then, von Däniken reduces the visionary to a kind of euhemeristic literalism. Doing so, he paradoxically engages in a form of mythologizing, one that replaces the ineffable visionary substratum of the biblical account with a myth of domination: God as an ancient astronaut. Accordingly, in *The Chariots of the Gods?* von Däniken views Ezekiel's *visio Dei* as an interplanetary "multipurpose vehicle," a technological wonder equipped with its own means of locomotion. Manned by remarkable astronauts who embody all the divine power and authority that any concept of deity is able to bestow, the chariot of these astronauts, von Däniken contends, is a vehicle exhibiting an advanced technology that is only in recent times being accorded the scientific scrutiny it deserves.[32] If we examine the various aspects of this vehicle, including its entry into the atmosphere, its movable parts, the means by which it is propelled, and the like, we shall immediately recognize it for what it is: a spacecraft of momentous proportions piloted by ancient astronauts who have since been worshiped as divine beings.

The inaugural vision of Ezekiel is everywhere present in the works of those who endorse the ancient astronaut hypothesis.[33] So widespread is the notion that it merits separate entries in both *The Encyclopedia of UFOs* and *The UFO Encyclopedia*. Books ranging from Morris K. Jessup's *UFO and the Bible* (1965) to W. Raymond Drake's *Gods and Spacemen in the Ancient East* (1968) and *Gods and Spacemen of the Ancient Past* (1974) suggest the currency of the idea, whereas Gerhard R. Steinhauser's *Jesus Christ: Heir to the Astronauts* (1976) indicates the direction it is bound to take.[34] The concept is of fundamental importance to the writings of George Adamski, one of the most famous of contactees. Originally a grill cook in a hamburger stand on the road to Mount Palomar, Adamski gained great notoriety after he described a face-to-face meeting with an astronaut from Venus on 20 November 1952.[35] Canonized by his disciples through the George Adamski Foundation, based in Vista, California, Adamski invokes Ezekiel's vision repeatedly in his books as evidence that such encounters are corroborated by the Bible itself.[36] Indeed, because of his encounters, Adamski has been looked on as a twentieth-century version of Ezekiel.[37] In fact, his acolytes include Dr. and Mrs. Thomas Armstrong, mentioned earlier in connection with Mrs. Keech, the contactee. The Armstrongs sought out and visited Adamski in order to benefit from his own profound experience of visitors from the other world.[38]

Among those who point to the vision of Ezekiel as evidence of ancient encounters with UFOS, however, none is more important than Josef F. Blumrich, whose *Da tat sich der Himmel auf* was published in 1973. Reissued in English as *The Spaceships of Ezekiel* in 1974, this book suggests how extensively Ezekiel's inaugural vision has been appropriated by those who espouse the ancient astronaut hypothesis.[39] Blumrich is interesting for a number of reasons. First, there is the matter of his background. Far from being a grill cook in a hamburger stand on the road to Mount Palomar, Blumrich (like Jacques Vallée) is a genuine scientist. Blumrich's particular credentials are in the area of aerospace engineering. In fact, Blumrich held a supervisory position as an engineer in NASA's George C. Marshall Space Flight Center, Huntsville, Alabama.[40] Second, there is what might be called the matter of his conversion. On reading von Däniken's statements concerning Ezekiel in *The Chariot of the Gods?* Blumrich set out to write a confutation, but a direct encounter with the biblical word caused him to see wisdom in von Däniken's conclusions.

Because of Blumrich's importance to the concept of *technē* in the interpretation of Ezekiel's vision, both his background and his work warrant serious attention. Blumrich has degrees in aeronautical and mechanical engineering. Born in Styr, Austria, in 1917, he trained at the Ingenieurschule in Weimar, Germany. During his years in Germany, he gained engineering and aeronautical expertise in various positions. Having worked on design and strength analysis of aircraft at Gothaer Waggonfabrick in Gotha, Germany, he became deputy chief of the Department of Hydraulic Structures at the United Austrian Iron and Steel Works in Lintz, Austria, from 1951 until 1959. Between 1944 and 1945, he served in the German army and was taken as a prisoner of war in 1945. After emigrating to the United States in 1959, he became a naturalized citizen in 1965. Joining the American space program in 1959, he was in charge of the Structural Engineering Branch of the Propulsion and Vehicle Engineering Laboratory, Marshall Space Flight Center at NASA. With publications on buckling loads, design of plywood shells, electrical spot welding of aluminum, and stage designs, he worked on the design research of the Saturn V rocket as well as various satellites, Skylab, and the shuttle. He became chief of NASA's Advanced Structural Development Branch before he retired in 1974. A member of the American Institute of Aeronautics and Astronautics and the American Association for the Advancement of Science, Blumrich was awarded NASA's Exceptional Service Medal in 1974. In short, both his standing and his qualifications as an aeronautical engineer are above reproach.[41]

Blumrich has paramount importance as an interpreter of Ezekiel's vision. It is here that his "other" life emerges and causes him to take his place among that host of visionary exegetes who, throughout the centuries, have imposed their own reading on Ezekiel's prophecy. As one might expect, that reading is of a distinctly "scientific" sort, an idea suggested by the use of the term *Technik* in the subtitle of the German edition. In Heideggerian terms, Blumrich's interpretation presents its own unique means by which *technē* causes that which is concealed to come into unconcealment or disclosure, to reveal itself as *alētheia*. As psychological event, this disclosure, one might argue, assumes the form of that which is customarily designated nonnormative, if not deranged. Its aesthetic is that of madness, its technology that of the bizarre. Because its *alētheia* springs from the reading of one whose credentials are of the highest sort, however, the particular "take" that Blumrich has on the vision assumes a significance, indeed, an urgency of the first order. Such significance and urgency only a genuine scientist can bestow.

In the assessment of Blumrich's other life, we must acknowledge that, as far as the scientist himself might conceive it, the concept of alterity would be considered nothing more than a fiction. In response to the notion of alterity, Blumrich would maintain that there is no "other life." For him, the occupations of aeronautical engineer and of biblically inspired ufologist are one and the same, a fact that renders him so fascinating a study. A belief in the concurrence of these occupations is discernible in his biographical entry in *Contemporary Authors*. The list of credentials there is cited as follows: "*Member:* American Institute of Aeronautics, American Association for the Advancement of Science, Ancient Astronaut Society."[42] It is the unself-conscious and matter-of-fact listing of all three affiliations *ad seriatim* that proves to be so revealing. The listing is so understated that anyone unfamiliar with the third affiliation might almost be inclined to pass it by unnoticed. In the context of our study, however, the reference to the Ancient Astronaut Society becomes a climactic moment in the sudden emergence of the other life that distinguishes our biblically inspired ufologist.

It is here that one is prompted to ask how the scientist-cum-ufologist assumed such an identity. What accounts for the emergence of the other life? Blumrich himself provides his own version of the answer in his book on Ezekiel's vision. The account is tantamount to a kind of conversion, a journey, as it were, on the technological road to Damascus. According to Blumrich, one day he received a call from his son Christoph, who had just read von Däniken's *Chariots of the Gods?* and recommended the book to his father. As a career engineer working on large rockets and spacecraft, Blumrich considered such books nothing more than a pleasant diversion. So, he says, "when the 'Däniken' arrived, I read, smiled, grinned, and laughed—until I found the passage in which von Däniken writes about the prophet Ezekiel." Encountering that passage, he discovered "technical statements and claims" directly in the fields of his own scientific expertise. As a result of encountering such statements, he resolved himself to become an exegete of the biblical text in order not to support but to confute von Däniken's theories. Little did he anticipate the outcome. Rather than confuting these theories, mirabile dictu, he became an avid convert to them. He soon abandoned his sense of superiority and scorn and underwent what he terms a "wonderful experience, unusual in every respect," that of discovering in the text of the vision the precise technical details necessary to support a scientific reading of the sort that he, as an aeronautical engineer, was accustomed to dealing with in his work at NASA.[43]

As a result of his conversion, Blumrich published *Da tat sich der Himmel auf.* What makes the book so fascinating is that, as a work of biblical exegesis, it adopts the kind of discourse one discovers in such technical "in-the-field" periodicals as *Astronautics and Aeronautics,* the official organ of the American Institute of Astronautics and Aeronautics. In fact, if one consults Blumrich's own publications in this journal, the correspondence between Blumrich the biblically inspired ufologist and Blumrich the NASA engineer is immediately apparent. The discourse of one venue is precisely what determines the discourse of the other. Both reflect a faith in (and indeed a celebration of) technology. Blumrich, for example, acknowledges his emphasis on the importance of technology: "I have," he says, "repeatedly mentioned technology" for, where "there is no technology, there is no design and no structure." Accordingly, he advocates the wisdom of working "closely together with the shop and materials people" to achieve a common goal.[44] Accompanying the article are photographs and illustrations that find their counterpart in those that grace the book on Ezekiel as well.

As might be expected, the stated purpose of this book is to provide "engineering proof [*ingenieurmässigen Beweis*] of the technical soundness and reality [*die technische Korrektheit und Realität*] of the spaceships described by Ezekiel." For Blumrich, the "key" to Ezekiel lies "in a very careful analysis of his description of the components of the spacecraft and of their function, carried out in light of today's knowledge of spacecraft and rocket technology." Blumrich observes that his technical interpretation of Ezekiel's writings has been made possible only as the result of recent findings. Despite all the progress of earlier decades, it is only since 1964 that rocket technology finally reached the point of discovering the kind of information that Blumrich could use in his own research on Ezekiel. To this end, Blumrich cites the work of Roger A. Anderson, a leading engineer at NASA's Langley Research Center. At issue is Anderson's pioneering research on "structures technology," which discloses the shape of a spacecraft design that Anderson had developed for "entry into planetary atmospheres." Appearing once again in the journal *Astronautics and Aeronautics*, the results of this discovery serve as the basis of Blumrich's own analysis of the central body of the spacecraft that Ezekiel beheld. Without the knowledge offered by Anderson's work, "a technical interpretation of the text," Blumrich avers, "would not be feasible even today."[45]

In venturing that interpretation, Blumrich distinguishes between his approach and that of the centuries of exegetes who preceded him. Whereas these exegetes were both learned and devout, they had no knowledge of space-age flying machines and rockets. By necessity, the interpretations of Ezekiel's enigmatic statements by the earlier exegetes "had to be sought in the only possible direction, which was that of religion and particularly mysticism." As important and well meaning as such interpretations are, they lack the scientific knowledge that only an aeronautical engineer is able to offer. An engineer, Blumrich avers, is concerned "with the assessment of structures or structural forms." As one who develops advanced structures and implements the conditions that determine their shape, the engineer (and in particular the design engineer) is "the most competent person to deduce the purpose and the use of a structure from its outward appearance."[46] It is for this reason that, in dealing with Ezekiel, Blumrich speaks in the language of engineers. Only in that way is it possible to understand what Ezekiel's vision is really about. In the light of such assertions, one need hardly emphasize the point that what Blumrich offers in his reading of Ezekiel's text is the new religion, that of science, with its particular system of beliefs and practices, indeed, its own "god," one that demands a form of worship all its own.

In his scientific reading of Ezekiel's text, Blumrich proceeds, however, in a manner that is strikingly like that of the earlier exegetes. Citing the text of the vision in full, Blumrich moves verse by verse to disclose the mysteries of the text. In this regard, his methodology has the ring of familiarity, especially for those acquainted with the traditions of biblical exegesis extending back to the early Middle Ages and up through the Renaissance. The difference, of course, lies in the results of Blumrich's *allēgoresis*. Rather than discovering typologies (Old Testament/New Testament figurations) of one sort or another, Blumrich discerns interplanetary space vehicles.

In Blumrich's reading, Ezekiel's vision is transformed into a spacecraft consisting of three major systems: the central main body, the four helicopters that

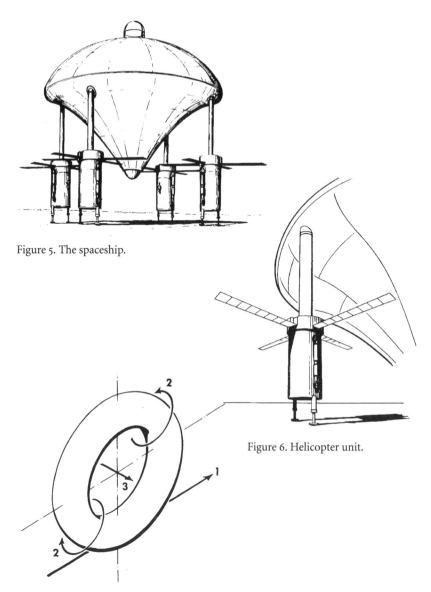

Figure 5. The spaceship.

Figure 6. Helicopter unit.

Figure 7. Wheel movement in any direction.
All figures reproduced from Josef F. Blumrich, *The Spaceships of Ezekiel*, pp. 14, 15, 37, by permission of Bantam Books.

support the main body, and the capsule for the crew, which is located on the upper side of the main body (figs. 5–6). So sophisticated is this spacecraft, suggests Blumrich, that NASA actually holds patents for a vehicle corresponding to the visionary objects Ezekiel beheld.[47] On the basis of his analysis, Blumrich claims that he even made a discovery concerning the rotation of wheels in opposite directions at the same time, a finding that has prompted him to file for a patent (fig. 7).[48] In fact, much of Blumrich's discourse is composed of scientific speculation concerning the nature of the wheels that Ezekiel beholds (wheels within wheels, wheels that move in any direction without the need for turning, wheels with eyes, and the like). An entire portion of an appendix to his book is devoted to a scientific, complex design investigation that has all the marks of an engineering manual. Accompanying this analysis are elaborate illustrations, formulas, equations, and graphs of one kind or another. In short, the visionary matter of the biblical original is transformed into an engineering manual.

Responding to the "construction" of the wheels as being "something like a wheel within a wheel" and to the movement as "in any of the four directions without veering as they moved" (Ezek. 1:16–17), Blumrich devises a theory of disks and outer rim surfaces that constitute the wheel work. By means of this theory, he explains how the system of paradoxical rotation operates. Within that system, he says, "multiple simultaneous movements can be observed on the wheel." An observer would discern, apart from the familiar rolling motion, the rotation of the tire segments, on the one hand, and the turning of the disks, on the other. As a result of the scientific explanation of the visionary that Blumrich provides, it is not unreasonable, he concludes, to envision a wheel within a wheel for such a device and to designate it *galgal* or "wheel work" (Ezek. 10:13).[49] If one has access to Blumrich's background and practical engineering expertise, that which apparently confounds in the text of the vision becomes perfectly clear when understood through the auspices of scientific explanation. It is from this perspective that Blumrich approaches the vision as a whole, as he proceeds verse by verse and feature by feature to disclose its mysteries.

Analyzing the opening verses of the prophecy, for example, Blumrich notes how the roar of the rocket engine exploding in the stillness prompts Ezekiel to look up in wonder. The engine's flames shoot from the center of a white cloud; from them emanate a blinding light and a mighty roar. This is the vision of God for the space age. In the manifestation of this vision, the sky itself seems to burst open, and the cloud reveals four elongated shapes above which something undefined moves. At the lower end of the shapes, it is possible to perceive straight legs and round feet (figs. 8–10). When the clouds are dispelled, these

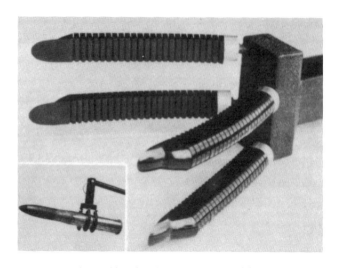

Figure 8. Mechanical hand and Figure 9. Legs and feet.

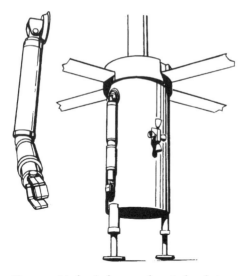

Figure 10. Mechanical arm and control rockets.

All figures reproduced from Josef F. Blumrich, *The Spaceships of Ezekiel,*
pp. 34 and 36, by permission of Bantam Books.

shapes emerge more clearly as possessing multiple wings. Below the wings, arms hang down alongside the body. Above them, Ezekiel discerns what appear to be faces (fig. 11). What he actually sees, however, is the spacecraft itself after its separation from the mother ship, at an altitude of about 220 nautical miles. As it flies through the atmosphere, the spacecraft reduces its speed because of aerodynamic drag, until eventually a brief firing of the rocket engine at low altitudes allows it to diminish its momentum enough to make use of its helicopters for the remainder of the descent. This final phase of the flight, which begins with the brief firing of the rocket engine, is what Ezekiel witnesses and describes at the outset of his remarkable account.

Like a medieval church father discoursing on the particulars of the text for the sake of absolute clarification, Blumrich continues in this manner of one-to-one correspondence. In the hands of the scientist, the vision is not mysterious or occlusive at all. Rather, it reveals itself as absolutely clear to those familiar with the technological components that constitute its various parts. The vision of ḥashmal is a case in point. Adopting the Lutheran Bible as the basis of his discourse, Blumrich interprets ḥashmal as *Glanzerz* (gleaming bronze), a rendering consistent with the Revised Standard Version.[50] In this respect, Ezek. 1:4 (the verse in which the first of the three appearances of ḥashmal occurs) represents the occasion by which Blumrich describes the process of igniting the rocket engine. Before that occurs, the entire system of suction lines, pumps, and the like must be cooled down to the low temperature of liquid hydrogen to achieve appropriate working conditions. "This is done," Blumrich observes, "by forcing liquid hydrogen under pressure from the tank through the system." When hydrogen is discharged into the atmosphere as a cold gas, the liquid residue freezes into ice crystals and resembles a cloud. In verse 4, this is the "great cloud, with brightness round about it." The main body of the spacecraft falling into this cloud forces it to flow radially outward. Blumrich compares the event to the experience of having witnessed the start of a Saturn rocket. This Blumrich describes as "the optical effect of the plume." Nobody will ever forget, he comments, the "fire flashing forth," the "brightness round about it," a shine "as it were gleaming bronze." He concludes with the following observation: "It is only in *this* Verse 4 . . . that Ezekiel describes the operation of the rocket engine."[51] Blumrich's discovery is that of the scientist who has been able to provide a workable solution to that which had previously been considered the most awesome secret of the "divine world," in this case, the secret of the *visio Dei* itself, centered in the ḥashmal.

Consistent with this discovery, Blumrich is able as well to disclose the meanings underlying the mystery of the enthroned figure (Ezek. 1:26–28) that culmi-

Figure 11. Gemini capsule.
Reproduced from Josef F. Blumrich, *The Spaceships of Ezekiel*,
p. 59, by permission of Bantam Books.

nates the vision. The nature of this figure can be explained entirely according to the schematics of Blumrich's technological discourse:

> Now Ezekiel looks above the rim of the arch and perceives an unreal, almost immaterial mixture of light and color in the midst of which he sees a man sitting on a throne. . . . Light and coloration of a transparent command capsule are optically much more impressive than its simple geometrical shape and therefore furnish a very plausible means for a description. The man whom he sees is the commander of the spaceship [*der Kommandant des Raumschiffes*], and his seat has an unmistakable resemblance to a throne. Ezekiel is overcome by the enormity of what he sees and sinks to the ground as a sign of submission.[52]

As articulated, the account makes a point of separating appearance from reality. That which the prophet is led to believe he perceives is really something else. The brilliant light emanating from the enthroned figure has its basis in the luminescence emitted by the command capsule, which, despite its dazzling appearance, is nothing more than a "simple geometrical shape." The awesomeness of the divine is reduced to the clarity of a scientific fact. Mystery is resolved in plausibility. Such is the nature of Blumrich's *alētheia*. So it is with the enthroned figure, whose godhead has now assumed the form of an interplanetary space traveler. God has been transformed into a spaceship commander.

Later in his analysis, Blumrich returns to this passage to observe that Ezekiel's *visio Dei* is such that the prophet "cannot reconcile the massive, real apparition of the spacecraft with a manifestation of the Lord, which he would not expect to be so utterly material." But, as Blumrich, the new prophet, would argue, the true scientist *is* able to reconcile appearance and reality, spirituality and materiality, for the God of the scientists is material: it is all that pertains to the spacecraft itself. If, like Blumrich, one is in possession of the truth, of *alētheia*, there is no need for metaphor. Trope is replaced by thing. Ezekiel "does not know what he is really seeing," Blumrich comments; "yet whatever it may be, it is colossal, mighty, and overpowering—it seems to be *comparable* to the glory of the Lord." In such a situation, the prophet is put to the expense of engaging in poetic discourse: "He does not identify, he compares [*er identifiziert nich, sondern er vergleicht*]."[53] As true scientist, Blumrich removes the necessity for comparison (*der Vergleich*). There is no need for poetry once we understand the reality behind the appearance, and that reality is the spacecraft itself. If Blumrich as a scientist eschews the prophetic or visionary, his writings nonetheless assume a prophetic or visionary quality: sacrificing all in the name of "science," they extol the machine. That is their truth, their *alētheia*. Collapsing vision into the world of thing, Blumrich becomes the prophet of the machine, the new *merkabah* mystic. Through him, *poiēsis* assumes renewed force as *technē*. How characteristic of the modern sensibility that the machine conceived by this *poiēsis* is nothing less than a UFO!

In response to the kind of perspective that Blumrich embraces, there are those among the scientific community determined to call into question the technologizing of the visionary experience as UFOs.[54] In the very process of doing so, however, they "invent" a new technology to accommodate their understanding of that experience. One such scientist is Donald H. Menzel, Paine Professor of Practical Astronomy and professor of astrophysics, emeritus, at Harvard University. His theories appear most fully articulated in his *Flying Saucers*. There he counters the idea of visionary experiences as UFOs by attributing UFO sightings to a meteorological phenomenon known as a *parhelion*. This phenomenon occurs when ice crystals in the sky create the appearance of a solar halo complete with mock suns (or sun dogs) and corresponding "glories." It is, in short, a "mirage" or "meteorological apparition" produced by the configuration of crystalline substances in the upper atmosphere. "The character of the display can range all the way from a simple unadorned halo or a single mock sun" to what Menzel calls a "super-deluxe model," which is much more elaborate.[55]

Invoking such a phenomenon to explain away UFO sightings is one thing, but Menzel does not stop there. In his zeal to do away with UFO speculations of all sorts, he applies his parhelion theory to the so-called flying saucers of the Bible, in particular, the vision of Ezekiel. No longer a flying saucer, that vision is now a meteorological apparition, which Menzel describes with as much care and technical precision as Blumrich devoted to his account of the spacecraft. No aspect of the inaugural vision is disregarded. The wheels, the living creatures, the eyes, the enthroned figure (Ezek. 1:26–28), and the sounds emitted by the vision: all become part of that grand parhelion that Menzel sees as underlying the biblical text. For purposes of comparison, Menzel provides detailed illustrations of a parhelion, on the one hand, and Ezekiel's vision, on the other. His discourse ends, moreover, with various extrabiblical references, including one to Milton's own Chariot of Paternal Deitie. An appendix titled "Theory of Mirages" constitutes a final flourish to the book as a whole. In this appendix, we appropriately find detailed graphs and equations comparable in complexity to those offered by Blumrich. Menzel concludes from all this that, whatever religious significance we wish to assign the inaugural vision, one fact is irrefutable: "Ezekiel proves to take top rank as an observing scientist and recorder of important meteorological phenomena."[56]

In later discussions of his ideas, Menzel not only reiterates his argument but authenticates its originality by maintaining that he "corresponded with the Vatican expert on Ezekiel, who replied that the idea was new to him" but that he did not object to Menzel's interpretation as long as the astrophysicist "did not imply that Ezekiel was not divinely inspired." In response to that proviso, one wonders what the phrase *divinely inspired* implied to Menzel since such "religious" concerns presumably lie beyond the purview of the scientist. Whatever the phrase implied, Menzel does not hesitate to place himself in the lineage of Ezekiel. Like Ezekiel, he says, he has himself witnessed parhelia during his lifetime, once in Colorado when he was a boy, and again in Alaska when he was an adult. They were, he allows, spectacular events, tantamount, one suspects, to those that accompany religious conversions. Although Menzel does not explicitly draw this conclusion, he is prompted to observe how "uninformed credulous people the world over and throughout history have regarded them [parhelia] with superstitious awe, as portents of some dreadful event." He discovered, he says, an account of a woman who had conceived during such an apparition and who nine months afterward gave birth to quadruplets. In the light of such clear evidence, he concludes ironically, "Who could doubt the malevolent influence of such a vision?"[57] As much as the astrophysicist might attempt to disassociate himself from the woman who gave birth to quadruplets, he shares something of the fervor of those "the world over and throughout history" who have regarded parhelia with awe and perhaps even dread. Doing so, Menzel becomes a true believer: in him the impulse to technologize the ineffable flourishes anew. Transforming the visionary experience into meteorological phenomena, he gives rise to his own creed, his own belief in the powers of *technē* to articulate that which transcends articulation.

As a final gesture of confidence in all that the worship of *technē* implies, any study of the impact of Ezekiel's vision on the phenomenon of the UFO must take into account the presence of that vision in the emerging world of cyberspace. There, the transformations undergone by the avatars of the great god *technē* assume renewed impetus.[58] As a manifestation of the culture of information technology, cyberspace is germane to our purposes, first because it is a phenomenon that exhibits compelling affinities with the visionary experience itself, and second because what might be called the *new romance* of cyberspace is grounded in a subculture that immediately suggests ties with the likes of Melchior Bauer and Josef Blumrich.

The subculture in question has its source in the dystopic realm of William Gibson's *Neuromancer*, published auspiciously in 1984. A work that has assumed canonical status in the discourse of cyberspace, this novel portrays a brutal technovisionary world of rebel hackers, corporate hegemony, and neural implants. Responding to the visionary implications of such a milieu, Gibson coined the term *cyberspace* as a "consensual hallucination" experienced throughout the world by a multitude of operators or "cowboys" who negotiate the Internet daily. As the phenomenon through which these cowboys operate, cyberspace is "a graphic representation of data abstracted from the banks of every computer in the human system." Inexpressible in its complexity, cyberspace is represented "by lines of light ranged in the nonspace of the mind, clusters and constellations of data" that recede into the darkness like city lights.[59] Within this brave new world (the "new romance" of the "neu/romancer"), a new reality (or "hyperreality") is forged as the supreme expression of what has come to be known as "cyberpunk."[60] In cyberpunk, we encounter the fusion of high technology and the modern pop underground. Two themes are characteristic: "the technological invasion of body and mind" (in the form of prosthetic limbs, implant surgery, genetic alteration, and artificial intelligence) and "the dislocation of time and space through the action of electronic global networks and the machinations of multinational corporations."[61] Along with a host of other fictions of the same kind, Gibson's *Neuromancer* is a prime example of the wedding of these themes.

For our purposes, the importance of this novel lies in the visionary dimensions it ascribes to cyberspace as a "consensual hallucination." Viewing the hallucinogenic aspects of cyberspace in decidedly sexual terms, Michael Heim maintains that what Gibson calls the "matrix," as the source of all cyberspatial communication, embodies the hallucinogenic wedding of the psychic and technological worlds. Heim compares that wedding to a kind of visionary, *unio mystica* discernible in such figures as John of the Cross and Teresa of Avila.[62] David Tomas views the spatial implications of the phenomenon known as cyberspace as a rite of passage from one realm to the next, indeed, from one spatial construct to another. Distinguishing between Euclidean space and cyberspace, Tomas sees the former as the space of Western "work-oriented cultures." The space of "measure and transport," it is also the "space of work," of the mason, the surveyor, and the architect. As such, it is the space of Western geometry: the geometry of perception, roads, buildings, and machines. Cyberspace or electronic space, on the other hand, is the "metasocial postindustrial work space." It is the space that is not a space, a "nonplace," a space in which "there are no shadows" for it is a medium void of "bodies" to cast them. It is founded on "a new hardwired communications interface that provides a direct and total sensorial *access* to a parallel world of potential work spaces." According to Tomas, this interface mediates "between the sensorial world of the organically human and a parallel virtual world of pure digitalized information."[63] Such a realm is both profane ("a metropolis of data") and sacred ("a cybernetic godhead"). In the rite of passage from one space to the next, the operator (cowboy, cybernaut) seeks through his special skills (known as "hotdogging") to gain access to the visionary realm of cyberspace.

Conceived either as a mystical union or as a rite of passage, the technovisionary experience embodied in the medium of cyberspace suggests a transport to a domain that is at once a "metropolis of data" and a "cybernetic godhead." The metaphor of cyberspatial transport is to the point. It is an idea that accords with the very notion of cyberspace as a "vehicle" of communication. Particularly from the perspective of the communication and control sciences, such a notion is apposite to the designation of *cybernetics* as a term that suggests the idea of a "steersman" or "pilot" (*kybernētes*) of a craft. It is used as such repeatedly by Philo, for example, in his references to the Logos as the "charioteer of the Powers," wielding "the reins of the Universe." As one "borne on the universe like a charioteer or pilot," the Logos "steers the common bark of the world, in which all things sail," and "guides that winged chariot, the whole heaven; exerting an absolute sovereignty which knows no authority but its own."[64]

In his pioneering work on control and communication in machine technology, Norbert Wiener appropriately coined the term *cybernetics* to designate the nature of his own investigations. Responding to the tumult of the war years that mark the publication of his 1948 *Cybernetics,* Wiener applies the notion of cybernetics to air transport and the deployment of missile technology in the conduct of flight combat and war games.[65] For Wiener, the advent of the "computing machine" in the realization of the goals of cybernetics was tantamount to what he calls the Second Industrial Revolution.[66] Here is where the "smart bomb," along with all the implications of mechanized warfare, was born. As much as he admired the possibilities engendered by the field of cybernetics in all areas of life, he was also sensitive to the crises that such a technology could precipitate. In *God and Golem, Inc.,* his reflections on the relation between cybernetics and religion are germane to the understanding of how the communication and control sciences foster an atmosphere in which the "machine" is liable to assume the aura of the supernatural in modern society. As Wiener observes, cybernetic mechanisms equipped to facilitate the transmission of information through the conversion of incoming messages to outgoing messages tend to be looked on as devices imbued with a knowledge and power tantamount to the divine.[67] Wiener warns that, if allowed to get out of hand, the powers bestowed on such machines in the age of "modern automatization" threaten to usurp those they are meant to serve.

Placing the dilemma in the context of game theory, he alludes to the "games" the creator (God) and his creation (Satan) play both in the biblical Book of Job and in Milton's *Paradise Lost*. Although in both works God appears to have the upper hand by virtue of his putative omnipotence and omniscience, Wiener argues that the game (or conflict) between the creator and his creation must be seen as more than merely one in which the "uprising of the rebel angels" is "foredoomed to failure." There must always be the possibility that the "adversary" might just win. In Milton's epic, this means that Satan must be conceived as possibly overcoming his creator in the War in Heaven, destroying the Chariot of Paternal Deitie, and assuming his place on the celestial throne. By implication, then, Satan's engine of warfare (*his* chariot) would defeat God's. Be careful, Wiener warns: the newly empowered machine might defeat its creator and prevail after all.[68]

In the world that Wiener envisions, the empowered machine is conceived as a golem. Recalling the figure of Rabbi Löwe of Prague, who claimed that his incantations inspired his creation with the breath of life to perform remarkable deeds, Wiener observes that "even now, if an inventor could prove to a computing-machine company that his magic could be of service to them, he could cast black spells from now till doomsday, without the least personal risk."[69] For our purposes, the metaphor of the golem is to the point. A manifestation of the mystical traditions of Judaism, it represents a staple of occult lore, especially that associated with theurgy and ecstatic spiritual experiences. Arising as a result of the incantatory pronouncement of permutations and combinations of the inscriptions in Jewish mystical documents, the golem is brought to life as a being with remarkable powers that are both daunting and dangerous. In its earliest form, the golem is associated with speculation on the *merkabah*. Its original milieu is that of chariot discourse (*ma'aseh merkabah*), especially in connection with the theurgic powers bestowed on the rider in the chariot in his journey to the *hekhalot* or celestial halls.[70]

In the modern parlance of cyberspeak, the golem assumes the form of the *cyborg*, the acronym for cybernetic organism, defined by Donna Haraway as "the awful apocalyptic telos of the West's escalating dominations of abstract individuation, an ultimate self untied at last from all dependency: a man in space."[71] Equipped with awesome powers, this "ultimate self," this machine self, is a figure dislodged from its ties with all that is human, all that binds it to the limitations of biological selfhood. In his analysis of the cyberspatial implications of this idea, William A. Covino moves from the apocalyptic bearing of Haraway's discourse to a tone that is at once domestic and quasi-comic but not for that any the less germane. Exploring the golem/cyborg relation, Covino views himself as such a "cybernetic organism" who embarks on a journey through the realms of cyberspace: "I enter the Net by typing the precise code required by my transmission software, and I materialize the parts of this immense cyberspatial body with additional codes. Thus, as I type 'ukanaix.cc.ukans.edu' for access to the hypertextual Internet domain known as the 'World Wide Web,' I mimic the invocatory power of the medieval adept, bringing to light the virtual world."[72] Logging on to the Internet, and processing our secret codes to access the Web, we all become cyborgs as we "interface" with the domain of limitless possibilities in the realm of cyberspace. In the process, we are imbued with theurgic powers of transmission, transcendence, and transport into the virtual world of hypertext. Such is the bearing that the cybernetics of control and communication assumes in the infinite domain of cyberspace.

If the cyberspatial phenomenon is to be viewed in cybernetic terms as a vehicle of access to the other world, this vehicle is one that has assumed a multitude of forms in modern culture. The most widely disseminated of these lies in the area of popular culture, particularly as manifested in New Age mysticism. According to Philip Hayward, "Euphoria over the revolutionary nature of cyberspace technology has often been expressed in decidedly 'mystical' tones echoing the 'spiritual' concerns of 60s hippie psychedelia and converging with aspects of those beliefs and practices which have come to be known as 'New Age.'"[73] Covino concurs: "A mystical enthusiasm for mind expansion in cyberspace—reminiscent of a more psychedelic age—is very much with us."[74] Reinforcing the conflation of the cyberspatial and the New Age mystical is the figure of Timothy Leary, for whom the technology of cyberspace surpasses in its visionary possibilities the power of LSD.[75] Reflecting this outlook, his paean to cyberpunk makes a point of celebrating the significance of heresy and madness, of revolutions against authority, and of the assertion of individuality. Leary accordingly hails the "cyber breed" of mavericks, inventors, freelancers, technologists, nonconformists, oddballs, troublemakers, kooks, visionaries, and iconoclasts as those who will lead us from the awful constraints of contemporary society into the freedom of the New Age.[76] In the terms that we have been exploring, these cyborgian cyberpunkers are the riders in the chariot of the Internet. Intrepid travelers through the infinite realm of cyberspace, they possess the powers of a new world order, one in keeping with their standing as harbingers of the New Age.[77]

It is within this context that the *merkabah* as New Age phenomenon is reconfigured. In that reconfiguration, cyberspace becomes the means by which the avatars of *technē* appropriate the Internet in their quest to bestow renewed significance on the work of the chariot both as a sublime vehicle of transcendence and as a UFO. Among those for whom the *merkabah* assumes such a significance, the figure of Drunvalo Melchizedek is particularly apposite.[78] Emerging from the infinite realm of cyberspace, this figure finds a home on several Web sites in which his teachings are communicated to all those for whom the New Age is a source of sustenance. All one need do is access such sites as those that deal with spirituality, *merkabah* meditation, alien cultures, UFO phenomena, "alternative technologies," or Drunvalo Melchizedek, and, like the Covino cyborg with his incantations, the guru himself will appear to greet his guest on the other side. One discovers from these close encounters an elaborate mythology of origins, contacts, communications, and strange life forms that result in consciousness expanding and a renewed sense of self.

The figure of Drunvalo Melchizedek, according to one Web site, originally existed for some ten billion earth years in the Melchizedek Order of unity consciousness in the thirteenth-dimensional realm until he was dispatched here to the realm of polarity consciousness. In this realm, he is what is known as a "walk-in," one whose body is occupied (or "channeled") by another until its true occupant is ready to use it. Drunvalo entered his present body in 1972. Before that, he manifested himself on earth in a variety of other forms, including that of a female born in 1850 into the Taos tribe of New Mexico. Having occupied his female body for forty years, he returned to the sixth-dimensional level in 1890, where he remained until 1972, when he "walked into" the body he presently occupies.[79] In an interview that appears at another Web site, Drunvalo maintains that he originally began life on this planet as a Catholic but, to the dismay of his parents, later converted to Judaism, after which he was attracted to Sufism, Hinduism, Buddhism, and Taoism. All this was before he became an American Indian. In his past lives, he was also very much given to visions, including that of two very tall angels, one of which emitted a chili-green color and the other an ultraviolet light.[80] In short, this is a New Age, multifaceted, multilived visionary who has occupied many bodies, both sexes, and has existed in a variety of dimensions.

Drunvalo's presence here is in response to the conflict between the forces of the Great White Brotherhood and those of the Great Dark Brotherhood, two bodies of consciousness opposed to each other in every possible way. Whereas the Great White Brotherhood does everything in its power to advance our evolution, the Great Dark Brotherhood does everything in its power to instill fear and delay evolution. Although the forces of these two brotherhoods balance each other out and therefore work ultimately toward the common good, they are viewed from the perspective of our polarity consciousness as a reflection of the battle of good against evil. Emanating from the forces of the Great White Brotherhood, Drunvalo has been sent here to help resolve the conflict.[81] At the heart of his "mythology," then, resides a kind of gnostic dualism, along with an implicit glance at racial difference, that is resolved at a higher level of consciousness. In his portrayal of this dualism/racialism, Drunvalo fashions a fable titled "A Love Story" that conceives of the war between white and dark in mythological terms. Within that mythology, the forces of Michael and the forces of Lucifer encounter each other on the field of battle to determine which will prevail in the consciousness of the individual.[82]

Underlying Drunvalo's teachings is his version of the *merkabah*.[83] Commissioned by higher powers to present his teachings in the Flower of Life Workshops, Drunvalo offers instruction in *merkabah* meditation techniques through which participants are able to ascend to superior levels of consciousness. As a vehicle of ascension, the *merkabah* transports participants from one dimension to the next during the course of training in the principles of meditation.[84] Recalling early meditative practices among the various schools of *merkabah* mysticism in medieval Judaism, the process of meditation that Drunvalo teaches correspondingly relies on prescribed bodily postures (reminiscent of those in yoga), patterns of breathing (with emphasis on the pace and number of inhalations and exhalations), and the recitation of codes (in keeping with the speed of ascent). Essential to the entire experience is the idea of love. So Drunvalo declares: "Feel Love. Feel Love for all life everywhere." When that feeling is part of one's being, one is ready to move toward the experience of the *merkabah*. It is not an *odium Dei* that drives this experience; if anything, it is an *amor Dei*. Moreover, this is a *merkabah* not of body, of "thing," but of mind, of "spirit." It is an internal phenomenon, to be sure. On the other hand, Drunvalo draws on technological discourse to delineate the experience. Thus, one must "cleanse" one's "electrical circuits" to reach the *merkabah;* these circuits, in turn, produce "rotating fields" of energy within and around the body. At the point of making contact with the "Higher Self," one is surrounded by energies of tremendous power that emanate from the *merkabah* in the form of spinning tetrahedrons. As manifestations of energy fields, these forms, in turn, are conceived alternately as male and female, a distinction that bestows on the *merkabah* a gendered presence. Entering into this vehicle, one is aware that its "motors" have begun to turn and that it is prepared to ascend. (At this point, Drunvalo observes, the experience is rather like that of starting an automobile motor with the transmission in neutral but ready to be shifted into gear.) In the final moments, "a disk about 55 feet in diameter forms around the body at the level of the base of the spine." The sphere of energy that encompasses the body fuses with the disk "to create a shape that looks like a FLYING SAUCER around the body." This "energy matrix" is called the *merkabah*. By means of it, one is ultimately transported through the speed of light into the fourth dimension. Within that dimension, one disappears from the present world and exists in a new home of ever-expanding consciousness.[85]

In the process described, the reference to the "flying saucer" suggests an essential dimension of the entire enterprise: that of UFOs. As an avatar of *technē*, Drunvalo generates his own UFO mythology, which he conflates with his delineation of the *merkabah*. This mythology has its roots in Lucifer's rebellion against God. Challenging God's authority, Lucifer constructs his own *merkabah* in an act of separation. Whereas God's *merkabah* as the product of love is "internal" and "spiritual," Lucifer's as the product of "the calculating mind" is "external" and "material." God's *merkabah* is the fulfillment of ascent to the higher spheres, Lucifer's the fulfillment of the desire to transform spirit into "technology." Resulting in the possible onset of chaos, Lucifer's rebellious act pits *merkabah* against *merkabah* in cosmic battle, reminiscent on a New Age scale of the War in Heaven in *Paradise Lost*, along with its antecedents.[86]

All this devolves into an extraterrestrial fantasy centered on a race of beings known as the Grays. Tracing their lineage to the Luciferians, the Grays, according to Drunvalo, have been transported here from Orion by spaceships in order to influence the course of history on this planet. In Drunvalo's New Age lore, the Grays are little people about four feet tall, with large heads, big eyes, and gray skin. The purpose of these creatures is to create a third race on earth that would enable them to regain control of their destiny and reascend to their former state. In their present state, they are incapable of accomplishing this objective because of their dependence on material, external technologies that hamper rather than aid them. Like the fallen Luciferians, they resort to the manufacture of *merkabah* vehicles devoid of spirit. Their *merkabah* vehicles are the prototypical UFOs that have left signs of their presence in the form of animal mutilations and accounts of abductions involving sexual experiments of various sorts.[87] The government is aware of these goings-on, of course. In fact, the government has been investigating the entire UFO-*merkabah* phenomenon all along. FBI agents have even been attending Drunvalo's various workshops in order to learn about *merkabah* energy fields and what bearing they have on the presence and power of UFOs for the government knows that UFOs are "real" and must be dealt with as responsibly as possible. As the government is well aware, earth itself is moving toward end time, toward possible collapse, and the government wishes to do all in its power to prevent this catastrophe. The presence of the UFOs represents a sign of "the end."[88]

As it moves from a concern with the *merkabah* as a vehicle of ascent to an emphasis on the technological implications of the phenomenon in the form of UFOs and alien creatures, Drunvalo's "Love Story" begins to assume some rather ominous overtones. Reminiscent of pulp fiction, the fantasies that Drunvalo conjures up are not far removed from the grim cyberspatial world of William Gibson. The New Age resolves itself into an apocalyptic mentality. But this is a cyberspatial eschatology, one in which the cyborgian mentality of dissociation, dislocation, and even anomie prevails. All is "virtual": nothing is "real." Who is Drunvalo Melchizedek, anyway? Beyond the cyberspatial commitment of the moment as one floats free of any responsibility to commitments in the "real" world, does anyone care?[89] One might pursue the Flower of Life Workshops, the course of meditations that Drunvalo advocates, the visions that he endorses, but to what end? To do so, in fact, would violate the cyberspatial "contract" of nonbeing entailed by the cyborgian flight to the electronic wonderland of virtuality. Here, *technē* resolves itself into a whole new configuration. As a true avatar of *technē*, Drunvalo Melchizedek fulfills his role most admirably.

The foregoing avatars of *technē* provide a visionary hermeneutic as complex and multifaceted as any that we have encountered in the interpretation of the *merkabah*. Reflected in their interpretive postures is an entire range of cultural assumptions concerning the place of Ezekiel's vision in the modern world. These assumptions incline us to view that vision as a moment of central importance to the understanding of technology (and pseudotechnology) as a manifestation of current modes of thought and discourse. By means of the so-called UFO experience that is seen to underlie Ezekiel's *visio Dei*, we are better able to assess the workings of the visionary imagination as an expression of *technē*. If that expression is viewed as inherently aberrant or pathological, this in no way lessens the fact of its significance in coming to terms with the nature of the modern.

3

Prophecy

Belief

and the

Politics of

End Time

In the ideologies embraced by the modern avatars of *technē*, one concept in particular assumes crucial import. Customarily designated *end time,* that concept is based on the belief that the world is approaching its final moment of existence. As Frank Kermode observes in *The Sense of an Ending,* it is a moment that is at once imminent (i.e., impending) and immanent (i.e., indwelling or inherent). As that which is about to occur, it is always already in the process of occurring. Throwing the weight of "end feeling" onto the moment, the pervasiveness of end time in modern religious thought causes the future to reside within the present sense that the end is now. Every moment contains within itself the crisis of its own eschatology. From that crisis emerge fictions of end time through which apocalypse is imagined in its various forms.[1] At the root of this outlook is biblical prophecy, conceived "as a complex, shifting, flexible, pervasive interpretive field, a wide range of narrative figures and frames, a living metanarrative that maps out history, geopolitics, current events, historical forces and trajectories, the telos of history, its agents, its benefactors, and its victims—all from 'a biblical point of view.' "[2]

Particularly from the perspectives that we have been exploring, that point of view is grounded in a mode of biblical interpretation that has been masterfully delineated by Paul Boyer. Focusing on the biblical prophets as the wellspring of end-time thinking, Boyer explores the impact of what he calls *prophecy belief* on modern American culture. Prophecy belief is a phenomenon that is the product of viewing the biblical text as a source of secret knowledge about the "signs" that constitute the events of one's own time. Pervasive throughout American culture, these signs are ones in which the number *666* and terms like *Gog, Magog, Antichrist, Mark of the Beast,* and *Armageddon* are as familiar as *Form 1040, VCR, MasterCard,* and *Hiroshima* are to the population as a whole. Those who inhabit the world of prophecy belief derive from the apocalyptic sections of the Bible "a detailed agenda of coming events." For such prophecy believers, the Bible becomes a road map into the future, one that culminates in the "promised end." He who can read the signs on that road map knows precisely what is going to transpire in the fulfillment of that end. Interpreted in this manner, the Bible assumes a distinctly predictive value: its language holds the secrets of all that will happen in the world to come. To know the hidden world of end time is to sort out the signs that constitute the language of the Bible. Underlying this knowledge is a methodology that corresponds to the eclectic approach advised by the prophet Isaiah: "For precept *must* be upon precept, precept upon precept; line upon line, line upon line; here a little, *and* there a little" (Isa. 28:10). Like artisans crafting a mosaic, prophecy interpreters "painstakingly build from hundreds of Bible verses a picture of the final days of human history—a picture strikingly similar to the world of today."[3] Indeed, not only is such a picture strikingly similar to the contemporary world: it veritably encodes the events of that world. Warfare among nations, large-scale political maneuverings, cataclysms of nature, the rise and demise of world leaders: all are present in the language of the biblical prophets.

Especially in the context of the evangelical and fundamentalist movements that represent a driving force in the history of American religious belief, end-time thinking looms large.[4] "Whatever the future holds for American evangelical religion," Timothy P. Weber observes, "one of its most noticeable elements is the interest in, even obsession with, biblical prophecy" as a source of apocalyptic speculation. Indeed, eschatology has come close to achieving "cult status" in American society.[5] An entire range of scholars confirms this observation.[6] In his psychological exploration of the subject, Charles B. Strozier avers that "the emergence in the last half century of a mass movement of Christian fundamentalism, energized by the apocalyptic, has moved the endist impulse into the center of our culture, where it works directly on large numbers of Christians and spills over in unpredictable ways into other cultural forms."[7] In today's world, the Jehovah's Witnesses, the Seventh-Day Adventists, and the Church of Jesus Christ of Latter Day Saints (Mormons), among other denominations, have "promulgated particular versions of end-time belief through their churches, door-to-door visitation, and ubiquitous magazines such as the Witnesses' *Watchtower* and the Worldwide Church of God's *Plain Truth,* distributed free in airports and elsewhere."[8] Although these evangelical movements would be viewed as anathema to many in the fundamentalist fold, they provide invaluable insight into the nature of end-time thinking in general and the primacy of Ezekiel's *visio Dei* in particular.[9]

Designated "one of the few truly American expressions of religion," the Jehovah's Witnesses is a case in point.[10] From its inception in Allegheny, Pennsylvania, in 1872, with only a handful of believers, it had already increased some hundred years later to a membership of over 2 million people around the globe. It now stands as a worldwide religious institution. Inspired by distinctive scriptural beliefs, made effective by an efficient social organization, and founded on the conviction that the end of the historical era is at hand, the Jehovah's Witnesses continues to exert a major influence in today's world.[11] In her illuminating study of the subject, Barbara Grizzuti Harrison establishes an appropriate setting for understanding the rise of the Jehovah's Witnesses as a potent religious force. Addressing the milieu that gave rise to the teachings of the movement, Harrison points to the emergence of apocalypticism in the 1840s and 1850s: "In 1843, William Miller had confidently announced the Second Coming of Christ, and his followers earnestly awaited their salvation and the end of the world." Although the world did not end, second adventists and other apocalyptic sects proliferated. This same period, moreover, witnessed the flourishing of science and secularism, industrialism and invention. This was an age of "mass movements," including Mormonism, mesmerism, and phrenology, Dunkers, Muggletonians, Baptists, Quakers, and abolitionists. Such events as "the revolution in technology, the reconstruction of the social order under the impact of the machine industry, [and] the advance of science into the domain of cosmogony" set the stage for what later came to be known as the Jehovah's Witnesses.[12] Because this movement represents the efflorescence of millennialist, adventist, and eschatological thought in America extending back to the period of the Great Awakening and beyond, a discussion of the founder and his heirs is apposite to our chronicle of those who most compellingly warrant the title *riders in the chariot.* Avatars of *technē*, they demonstrate the extent to which the transformation of the visionary experience into the wonders of technology underscores an essentially eschatological point of view.

At issue is the figure of Charles Taze Russell (1852–1916), known by his flock as "Pastor" Russell. Born in Pittsburgh, Pennsylvania, the son of Scottish-Irish parents, Russell was raised first as a Presbyterian and later joined the Congregational Church. Coming under the influence of the adventist teachings of William Miller (founder of the Millerites), among others, Russell formed a small group to study the Bible from the perspective of the Second Coming of Christ. In time, Russell's teachings spawned a much larger following and prompted the publication of *Zion's Watch Tower and Herald of Christ's Presence* (later called simply the *Watchtower*). By 1884, the Zion's Watch Tower Bible and Tract Society was organized, later to become the Watch Tower Bible and Tract Society. Eventually, groups known as Russellites, among other names, arose throughout the United States and in several other countries. In 1909, the organization moved its headquarters to Brooklyn, New York. There, its international operations have been directed since that time.[13]

As Harrison depicts him, Russell was a child of his time. A staunch believer in "progress," he declared that "the old order of things must pass away" and that "the new must supersede it." Although the change would be violently opposed by the entrenched classes who sought to halt this progress, it would occur nonetheless. In the third volume of his *Studies in Scriptures* (1886–1917), he wrote that the "Time of the End," a period between A.D. 1799 and 1914, is particularly marked in the Scriptures. Paving the way to the coming millennium, discoveries and inventions will be realized in "mechanical devices" that will "economize labor" and provide the world with "time and conveniences." There will be an uprising of the masses and the overthrow of big businesses. "All the discoveries, inventions and advantages which make our day the superior of every other day are but so many elements working together in this day of preparation for the incoming millennial age" of reform and progress. Worldwide revolution will result in the final destruction of the old order and the introduction and establishment of the new. Drawing on Dan. 12:4 ("But thou, O Daniel, shut up the words, and seal the book, even to the time of the end: many shall run to and fro, and knowledge shall be increased"), he was accordingly convinced that "a wonderful new world order was at hand." In this context, he recalled the time when he was a child contemplating his first toy railroad: "The predicted running to and fro—much rapid travelling—also confirms it," he declared. The "steamboat, steam-car, telegraph," all "belong to the Time of the End." Elaborating further on the visionary "to-and-fro" motion in

Daniel, he therefore celebrated the rise of the machine: "Today thousands of mammoth cars and steamships are carrying multitudes hither and thither, 'to and fro.' "[14]

Such pronouncements were characteristic of one who transformed the biblical text into a program for the enactment of millennial prognostications. His prophecies of technological progress as a sign of the coming new age were infused with adventist fervor. A "fevered visionary" who would not permit the world to confound him or his reading of the events of history through biblical chronology, Russell conflated "class conflict and steam cars" to forge a brave new world of speculation. Harrison observes that this outlook was further bolstered by extrabiblical speculations drawn from Madame Blavatsky's ideas concerning such phenomena as the "inner meaning" of the Great Pyramid of Egypt. By means of these and other sources, Russell founded a new millennialist religion determined to conceive American technology and class struggle in "quasi-mystical terms."[15]

Complementing Russell's theories was his behavior. He was to become notorious during his ministry for various scandals as the result of which he was accused, both in and out of court, of being "money-mad, power-mad, and sex-mad." His estranged wife alleged that he defended his putative marital excesses by maintaining, "I am like a jellyfish; I float around here and there. I touch this one and that one, and if she responds I take her to me, and if not I float on to others." Marked by behavior that was at times nothing short of bizarre, his escapades were apparently widely publicized. Pittsburgh newspapers reported that, in 1878, Russell (attended by his "saints") commemorated the occasion of Christ's death by waiting to be wafted to heaven as he stood attired in a white robe on the Sixth Street Bridge. (Russell later confided to reporters that, on that night of "glory-be that was not to be," he was at home in bed.) When the miracle of his hoped-for transumption failed to occur, his saints remained undaunted—although they were quite disappointed—and continued to adhere steadfastly to the faith. As one warranting the kind of transumption befitting a prophet of his stature, he is still regarded by the Jehovah's Witnesses as a modern-day "Elias," perhaps the first true Christian since "apostasy came to full bloom."[16] Russell's final days are legendary. In bad health, Russell embarked on a preaching tour in the fall of 1916. Stopping in Detroit, Chicago, Houston, and San Antonio, he gave his last public lecture in Los Angeles on 29 October 1916. Too ill to go on beyond that, he decided to cancel the remainder of his tour and return home. Journeying homeward on the Atcheson, Topeka, and Santa Fe railway, the dying Russell made a final request of his traveling companion, Menta Sturgeon: "Make me a Roman toga." Sturgeon did the best that he could with the Pullman sheets available. Calling the conductor and porter, Sturgeon declared: "We want you to see how a great man of God can die." Russell gathered the sheets around him, drew his feet up on the bed, and expired.[17]

More than anything else, Russell can be looked on as the heir of Ezekiel's vision. In fact, he was later identified by the Witnesses with "the man clothed with linen, which *had* the writer's inkhorn by his side," as described in Ezek. 9:1–11.[18] In his own works, Russell's allegiance to the vision of Ezekiel is most pronounced in *The Finished Mystery* (1917), the seventh and final volume of his *Studies in Scriptures*.[19] Published posthumously, and prepared by George Fisher and Clayton Woodworth shortly after Russell's death, this work is based largely on Russell's notes and writings.[20] As is widely recognized, *The Finished Mystery* provoked a good deal of controversy when it was first issued under the auspices of "Judge" Joseph Franklin Rutherford (1869–1942), who assumed control of the Jehovah's Witnesses on Russell's death.[21]

In *The Finished Mystery,* the figure of Charles Taze Russell is conceived as the direct descendant of a line of prophets and reformers beginning with Ezekiel and moving through Saint Paul and Saint John the Divine down to Arius, Peter Waldo, John Wycliffe, and Martin Luther. Russell thus becomes imbued not only with the prophetic and apocalyptic aura of the biblical stalwarts but also with the zeal of those determined to purify the church, even at the expense of being viewed as possibly subversive and even heretical. This prophet of the new order reflects those characteristics embodied in the name *Ezekiel ben buzi* as both "strength of God" and "contemned of God."[22] From the second point of view, he is looked on as one born of a system "unfaithful to Jehovah" and made to endure the aspersions of the unbelievers who accost him. This act of self-representation places Russell within the context of those who define their own mission as a response to the paradoxical nature of Ezekiel's presence in his prophecy. This presence is one that at once revels in God's power but is aware of the extent to which the bearer of that power will be viewed as suspect by an alien and faithless world.[23]

Particularly interesting is the way in which biblical texts are conflated to illuminate present concerns. Like Ezekiel, Russell "watches" for the vision to arrive. This act of watching is confirmed by Hab. 2:1: "I will stand upon my watch, and set me upon the tower." On the basis of this conflation of Ezekiel and Habakkuk, the former prophet is made to anticipate, if not to become, the source of the very idea of the witnesser and watcher that underlies the movement that Russell represents. The coming forth of the vision beheld by Ezekiel as watcher is that of "Jehovah" conceived as the very God of the Jehovah's Witnesses. As in the vision of Ezekiel, this Jehovah comes forth from the north to be beheld by Russell as a sign of the approach of "a great Time of Trouble, a whirlwind of warfare, revolution, and anarchy" (cf. Jer. 25:32). In his Ezekiel-like posture, Russell beholds what will be the conflagration of a world war on the earth. This, of course, is the First World War as a kind of Armageddon all its own. For Russell, Ezekiel's vision becomes a harbinger of this event.[24] It is within such a context that Russell analyzes the various dimensions of the vision itself. Focusing on the wheels that the prophet beholds, Russell maintains that they embody in their motion the cycles of history as applied to epochs or ages. For Russell, such cycles are evident in the cyclic motions of machines during his own time. On a historical scale, this is the divinely appointed "mechanism of ages" through which the course of history runs. "The Divine operations are not in one simple age, cycle, or manner of operation," he observes, "but cycle within cycle, age within age, many operations working together 'manifold' (Ephes. 3:10), like a vast and complicated machine." Made evident in figures like Joachim of Fiore, the notion of such cycles had gained wide currency in the apocalyptic thought of the Middle Ages.[25] In Russell, the notion gains new currency as it is assimilated into his distinctly "machine-based" view of the visionary substratum of Ezekiel's *visio Dei*. That view is applied with renewed vigor to what Russell perceives as the forces of historical change operating in his own time. In one of the two illustrations that accompany the analysis of the text of Ezekiel, the wheels within wheels are depicted as the cycles of ages progressing from the earliest periods through the "Messianic Cycle" and beyond. Along with these wheellike cycles, the fourfold creatures of Ezekiel become a means of understanding the "Plan of the Ages," comprised of the "lengths, breadths, heights, and depths" of Jehovah's wise dispensations. Through these dispensations, we are able to behold the character or face of Jehovah.[26]

The kind of analysis made evident in *The Finished Mystery* can be discerned in the publications that succeeded it. Joseph Franklin Rutherford's own writings are apposite. His tract *Vindication* (1931) ably carries forward the interpretive practices of his predecessor.[27] As indicated by the subtitle of the tract, Rutherford's purpose is to examine Ezekiel's prophecy as a means of disclosing "what must speedily come to pass upon the nations of the world." Viewing the prophecy from the perspective of his own times, Rutherford finds correspondences between the events surrounding the Babylonian Captivity and the travails faced by the Witnesses. Specifically, Rutherford places the period between the Captivity and the receipt of Ezekiel's vision within the context of the First World War.[28] Prefigured in the vision and its aftermath, this war becomes a sign that God has set Christ on his celestial throne in anticipation of the act of driving Satan out of heaven.[29] Accomplished with the completion of the First

World War, that event was anticipated by the destruction of the city in Ezek. 8–11. In Rutherford's interpretation of the prophet, the act of destroying the city assumes apocalyptic overtones consistent with his emphasis on Armageddon as that which is realized by the warfare his own generation has faced. By means of this warfare, the name of Jehovah and all who witness his triumph are vindicated.[30]

Underlying Rutherford's rhetoric throughout the tract is an idea ultimately derived from Russell but here brought to the fore even more compellingly. This is the idea of the *organization*. A distinctly bureaucratic notion, it becomes for Rutherford and those after him a unifying motif through which to understand the calling and struggle of the Jehovah's Witnesses. As conceived by Rutherford, the organization is either good or evil, either that of Jehovah or that of Satan. The two are fundamentally in conflict. The purpose of the Jehovah's Witnesses as the divinely sanctified organization is to do battle against Satan's organization. This battle of corporate entities is epitomized by Ezekiel's vision as an emblem of the warfare through which Jehovah's organization overcomes all that Satan's organization represents.[31]

In its most essential form, the Satanic organization is what the Witnesses refer to as *Christendom*, that is, organized Christianity. Typified in the destruction of the city in Ezekiel, it is this entity that is under siege by the vision. Although Satan's organization may assume many forms, Christendom or organized Christianity is for Rutherford most immediately to be understood as the Roman Catholic Church or what Rutherford in *Vindication* calls "the Roman wing of Christianity." In terms of organized religion, Roman Catholicism is what Rutherford finds most characteristic of the workings of Satan in the modern world. It is this, among other entities, that the organization known as the Jehovah's Witnesses seeks to overcome. Adopting this organizational point of view, Rutherford's discourse engages in what might be called the *bureaucratization of vision*. In that bureaucratization, the elements of Ezekiel's *visio Dei* and all that it implies assume a corporate status, one in which the Lord God Almighty sits on his throne over and above all the forces of his "organization" as he moves on to the destruction of the city.[32]

Particularly within the apocalyptic context to which this event gives rise in the thinking of Rutherford and his heirs, the zeal that underlies his determination to overcome the organization of Satan in all its forms has prompted some observers to view the Witnesses as an organization that has made "hate a religion."[33] So Herbert Hewitt Stroup observed in 1945 that, "whatever else the Witnesses may believe, they do somehow feel that the whole world is arrayed against them, and respond with resentment, hatred, and bitterness." The Witnesses "believe that they owe the world nothing, and from the world they want nothing, for it is evil and not in any sense a positive part of the divine purpose." The list of resentments they harbor is long: it encompasses all of "Satan's organization," which in turn may be defined as any institution or individual that stands opposed to "the Lord's organization."[34] Whether such an outlook obtains at present remains to be seen. Certainly, it was true of the outlook that Rutherford embraced. In his zeal to uphold the cause of Jehovah's organization, he made a point of distinguishing those who supported him and those who opposed him. The former were considered "saved," the latter "unsaved." The idea reached its peak in 1937 with the publication of Rutherford's *Enemies*, which clarified his stand on the nature of Satan's organization even further. Ultimately, however, the Witnesses felt that their battle was not of a mundane but rather of a celestial character, its proportions cosmic, as they waged war against the principalities and powers of Satan's forces.[35]

Underlying their beliefs throughout their history has been the determination to forge an organization capable of withstanding all the wiles of Satan's organization. What has emerged from this determination is a complex corporate entity founded on the concept of a theocracy, one that functions with its own structured form of government, its own polity and chain of command, and its own hierarchy. Within the chain of command, the Governing Body oversees the organization as a whole. Beneath the Governing Body are the Committees (Personnel, Publishing, Service, and the like), beneath them the Watch Tower Societies, and beneath them the Branch Committees, the District and Circuit Overseers, the Congregation Elders, and the Ministerial Servants. Over the theocracy as a whole, Jehovah himself presides through the Lord Jesus Christ as chief executive officer. For the Witnesses, this is the government of God on earth.[36]

In the publications by the Jehovah's Witnesses that have proliferated over the years, this theocratic point of view prevails. It is particularly apparent in the literature that the Witnesses have produced within the past twenty-five years concerning the vision of Ezekiel. In this respect, their *"The Nations Shall Know That I Am Jehovah"—How?* (1971) is germane. A detailed and extensive reading of Ezekiel's prophecy as a whole, this work suggests the extent to which the original teachings of the Witnesses find expression in the current literature. Recalling the Heideggerian notion of bringing forth into unconcealment, the work immediately aligns itself with those avatars of *technē* determined to re-inscribe the vision of Ezekiel in technological terms. In a chapter titled "God's Chariot Is on the Move," the work declares: "Two thousand five hundred and seven years before the gasoline automobile began to be manufactured industrially (in 1895 C.E.), a self-moving or self-propelled chariot was seen in south-western Asia." This vehicle "was not of man's invention." It was no mere human contrivance. Rather, the industrialist was Jehovah himself. What he manufactured was the original "horseless chariot" beheld by Ezekiel, the son of Buzi the priest, who saw it moving from outer space to an earthly destination. If Jehovah was the inventor of this chariot, he was also its driver, one adept at setting the vehicle on a proper course. It would be impossible for modern automobile manufacturers to duplicate such a vehicle, let alone drive it. Only Jehovah himself could maneuver this chariot and do so, moreover, "without any steering wheel or shifting of gears." The work assures us that the prophet reported accurately and in detail what he saw. It "was no hallucination caused by the taking of some drug like LSD, but was presented to him in vision by the 'hand' or applied power of Jehovah." Does this chariot still exist? Is it still "on the move"? These are the questions that the work poses and seeks to answer in the context of our mechanized, industrialized, "locomotive twentieth century."[37]

What is clearly at work in this analysis is not only a reaffirmation of Russell's belief in the power of progress but also a conscious adoption of what might be called *chariot discourse* for the purpose of furthering the argument. This discourse reconceptualizes the visionary substratum in the language of the machine as a divine vehicle that has no need of steering wheels and gear mechanisms. Manufactured and driven by Jehovah, this is a masterpiece of technology indeed. Its purpose, *"The Nations Shall Know"* makes clear, was not simply for the sake of undertaking a holiday drive in the country. Rather, this is a mighty and fierce vehicle of visionary combat. Accompanied by the tumultuous sound of an armed camp or an encampment of soldiers, what Ezekiel beholds is God's war chariot. In this respect, Ezekiel's chariot becomes a vehicle with a distinctly military purpose.[38]

This fusion of the automotive and military aspects sets the stage for the argument that follows. Reflecting the discourse of its founders, that argument is one in which the vision is portrayed through the ongoing struggle between Jehovah's organization and Satan's organization. Imbued with the spirit of Ezekiel's vision, the former is clearly a "chariotlike organization." Drawing on the impulse to bureaucratize the vision, *"The Nations Shall Know"* (itself the "corporate" product of unnamed individuals coming together for a larger cause) incorporates into its reading what we have seen as the organizational outlook of the Jehovah's Witnesses as a theocratic entity. In this way, Ezekiel's vision as a spiritual manifestation of God's presence in this world is at once a remarkable work of technology and a fully realized emblem of the divine theocracy. So the work declares of the vision: "All cherubs, seraphs and angels, together, make up this united, coordinated, harmonious, obedient heavenly organization." As a "superhuman, spirit organization," the Jehovah's Witnesses is itself "the celestial chariot seen in Ezekiel's vision and is pictured by it." Propelling the chariot forward, Jehovah "rides this organization, causing it to move to wherever his spirit causes it to move."[39] It was on the move in the tumultuous wars endured in the twentieth century, and its "wheels of progress" are now "turning faster than ever": "Jehovah rides again!"[40] That which he rides out against is the enemy. In keeping with the delineation of Satan's organization in the ideology of Witness polemic, that enemy is "Christendom." Guilty of the misuse of "religious things that she adopted from the idolatrous pagans in order to carry on her spiritual immorality," Christendom has no right to claim the name *Christian*. Both its "standing" and its "name" will be destroyed by Jehovah, just as the city is destroyed in the prophecy of Ezekiel. As with the destruction of the city, so will be the destruction of Christendom. Dead bodies of the "spiritual fornicators" will be seen surrounding their "idolatrous altars," with their incense stands through which they worshiped false gods at their "high places" on hilltop and mountaintop.[41] Such will be the day of reckoning when Jehovah rides forth on his chariot to overwhelm the heathen hordes. It is the day of Armageddon, when the forces of Satan's organization embodied in Christendom meet their final end.

If the Jehovah's Witnesses represents a strand of end-time thinking that resorts to the prophecy of Ezekiel as a wellspring of technological speculation, corresponding forms of evangelical fervor are at least as apposite.[42] These forms are particularly fascinating because they result in what Andrew Lang aptly calls the "nuclearization of the Bible," that is, the act of looking on the Bible, in general, and biblical prophecy, in particular, as a source of discourse concerning the annihilation of the world through the use of nuclear armaments.[43] As Paul Boyer makes clear, such an idea became current with the advent of the atomic bomb, a turning point in the history of prophecy belief. Tracing the roots of the eschatological outlook from its biblical origins through the first half of the twentieth century, Boyer establishes the extent to which the world of prophecy belief assumes a new form and a renewed urgency with the awareness of the powers unleashed in the August 1945 bombing of Hiroshima and Nagasaki. "With the coming of the atomic bomb," Boyer observes, "everything changed: it seemed that man himself had, in the throes of war, stumbled on the means of his own prophesied doom."[44] As a result of the new "nuclear awareness" that pervaded the "Atomic Age," end-time prophecy flourished. "Today the whole world lives in fear of annihilation," proclaims the manifesto of a prophecy conference held in New York City in 1952.[45] The mid-1950s and 1960s saw an even more pronounced increase in the prophetic fervor reflected in such proclamations. This was true particularly among the end-time pronouncements of the evangelical seers whose prophecies were rooted in the biblical text as a source of ultimate inspiration and authority. Such end-time seers interlaced their prophecies of nuclear holocaust with a stock set of biblical proof texts: "The vision of a melting earth in II Peter; the crescendo of catastrophes in John's Apocalypse; the all-consuming conflagration and terrifying astronomical events woven through the book of Joel's three short chapters . . . ; and the prophet Zechariah's terrifying description of Jehovah's judgment on Israel's enemies." Influential prophecy interpreters like M. R. DeHaan of Grand Rapids, Michigan, and J. Vernon McGee of Los Angeles adopted proof texts of this sort to confirm that the holocaust of atomic annihilation would fulfill the biblical prophecies and that "the Bible and science go right down the line together on the forecasting of future events for earth."[46]

Among the biblical prophets invoked to demonstrate the confluence of technology and end-time belief, Ezekiel has become the veritable centerpiece of end-time speculation throughout the course of evangelical history from the very beginnings up to the present time. So significant has Ezekiel been, in fact, that Boyer designates him the "first cold warrior," an ascription that suggests the essentially political bearing that his prophecy assumes as the oracle of end time.[47] In the articulation of the oracle's message as eschatological proclamation, a new dimension arises to complement and augment the chariot discourse that has been the center of our discussion throughout. That new dimension causes the interpretive locus to shift from the first chapter of Ezekiel (the "chariot" chapter) to the thirty-eighth (what might be called the "Gog and Magog" chapter).[48] Focusing on Gog and Magog, this oracle is part of a complex of oracles that anticipate the return of the glory of God in the form of the *merkabah* to the restored, visionary temple in the chapters (40–48) that conclude the prophecy as a whole.

According to the Gog and Magog oracle, Ezekiel proclaims that the word of the Lord came unto him, saying,

> Son of man, set thy face against Gog, the land of Magog, the chief prince of Meshech and Tubal, and prophesy against him, And say, Thus saith the Lord God; Behold, I *am* against thee, O Gog, the chief prince of Meshech and Tubal: And I will turn thee back, and put hooks into thy jaws, and I will bring thee forth, and all thine army, horses and horsemen, all of them handling swords: Persia, Ethiopia, and Libya with them; all of them with

shield and helmet: Gomer, and all his bands; the house of Togarmah of the north quarters, and all his bands: *and* many people with thee. . . . And thou shalt come from thy place out of the north parts, thou, and many people with thee, all of them riding upon horses, a great company, and a mighty army: And thou shalt come up against *my* people of Israel, as a cloud to cover the land; it shall be in the later days, and I will bring thee against my land, that the heathen may know me, when I shall be sanctified in thee, O Gog, before their eyes. (Ezek. 38:1–6, 15–16)

In order to appreciate the import of this crucial passage to end-time thinking, one must account for the significance of the mysterious terms *Gog* and *Magog* in the context of the other names registered in the oracle. As scholars generally agree, neither term can be precisely defined. A ruler whose domain is known as Magog ("place of Gog"?) and whose office is that of chief prince of the realms of Meshech and Tubal, Gog comes from the northern regions "in the later days" to destroy Israel but is instead overcome by Yahweh in a cataclysmic battle. Speculation about the meaning of Gog's name abounds: it is an apocalyptic pseudonym for the Babylonians; the Scythians; the Lydian king Gyges; Gagaia, one of Cyrus's generals; Alexander the Great; Antiochus Eupator; Mithridates VI; Artaxerxes Ochus. The point is that no one really knows.[49] The particular relation between *Gog* and the other names (*Meshech, Tubal,* and the like) invoked in the oracle is correspondingly difficult to understand. Appearing elsewhere in Hebrew Scriptures, these names lend a certain "historical aura" to the oracle as a whole.[50]

One issue remains clear: the complex of narratives that surround the figure of Gog bestows on him and his forces a bearing that suggests the larger "mythic" import of the oracle.[51] That Gog comes from the northern regions is a case in point. In the biblical traditions, an entire range of associations has accrued to the northern regions. These associations underscore the mythic quality of the idea. Thus, the very word *north* (*ṣaphon*) implies that which is hidden, dark, and unknown.[52] These qualities reveal themselves both in a positive and in a negative way. As Ezekiel articulates them throughout his prophecy, an ambivalence arises concerning the signification of the northern regions. The positive perspective is delineated at the very outset of Ezekiel's prophecy for, after all, the north is the region from which God's own throne-chariot emerges: "And I looked, and, behold, a whirlwind came out of the north [*min-haṣaphon*], a great cloud, and a fire infolding itself" (Ezek. 1:4). The idea of God's coming from the north has caused a good deal of consternation among the scholars. Such a notion, declares Walther Zimmerli, "is in opposition to all

Israelite tradition."[53] Zimmerli's statement is based on the idea that it is the Babylonian pantheon (which Ezekiel himself opposes) that is customarily assigned to the northern regions.[54] For this reason, as Moshe Greenberg argues, it is unlikely that Ezekiel would adopt such a location for Israel's God.[55] Nonetheless, an association of this sort is already present in the Psalm literature: "Great *is* the Lord, and greatly to be praised in the city of our God, *in* the mountain of our holiness. Beautiful for situation, the joy of the whole earth, *is* mount Zion, *on* the sides of the north [*yarkitei ṣaphon*], the city of the great King" (Ps. 48:1–2).[56] Present in Ezekiel's ascription of the northern regions to his deity, then, is a problematic that is fundamental to the inaugural vision (a phenomenon itself deeply conflicted) as a whole. What David J. Halperin aptly calls "the dark side of the *merkabah*" is already present in the notion of God's emergence from the northern regions.[57] This "dark side" is even further intensified in the oracle concerning Gog.

However one interprets the inaugural vision of Ezekiel, it is clear that the oracle concerning Gog represents an implicit pairing of what we have called the chariot chapter with the Gog and Magog chapter. Reinscribing the one in the context of the other, Ezekiel establishes a dialectic through which the chariot chapter and the Gog and Magog chapter are counterpointed. If the positive implications of ṣaphon appear in the chariot chapter, the negative implications (already present at the very outset of the inaugural vision) appear in the Gog and Magog chapter. There, the expressions "the north quarters [*yarkitei ṣaphon*]" and "thy place out of the north parts [*miyarkitei ṣaphon*]" (Ezek. 38:6, 15; 39:2) imply not only that which is hidden and dark but also that which is evil. So Jeremiah proclaims that the word of the Lord came unto him, declaring that "out of the north an evil shall break forth upon all the inhabitants of the land. For, lo, I will call all the families of the kingdoms of the north, saith the Lord; and they shall come, and they shall set every one his throne at the entering of the gates of Jerusalem, and against all the walls thereof round about, and against all the cities of Judah" (Jer. 1:14–15; cf. Jer. 4:6, 6:1, 22–23; Joel 2:20). As the literature surrounding ṣaphon in this context suggests, "the evil powers that are hostile to God reside in the north, whence they are set loose."[58] The mode of their being set loose, moreover, is through the kind of discourse that has already been encountered in Ezekiel's own *visio Dei:* "Behold, he shall come up as clouds, and his chariots *shall be* as a whirlwind" (Jer. 4:13). Whirlwinds, clouds, chariots: these are the very signatures of the inaugural vision (Ezek. 1:4 and passim). So with Gog: "Thou shalt ascend and come like a storm, thou shalt be like a cloud to cover the land, thou, and all thy bands" (Ezek. 38:9, 16).

In keeping with the mythic import of Ezekiel's oracle, the motifs that constitute its articulation move inextricably toward the sense of end time. "It shall be in the later days [*b'aharit hayamim*]," the prophet declares, when Yahweh will bring Gog against his land (Ezek. 38:16). If the meaning of the phrase *b'aharit hayamim* remains deliberately open to question (exactly when will these "later days" or "after days" occur?), the thrust of its bearing does not: the coming of Gog and his forces, followed by the triumph and sanctification of Yahweh in the eyes of his enemies at some future time, is inevitable. Whenever that future time occurs, the world will reel in the tumultuous throes of its effects. The earth will shake, and mountains will fall. There will be pestilence and blood, hailstones, fire, and brimstones (Ezek. 38:22). This will indeed be the promised end.

I devote so much attention to these dimensions of the Gog oracle in Ezekiel to provide a context for the evangelical rearticulation of this seminal text not as some "mythic" conception of the attack on Israel by mysterious hordes from the north but as the prophecy of what will be an actual event in the world of today. That is, the evangelical impulse is to demythologize (some would argue remythologize) the oracle and to place it in a distinctly political framework that our world is able to understand. In order to appreciate the nature of this hermeneutic, we must attend to the foundations on which it is constructed. Those foundations are canonized in one of the most authoritative texts (indeed, *the* most authoritative text) in the interpretation of Hebrew Scriptures from the nineteenth century to the present time. That text is the *Hebräisches und chäldaisches Handwörterbuch über das Alte Testament* (Leipzig, 1828) of Heinrich Friedrich Wilhelm Gesenius (1786–1842), the father of modern Hebrew lexicography.[59] For the purposes of the Gog oracle, Gesenius's *Lexicon* provides a definition of the term *ro'sh* (Ezek. 38:2, 3; 39:1) in the phrase *nesh'ey ro'sh meshekh* (which the Authorized Version translates "the chief prince of Meshech") as a proper noun that refers to "a northern nation, mentioned with Tubal and Meshech; undoubtedly the *Russians,* who are mentioned by Byzantine writers of the tenth century, under the title of *hoi Rōs,* dwelling to the north of Taurus."[60] At the same time, Gesenius also claims that "Tubal" and "Meshech" are Moscow and the Siberian city of Tobolsk.[61]

As it influenced early fundamentalist thinking in the nineteenth century, Gesenius's insistence on the *ro'sh*/Russia correspondence became commonplace.[62] From the twentieth century on, it was a staple of end-time interpretation. So widespread was this reading that it found its way into the so-called Scofield Reference Bible (1909), the work of Cyrus Scofield (1843–1921), a towering figure in fundamentalist thought. A Tennessean plagued by scandal, drinking, and marital problems in his early years, Scofield fought as a Confederate in the

Civil War, later practicing law in Kansas, which he fled in 1877 (abandoning his family in the process) under accusations of theft. Jailed in St. Louis on forgery charges, he underwent a religious conversion in prison. In 1882, he became a pastor and, in 1895, joined the faculty of Dwight L. Moody's Northfield Bible School in Massachusetts. After 1902, he devoted his time to lecturing and Bible research.[63] The major achievement of this reformed, born-again sinner is his Reference Bible. So influential is this work that it has been termed "the most important single document in all fundamentalist literature."[64] Among those who adhere to its teachings, it is viewed as gospel, a view reinforced by the format of the Scofield Reference Bible itself. Because Scofield as editor and annotator combined his notes and the biblical text on the same page, the word of Cyrus Scofield assumed just about the same authority as the word of God: "Readers often could not remember whether they had encountered a particular thought in the notes or in the text."[65] In effect, the two flowed into one another.

Such is particularly true of Scofield's reading of Ezekiel's prophecy. The words of the born-again sinner are imbued with the prophetic fervor of the prophet himself. In his notes to the Gog oracle, Scofield mounts an end-time argument by grounding his interpretation in the correspondences made evident in Gesenius's *Lexicon*. Drawing on the relation between *ro'sh* and Russia, Scofield's apocalyptic speculations are eloquent in their politicizing of the text: "That the primary reference is to the northern (European) powers, headed up by Russia, all agree. . . . The reference to Meshech and Tubal (Moscow and Tobolsk) is a clear mark of identification. Russia and the northern powers have been the latest persecutors of dispersed Israel, and it is congruous both with divine justice and with the covenants . . . that destruction should fall at the climax of the last mad attempt to exterminate the remnant of Israel in Jerusalem. The whole prophecy belongs to the yet future 'day of Jehovah' (Isa. 2. 10–22; Rev. 19. 11–21) and to the battle of Armageddon (Rev. 16. 14), but includes also the final revolt of the nations at the close of the kingdom-age (Rev. 20. 7–9)."[66]

Throughout his reading of Ezekiel 38, Scofield weaves an entire series of what he considers to be corresponding texts from both Hebrew Scriptures and the New Testament. Among those texts, references to the Book of Revelation are paramount. Especially important are the portrayals of end-time battles that underscore Rev. 16:14–16, 19:11–21, and 20:7–10. These speak of the confrontations in which the powers of Satan are finally destroyed. Portraying the penultimate confrontation as that in which "the spirits of devils" go forth "unto the kings of the earth and of the whole world, to gather them to the battle of that great day of God Almighty," Saint John the Divine calls the place of this battle "Armageddon," that is, *har-megiddo* or mountain of Megiddo, the ancient hill and valley on the southern rim of the plain of Jezreel (Rev. 16:14–16).[67] That battle is conceived as an event in which heaven opens and the Word of God, accompanied by his hosts, goes forth on his white horse to overcome the enemy with the forces of Satan, who is, in turn, cast into the bottomless pit and bound a thousand years (Rev. 19:11–21, 20:1–3). After this millennial period, there occurs what Scofield calls "the final revolt of the nations at the close of the kingdom-age [i.e., the millennium]." Here, Gog and Magog once again come into play. Speaking of the millennial reign of Christ, Saint John the Divine declares that, "when the thousand years are expired, Satan shall be loosed out of his prison, And shall go out to deceive the nations which are in the four quarters of the earth, Gog and Magog, to gather them together to battle." So the wars of Gog and Magog are reenacted against God's faithful, who once again prevail as a result of the unleashing of divine power: "And they [Gog and Magog] went up on the breadth of the earth, and compassed the camp of the saints about, and the beloved city: and fire came down from God out of heaven, and devoured them. And the devil that deceived them was cast into the lake of fire and brimstone" (Rev. 20:7–10).

It is within such contexts that Scofield politicizes the Gog oracle of Ezekiel 38. In his response to this text, Scofield conceives of Russia as an "evil empire" that will marshal the northern European nations in order to persecute and ultimately attack in all-out war the remnant of Israel in Jerusalem. Focusing on a pre-Communist Russia and a "dispersed Israel" well before the actual founding of the Jewish state, Scofield's remarks both call on and look forward to those "wars and rumours of wars," when "nation shall rise against nation, and kingdom against kingdom" (Matt. 24:6–7). From Scofield's comments, it is unclear what might motivate Russia to undertake such an all-out war against the remnant of Israel. During the period of the Scofield Reference Bible's initial publication (1909–17), it was generally felt that, given the extent of Jewish persecution within Russia, the attack would arise from a history of anti-Semitism.[68] Whatever the motivating factor, the speculations of the born-again sinner involve power politics of the most cataclysmic sort. At the center of the politics are a people who must be protected at all costs: the Jews. The enemy is Russia. If the warfare that will ultimately ensue lies in the indefinite future, clearly the context is that of Scofield's world, the world of his day. It is a day that expectantly awaits the battle of Armageddon and encompasses the final revolt of the nations prophesied in the Book of Revelation.

With the institutionalizing of the *ro'sh*/Russia correspondence provided by the Gesenius *Lexicon* and the Scofield Reference Bible, the politicizing of the Gog oracle among the prophecy believers flourished in the years extending from the two world wars up through the present time. As a counterpart to the inaugural vision, the Gog oracle provided generations of evangelical prophets a means of interpreting the Bible according to their own notion of how scriptural prophecy must be viewed as a harbinger of current events. Especially as those events reflect an overriding concern with technology and its role in hastening the end of the world, prophecy believers were determined to look on the Bible as a wellspring of technological innovation.[69]

It is perhaps not coincidental, Paul Boyer observes, that an impressive number of post-1945 prophecy writers had a background in science and engineering. Such end-timers as Edgar Whisenant was an electrical engineer and NASA scientist; Leon Bates had a long career in electronics before he established the Bible Believers' Evangelistic Association; Robert Faid was a nuclear engineer; Gerald Stanton was preparing to become a chemist with the Dupont Corporation when he decided to attend Dallas Seminary instead; and Texe Marrs listed his background as "U.S. Air Force officer, professor of American defense policy and international affairs, and high tech consultant."[70]

Whether trained in science or not, they often established their credentials by referring to themselves as "Dr." At the same time, they endlessly invoked "famous scientists" and "well-known experts" and alluded in their works to DNA, gene splicing, the neutron bomb, laser technology, and computer electronics. A prophecy film produced by end-timer Chuck Smith announced that, within the next hour, viewers of his film would be meeting with "doctors, professors, scientists, world leaders, and Chuck Smith." Albert Einstein, Paul Harvey, the *New York Times*, the *Reader's Digest,* and papers such as the *South Bay Daily Breeze* were "on an equal footing in prophecyland." In that kingdom one discovered a plethora of illiteracies and howlers. (For example, one seer referred to the speed of an F15 jet as "mock 2.2.") For the millions of those faithful to the beliefs of the end-timers, such faux pas did not detract in the least from the urgency or the accuracy of their message.[71] In fact, among the prophets of end time, the beliefs they fostered have become in the last three decades or so an essential part of the American psyche.

So extensive has the influence of prophecy belief become in America that Grace Halsell provides a fascinating firsthand account of her experience among the end-timers in the early 1980s. In 1983, she participated in a Jerry Falwell–sponsored tour of Israel. One of 630 Christians who flew from New York to Tel Aviv, she traveled by bus to Megiddo. Arriving at a site that lies twenty miles south-southeast of Haifa and about fifteen miles inland from the Mediterranean, she falls in with Clyde, a retired business executive from Minneapolis. Having served as an army captain in North Africa and Europe during World

War II, this confirmed prophecy believer is eager to survey the area where the final battle is to be fought. Enumerating for Halsell all the battles waged here in antiquity, he moves to the twentieth century to mention the crucial victory over the Turks that the British general Allenby won in 1918, near the end of World War I. "At last," Clyde declares with complete conviction, "I am viewing the site of the last great battle!" Recounting the evidence for that battle in the reference to Armageddon in Revelation, Clyde proceeds to describe the battle plan as indicated in Ezekiel's oracle concerning Gog. In response to the speculations of this end-timer, Halsell remarks that the entire valley where the battle is to be waged is so small that "it would fit into a Nebraska farm and be lost if placed in a big Texas ranch." In short, it looks too small to accommodate the last, decisive battle. "No," Clyde protests. "You can get a lot of tanks in here." When Jesus as supreme commander marches forth to overwhelm the enemy, he will use a new weapon, one that, according to Clyde, "will have the same effects as those caused by a neutron bomb." In fact, it will be a neutron bomb of the most cataclysmic sort.[72]

For Clyde, the nuclearization of the Bible represents an actual event. A true avatar of *technē*, Clyde grounds his faith on a biblical inerrancy that at once demythologizes Scripture and reconceives it as "fact," or at least fact as he knows it. To undertake such an act of demythologizing is, of course, to reconceive the Bible through a new mythology, one driven by a belief in the power of technology to fulfill all that is contained in biblical prophecy. This is a comforting belief for Clyde: it means that biblical figuration finds its counterpart in the world of "place" and the world of "thing." He can travel to it, look about it, and touch it. It is part of his experience. The relation between type and antitype assumes a resonance entirely consistent with the world Clyde knows. Nothing will shake his confidence in the placeness and thingness that empower his faith. A true believer, he is determined that end time will be realized precisely through the terms set down by the prophets in general and Ezekiel in particular.

The ideas that inspire Clyde's end-time belief have been recorded in Jerry Falwell's prophecy sermons. In one such sermon (issued by the Old Time Gospel Hour) that Halsell recorded in 1979, Falwell declares that Armageddon is "a reality, a horrible reality" for those who are part of "the terminal generation." Invoking the prophet Ezekiel as his source, Falwell alludes to the Russians as the Communist "God-haters" and "Christ-rejecters," whose ultimate goal is to conquer the world. Ezekiel, he declares, prophesied that "such a nation would rise to the north of Palestine just prior to the Second Coming of Christ." Reiterating the association of *ro'sh* and Russia, he proceeds to analyze Ezekiel's oracle against Gog verse by verse. The allies of Russia in Ezekiel are modern-day Iran, Ethiopia, Libya, Eastern Europe, and the Cossacks of southern Russia. The purpose of Russia's invasion of Israel, according to Ezek. 38:12, is to take "spoil." "If one but removes the first two letters from this word," Falwell cleverly observes, "he soon realizes what Russia will really be after—obviously, oil. And that is where we find ourselves today." Such is Ezekiel's prophecy concerning Russia. Despite all the attempts to reach an accord in the Middle East, Falwell concludes, there will never be any real peace "until the Lord Jesus sits down upon the Throne of David in Jerusalem." Such a day is certainly coming, as the aftermath of the nuclear war that the Middle East will undergo in the battle of Armageddon.[73]

It is this kind of preaching that underscores the end-time perspectives of the prophecy believers of Amarillo, Texas, about whom the novelist A. G. Mojtabai writes in her study of evangelical movements.[74] During the four years that Mojtabai spent in Amarillo, she became immersed in the religious life of the city, which is the site of Pantex, originally an army ordnance facility established in 1942 for the manufacture of conventional armaments for use in the war but in 1952 converted by the Atomic Energy Commission into a plant for the assembly of nuclear warheads.[75] At the center of a vast network of the military-industrial complex, Pantex received plutonium and tritium for nuclear warheads, electronic arming, fusing, and firing switches, detonators and timers, neutron generators, and uranium and testing devices from various production sources throughout the nation. After the warheads were assembled at Pantex, they were sent to military bases in California, Georgia, Michigan, Montana, Nevada, New Mexico, and New York, among other states. The white locomotive bearing the warheads "moved slowly in and out of Pantex under cover of night." With its nuclear payload, the locomotive remained "invisible" for more than twenty years as it made its way slowly back and forth across the nation. Once the purpose of the locomotive was discovered, the antinuclear protestors who kept vigil along the tracks looked on its "whiteness" in Melvillean terms as an ominous and disturbing symbol of the "blinding white night of annihilation."[76]

For the prophecy believers who lived in Amarillo and helped assemble the warheads, Pantex became a consummate symbol of end-time belief. It was by means of this facility that the events moving toward Armageddon would be hastened. Those committed to prophecy belief looked on themselves as facilitators of the promised end. In the sermons that were preached to them in their churches, these avatars of *technē* delighted in knowing that they were doing God's work. Their belief was in an end-time theology through which they were viewed as instruments of Providence as they helped hasten the onset of the millennium. For them, the world was divided between absolute good and absolute evil, between those who followed God and those who followed Satan. Under such circumstances, Mojtabai found, peace became impossible and war inevitable. At the forefront of these beliefs were the biblical prophets, especially Ezekiel. For those who saw in his oracles that a Soviet attack on Israel has already been preordained as a prelude to the millennium, any provocation and nuclear belligerence were to be viewed quite simply as the "implementation of God's own design for creation."[77] That is truly how one can come to love the bomb, particularly as a construct sanctioned by Ezekiel himself.

From this perspective, perhaps the one end-timer who has been most successful in capturing the apocalyptic imagination of those committed to the concepts that inform prophecy belief is Hal Lindsey. To address the nature of end time in the modern world is to concern oneself with his works. Educated at the University of Houston and the Dallas Theological Seminary, in the 1960s Lindsey devoted his time as a minister to students at UCLA under the auspices of the Campus Crusade for Christ. Lecturing on prophetic themes to college and church groups, he published *The Late Great Planet Earth* in 1970.[78] With sales of more than 18 million copies since its publication, this book has established Lindsey as one of the most widely read authors on prophetic themes in history.[79] Beyond *The Late Great Planet Earth,* over the past quarter century Lindsey has published an entire series of studies that address related end-time themes and update his proclamations according to the changing political climate.[80] A paramount prophet of end time, Lindsey has been thrust on the international stage of power politics. In the capacity of strategist and truth bearer, he has apparently been called on to address not only the military planners at the American War College but also officials at the Pentagon. Among the military and officials in high office, his books have circulated widely.[81]

Reflecting and consolidating the views of end-timers before him, Lindsey incorporates in his works the prevailing themes of prophecy belief. In brief, Israel is looked on as the most important sign of the coming arrival of end time, which could not be realized until the reestablishment of Israel as a Jewish state. The occurrence of that event in 1948 set in motion the other prophetic fulfillments that would, in turn, pave the way for Armageddon and the Second Coming. For Lindsey, these included his view that the European Common Market was on the verge of assuming the role of a revived Roman Empire under an Antichrist whose identity would eventually become known. Along with its Arab allies, Russia was the sworn enemy of Israel and an ever-present threat to its security. With their ability to raise an army of 200 million troops, the Communist Chinese had become a military threat as well. Modern technology had reached the point where the biblical prophecies could be fully realized. In keeping with these ideas, Lindsey formulated an elaborate end-time scenario in which the Israelis will sign a treaty with the leader of the ten-nation Common Market. Unknown to the Israelis, he will be the Antichrist who will momentarily guarantee Israel's security and allow the Jews to rebuild their temple in Jerusalem. Soon after, the Antichrist will enter the new temple and declare himself to be God. Israel would then be the subject of attacks from an Arab-African confederacy and a Russian and eastern European alliance. Sweeping into Israel from the north, the Red Army will double-cross its Arab and African allies, leaving Russia in control of the Middle East. Attempting to destroy the remaining Jewish populations, the Russians would then be overcome through divine intervention. At that point, the Chinese army and the forces of the Antichrist will do battle at Armageddon. This cataclysm will assume the form of a nuclear confrontation with the obliteration not only of the Middle East but of the world's major cities as well. Just as the nations of the world are on the point of complete annihilation, Jesus Christ will return, destroy the remaining forces, subdue the Antichrist, and establish his millennial kingdom on earth.[82]

In one form or another, Lindsey's corpus embodies such themes. Throughout that corpus, major emphasis is placed on the technologizing of biblical prophecy.[83] Already present in the various references that *The Late Great Planet Earth* makes to nuclear weapons and the deployment of ballistic missiles, the cinematic version of that book is even more dramatic.[84] Produced as a film in 1977, *The Late Great Planet Earth* is narrated by Orson Welles, whose deep, resonant voice lends an air of authority and verisimilitude to the entire affair.[85] After reenacting the circumstances surrounding the biblical prophets, the film moves to the twentieth century. There, Hal Lindsey addresses the audience from the "actual" battlefield of Megiddo on the southern rim of the plain of Jezreel, where Armageddon is to take place. Pointing to natural disasters (earthquakes, tornadoes, floods, and pollution), the growth of atheism, New Age beliefs, and the reliance on the occult sciences, Lindsey invokes Ezekiel 38 as the source of the prophecy that Russia and its confederates will attack Israel from the north in preparation for the major conflicts to follow. It is all in Ezekiel. Indeed, "Ezekiel's prophecy," Lindsey declares, "reads like a blueprint of the present situation." This is a situation in which the final battle will be waged by nuclear armaments, tanks, ICBMS, and bombs of the most awesome sort. At the center of the struggle is Israel, which, as an independent nation, must protect itself from being attacked by the brutal and unrelenting enemy.[86]

To lend even further credence to Lindsey's focus on the "Holy Land" in the context of his nuclearization of Ezekiel's prophecy, Major General Chaim Herzog of Israel offers his view that the prophetic warnings encoded in Ezekiel will no doubt come about.[87] Israel is indeed as much at the center of today's world as it was in Ezekiel's day. Be forewarned, General Herzog declares: apocalyptic history will be realized in Israel. The dispensationalists are entirely correct in their reading of the biblical text. If we are living in the era of what will ultimately be known as "the late great planet earth," it is because of the coming together of end-time events that all revolve around Israel. Those events represent the nuclear fulfillment of all that is prophesied in the oracle of Ezekiel and now made entirely clear by the interpretations of Hal Lindsey. General Herzog is only one among several other "authorities" in the film version of *The Late Great Planet Earth* called on to lend credence not only to Lindsey's predictions but also to the idea that biblical prophecy invites the act of technologizing the oracular. Here, an allegiance to *technē* as the true source of gnosis assumes an urgency reinforced by a pronounced belief in the power of the biblical text to determine the conduct of international politics.

If Israel assumes center stage in these politics, those who are determined to attack it are demonized in the most extreme form. As Paul Boyer has brilliantly demonstrated, this act of demonization is founded on a politics of racial difference that pits Israel against its enemies. Prophesied in Ezekiel, end time is a matter not simply of nation against nation but of race against race. Beware the Red Peril (the Russians), the Yellow Peril (the Chinese), and the Black Peril (the African nations), Lindsey warns: all conspire to destroy the true people of the Holy Land. The vast hordes of the perilous races will overrun not only Israel but the planet itself in a final conflagration. Among the forces of evil foreordained to shape the course of history in the last days, Russia and its allies, the Arabs and the "darker-skinned peoples of Africa and Asia," are for Lindsey especially to be feared. They are at the source of anxieties about end time. As much as one might prattle about the importance of racial and international harmony, the end-timers know better: dangers lurk in foreign nations and alien people of all stripes. Eventually, those dangers will coalesce in an overwhelming cataclysm of global proportions.[88]

Such attitudes are discernible not only in Lindsey but in a host of prophecy believers who espouse corresponding views. Expanding the coming attack on Israel from the northern front to the four corners of the globe, end-timer Robert Glenn Gromacki, for example, declares that four major coalitions will confront Israel. The western powers will be led by the Antichrist, whose forces include "the nominal Christian white segment"; the northern powers are the atheistic, communistic, materialistic peoples of Red Russia and its allies; the southern forces are composed of the Muslim, Pan-Arab alliance led by Egypt and the "nationalist Black African peoples," both of which have always hated the Jews and have a "death pact" against them; and, finally, "the eastern king-doms can be seen in the yellow, Oriental peoples of Red China and her neigh-bors," united by race, religion, and years of oppression. "The stage is prepared; the players are costumed; and the time is near. In the fullness of God's time, the curtain will go up, and the tribulation events, including Armageddon, will take place." Armageddon is "just around the corner."[89] Underlying prophecy belief, then, is a set of assumptions based on categories of race. At one extreme are the Israelis as the chosen people of God whose cause must be upheld at all costs. At the other extreme are the alien hordes, composed of reds, yellows, blacks, and even misled whites (nominal Christians), that threaten the very foundations of existence and that will ultimately be obliterated by a righteous God in the battle to end all battles. All this has been prophesied: it may be found perfectly delineated in biblical prophecy and in Ezekiel with his oracle against Gog.[90]

The demonization of the "other" that this oracle portrays strikes at the heart of Lindsey's discourses on UFOs, the very embodiment of otherness. So in *The 1980's: Countdown to Armageddon* (1981), Lindsey professes his belief in the ex-istence of UFOs and maintains that there will be a "close encounter of the third kind" shortly. The driving force of this phenomenon is "some type of alien being of great intelligence and power." In fact, it is a "demon" in a state of war with God. Such demons will be empowered to exercise their influence during the end-time period by staging a spacecraft landing on earth. The chief of the demons is Satan, who will tempt the world into total error in the final days.[91] Over a decade later, Lindsey reiterates his stance in *Planet Earth—2000 A.D.* (1994). In an extended discourse on UFOs, he looks on their presence as evi-dence that Jesus' own admonition to expect "terrors and great signs from heaven" (Luke 21:11) will be realized in the form of UFOs. Over the years, he once again asserts, he has become thoroughly convinced that UFOs are "real." Repeat-ing his observations from *The 1980's: Countdown to Armageddon,* he maintains that UFOs are spacecraft operated by alien beings, extraterrestrial and super-natural in origin. Filthy, ugly, repulsive, dark, and evil, they are frightening creatures whose intentions are entirely sinister. The very embodiment of anx-ieties directed against that which is "other," these aliens from space are the counterpart of those descending to attack the "true believer" from the outer-most regions. The object of the true believer's fears and, indeed, his hatred as well, they are the demons come to make their presence known in the last days.[92]

The purpose for which these demons descend from the heavens is what Lindsey calls "the Great Deception." This event for Lindsey has a biblical basis in 2 Thess. 2:8–12, which speaks of a time when the "Wicked shall be revealed" under the auspices of Satan, who shall bestow on them "all powers and signs and lying wonders." Those who fail to receive the "love of the truth" will be deceived by these wicked demons, "and for this cause God shall send them strong delusion, that they should believe a lie: that they all might be damned who believed not the truth, but had pleasure in unrighteousness." In the context of Lindsey's prophecy belief, the Great Deception will be perpetrated during the period of the tribulation that accompanies the attack of Gog and his alien forces on Israel (cf. Rev. 20:3, 8). Those who fail to realize the full import of this event will be deceived and ultimately destroyed by the warfare and suffering that ensue. Because of the extent to which the Russians in particular are demonized as agents of Satan in Lindsey's outlook, they, more than any other people, are associated with the UFO phenomenon. So Lindsey observes that "Russia and the former Soviet Union have been experiencing a dramatic increase in encounters with unidentified flying objects." Citing numerous examples of such encounters, Lindsey in effect conjoins the demonization made evident in his portrayal of the space aliens with the demonization made evident in his portrayal of the Russians. Agents of Satan both, they descend on the unsuspecting to deceive, to exact suffering, and ultimately to destroy.[93]

In the context of this coming forth of the demons (whether as agents of Satan or as agents of Gog) during the terrible period of tribulation, Lindsey assures us that those who are adherents of the "True Way" will escape the UFOs. Just at the point that the UFOs are landing, the redeemed "will be snatched up to be with the Lord."[94] With its counterpart in the descent of space vehicles to destroy the world, the experience of being snatched up is known in dispensationalist circles as the Rapture. This event finds its source in 1 Thess. 4:16–17, which declares that the Lord "shall descend from heaven with a shout" to call up both the saved who have died and the saved who are alive: "Then we which are alive *and* remain shall be caught up together with them in the clouds to meet the Lord in the air: and so shall we ever be with the Lord."[95]

For those like Lindsey who believe that the Rapture is to occur before the onset of the millennium (Rev. 20:2–6), such an experience is looked on as a great source of comfort. More than that, it is one of the profound mysteries that define premillennialism.[96] Comparable in import to the transport and transcendence that Enoch (Gen. 5:24) and Elijah (2 Kings 2:11) experienced in being taken up to God, the Rapture becomes the premillennialist answer to the experience of ascent. If Enoch defies his mortal bounds by "walking with God," Elijah undergoes his own chariot ascent: "Behold, *there appeared* a chariot of fire, and horses of fire," and "Elijah went up by a whirlwind into heaven." As types of the Rapture, these ascents represent the means by which premillennialists view themselves as riders in the chariot.[97]

Conceiving himself as chariot rider, Lindsey, like his visionary forebears, does not hesitate to transform visionary matter into technological wonder. In fact, he makes a point of associating the Rapture with the launching of rocket ships. In his book on the Rapture, he comments that he "recently watched in awe as the space shuttle blasted off into space. Within a matter of a few minutes it was out of sight and traveling at more than six times the speed of sound." In comparison to this event, the Rapture will surpass all such marvels of technology. As believers, we, too, "will suddenly one day just blast off into space." As a result of this blastoff, the world will no doubt hear "a great sonic boom from all our transformed bodies cracking the sound barrier."[98] Those committed to such beliefs among the end-timers involved in the Pantex nuclear power plant, discussed earlier, did not hesitate to look on "blast-off for the Rapture" as "God's space program." Transported into outer space, they knew that they would be "traveling 186,000 miles a second," as they sang on their way like a meteor, "right up into the firmament above Amarillo, Texas."[99] So widespread is this belief that bumper stickers have appeared that bear the message "In Case of Rapture, This Car Will Be Driverless" and "Beam Me Up, Jesus."[100] In some circles, such an experience is not only technologized but sexualized as well. It is the union of bride and bridegroom (cf. Rev. 21:2), indeed, "a kind of heavenly sexual intercourse," one in which "the passively waiting, virginal, and submissive female church" is "snatched up" by "Jesus, standing erect in the clouds."[101]

The technomysticism with which the Rapture is imbued in premillennialist thought distinguishes the experience as an event of utmost significance to those determined to escape the ravages brought on by the tribulations promised in Ezekiel's oracle against Gog. Particularly as it is portrayed through a language of sexuality, the Rapture becomes an experience in which technology is divinized to the point of mystical transcendence. Rocket ships, ballistic missiles, and nuclear armaments of one sort or another represent the "*object*ification" of realizing union with "the One." The rhetoric underlying such "*object*ification" amounts to what has been called "a kind of madness."[102] If this is the case, it is a madness that we have encountered before in various contexts associated with the figure of Ezekiel and his fascinating visions and oracles. With its belief in technomystical ascents and transcendent unions, the Rapture becomes the counterpart to the experience of the *merkabah* itself. Fully technologized into a transport of the most sublime nature, it renders the premillennialists consummate riders in the chariot.

Arming

the

Heavens

4

As suggested by the foregoing chapter, the prophet Ezekiel has had a profound impact on the traditions of fundamentalism and evangelicalism during the course of the past century and a half both in this country and abroad. Whether in the form of the inaugural vision or of its counterpart, the oracle concerning Gog, the Book of Ezekiel has prompted fundamentalist and evangelical alike to resort to biblical prophecy as the essential source of an abiding concern with end time, especially as that concern represents a crucial aspect of prophecy belief. More than anything else, this belief is political in its bearing.[1] Apocalyptic at its core, prophecy belief transforms the Bible into politics. Nor is the political dimension limited to closed circles or conventicles of prophecy believers whose influence rarely extends beyond the margins of their immediate social and religious spheres. As we have seen, the profound effect that prophecy belief has had on an entire cross section of American life is well documented. Reaching millions of viewers, television evangelism alone has assumed an institutional status in the United States.[2] The influence of end-timers has been enormous not only among the not-so-silent "moral majority" of the American populace but within the very corridors of power where national decisions are made. According to Susan Harding, U.S. foreign policy debates since World War II have been significantly attuned to the kinds of thinking associated with prophecy belief. What Harding calls a "dispensational narrative framework" has underscored the manner in which the entire foreign policy establishment, including U.S. presidents, responded to both national and international crises and to victories and defeats at home and abroad.[3]

Harding is prompted to ask rhetorically: "What difference does apocalyptic belief make?" (Implicit in her question is the impact of apocalyptic belief at the highest political levels.) Her response is telling. James Watt, secretary of the interior under Ronald Reagan, declared publicly "The End" was near, "and he turned nature over to business at an alarming pace." Did he behave in this manner, Harding asks, "because he knew this era of human dominion was almost over and there was no point in protecting the environment?" This question is followed by others. President George Bush is known to have consulted dispensationalist preacher Billy Graham during the Persian Gulf Crisis, including the day the commander in chief ordered the bombing of Baghdad and started the war. Did Graham inform Bush that, in undertaking this war, he might be furthering events that were part of God's providential plans? In keeping with such beliefs, Harding is moved to question whether born-again Christians supported the Persian Gulf War because of their millennial outlook.[4] Although cause-and-effect relations of this sort are difficult to establish, they do raise interesting questions about the effect of end-time thinking on matters of public policy. Within this context, Ezekiel's visions and oracles assume a crucial role.

The extent to which end-time thinking has influenced public policy in recent times is especially evident in presidential politics. This dimension can be approached from two perspectives. The first might be called the incorporation of end-time thinking into fictions of power, the second the implementation of end-time thinking in the exercise of power. Both perspectives are the products of a prophecy belief grounded in Ezekiel's visions and oracles, and both involve the nuclearization of the Bible by individuals who have either aspired to or held high office. Imbued with a perspective that is distinctly apocalyptic, these individuals are consummate avatars of *technē*, figures for whom eschatology and technology have their source in the ineffable and for whom the sacred text of biblical prophecy yields concrete answers to its otherwise inscrutable questions.

As a figure who at one time moved at ease in the corridors of power, Charles W. Colson is an excellent representative of the first perspective, that which arises from the act of incorporating end-time thinking into fictions of power. A textbook case of religious conversion (of being "born again"), Colson's life has been fashioned and refashioned in his own self-portrayals of the movement from sinner to saint on the road to Damascus.[5] Having served as special counsel to President Richard M. Nixon from 1969 to 1973, Colson was found guilty of offenses related to Watergate in 1974, following which he served seven months in prison. During the period of this personal tribulation, Colson underwent a conversion that put him on the true path of undertaking a Christian ministry as a spokesman for criminal justice reform.[6] However one might view Colson's conversion and the path that he has followed since his Watergate days, one fact is clear: his White House experience placed him at the very seat of power. For this reason, his act of incorporating end-time scenarios into fictions of power at the highest reaches of political office bears the stamp of verisimilitude as well as exhibiting a certain urgency. These fictions translate prophecy belief into the eerie possibility of realization. As such, they become warnings of what (under the right circumstances) might well come to pass.

The force of this observation is made evident in an "Ezekiel drama" that Colson portrays in the prologue to his *Kingdoms in Conflict.* It is a drama delineated not to engender prophecy belief but to serve as a reminder of its potential for influencing political decisions even within the presidency. Colson is an able portrayer of end-time scenarios. Occurring in 1998, shortly before the onset of the second millennium, this particular scenario assumes the form of a political parable of the promised end. The parable is founded squarely on Ezekiel's oracle against Gog. In addition to the figure of one Shelby Hopkins, president of the United States, six actors round out the parable: General Brent Slocum, chair of the Joint Chiefs of Staff; Larry Parrish, White House chief of staff; Alexander Hartwell, secretary of defense; Henry Lovelace, secretary of state; Alan Davies, national security adviser; and Hyman Levin, attorney general. Each has his role in either questioning or supporting the president, a national leader whose entire administration is founded on the Bible and its inerrancy in guiding policy decisions.[7]

Hurriedly called to an emergency meeting by President Hopkins, the advisers gather in the Lincoln Sitting Room of the White House on this most important of occasions. At this meeting, the president discloses a confidential intelligence communiqué indicating that a crisis has emerged in Israel as the result of the activities of the Tehiya, a small fringe party that favors driving all the Arabs out of the occupied territory. Under the leadership of one Yosef Tzuria, Tehiya is committed to blowing up the Dome of the Rock, the sacred Muslim shrine in Jerusalem, and to building a temple in its place. This act of destruction and reconstruction will be undertaken to restore the Temple to its true place as the original site of worship and as the foundation of Israel's ancient heritage. As mad as this scheme sounds, it has the financial backing of not only "big industrialists in Israel" but also fundamentalist groups in the United States. More alarming still is that Tehiya has arrived at a tacit agreement with the Likud Party under Moshe Arens to proceed without intervention. If allowed to pursue these plans, Tehiya might well precipitate a major international confrontation.[8]

In the light of this crisis, President Hopkins must decide how to proceed. Guided by his own strong evangelical convictions, the president acts in the only manner he deems appropriate: he consults the book of Ezekiel, the one document that underlies all high-level international policy decisions. Putting on his reading glasses, and picking up his "well-worn brown-leather Bible," he says to his assembled cabinet, "At the risk of appearing fanatical, I'd like to read you all a passage from Ezekiel," and then cites what is essentially a conflation of Ezek. 37:26–27: "My dwelling place will be with them: I will be their God, and they will be my people. Then the nations will know that I the Lord make Israel holy, *when my sanctuary is among them forever.*" A prologue to the Gog oracle, this passage in effect justifies for the president the notion (represented in the italicized verses) that, whatever happens to the Temple, it will ultimately endure. Although currently displaced under the Muslims, the Temple must reassume its rightful place as the dwelling place of the true God. With this idea in mind, President Hopkins recites the mantra of Gog: "Ezekiel tells us that Gog, the nation that will lead all the other powers of darkness against Israel, will come out of the north. Biblical scholars have been saying for generations that Gog must be Russia. What other powerful nation is to the north of Israel? None. But it didn't seem to make sense before the Russian revolution, when Russia was a Christian country. Now that Russia has become communistic and atheistic, it does. Now that Russia has set itself against God, it fits the description of Gog perfectly."[9]

This is all the justification the president requires to determine his course of action. Although he knows full well the consequences of allowing Yosef Tzuria, with his Tehiya party, to sabotage the Dome of the Rock, the president will do nothing to prevent the "inevitable," which has been fully delineated in Ezekiel. The president knows that blowing up the Muslim holy site is the only means by which the Temple can be restored to its rightful place. This restoration, in turn, provides the way for the onset of Armageddon, followed by the coming of Christ. In response to the foreign policy revelation afforded by Ezekiel, the president obliges his advisers to follow him in the act of kneeling in prayer, as he asks for continued divine guidance. By phone, he later consults his evangelical guide to refresh his memory on the significance of the Rapture and the Tribulation. Armed with such knowledge, the president is confident that he is in possession of the true path: all is in order, all as it should be. The Dome of the Rock is destroyed by Israeli commandos; the Russians are in a state of alert to retaliate; and the United States is on the verge of war to counterretaliate. Armageddon may now proceed as planned.[10]

As an addendum to this parable, Colson notes, almost in passing, that, "although [his] story is fictional, certain quotations attributed to Israeli and U.S. political and religious leaders have been taken from actual public statements; material regarding the takeover of the Temple Mount is also taken from public records."[11] This is a fascinating aside for it suggests the extent to which fact and fiction are liable to converge. If Colson does not specify the sources of the "actual public statements" and the "actual public records" he has in mind, one is inclined to accord him the benefit of the doubt. Given the impact of end-time thinking on the traditions of prophecy belief explored thus far, the Ezekiel scenario that Colson portrays has the ring of "truth" about it.

A cautionary tale by one who is himself "born again," Colson's parable concerning the potential excesses of end-time thinking at the highest levels transforms what might otherwise be deemed mere fantasy into a drama that arises from Colson's own personal experiences during his White House years. One is prompted to see in this drama a residue of past meetings, of intrigues, of conversations in high places. The suggestion of public statements and public records might well be bolstered by that which is not public, that which remains undisclosed. At the very least, the names that Colson bestows on the characters in his drama recall those who served in past administrations. Henry Lovelace brings to mind Henry Kissinger, Alexander Hartwell Alexander M. Haig Jr., and Hyman Levin Edward Levi. The figure of Shelby Hopkins, in turn, is a composite of the presidents between the time of Nixon and Reagan. One need not push the analogies too far. The point is that the specter of end time may haunt any administration. As the crucial source of prophecy belief in its most extreme form, Ezekiel is an all-pervasive presence in the political arena.

Such an idea assumes a particular cogency in the figure of evangelist Pat Robertson, in some respects the living embodiment of Colson's Shelby Hopkins. In the possible playing out of an end-time scenario, Robertson is apposite as one who aspired to the highest political office in the land. His aspirations might well have been realized: in 1988, Robertson marshaled enough support to place third behind George Bush and Robert Dole in the Republican nomination for the presidency of the United States.

Grace Halsell notes that, on 9 June 1982, three days after Israel commenced its invasion of Lebanon, Robertson appeared on television to explain to his viewers the horrors of a forthcoming battle of Armageddon. Beginning his program by reiterating a prediction he had made some months earlier, he said, "I guarantee you by the fall of 1982, that there is going to be a judgment on the world, and the ultimate judgment is going to come on the Soviet Union." With television cameras following him, Robertson moved to the blackboard, and, with his pointer indicating the Middle East, he paraphrased Ezekiel's oracle against Gog. Then, explaining how the oracle includes Russia, Armenia, Libya, South Yemen, and Iran, he declared that "this whole thing is now in place" and "can happen at any time." By the fall of 1982, the Ezekiel prophecy will be fulfilled. "The United States," he said, "is in that Ezekiel passage," and we are waiting for the final battle.[12] From the perspective of such a prediction, one is invited to conjecture what might have occurred had Robertson's aspirations to high office ultimately been fulfilled. As a reflection of deeply held end-time beliefs, the kinds of statements that Pat Robertson ventured some six years before he sought his party's nomination provide a clear indication of the principles on which his own foreign policies would have been based. The fictional Shelby Hopkins does indeed find his "real-life" counterpart in Pat Robertson.

Under the circumstances, it is perhaps something of an irony that Robertson has come forth with his own end-time fiction, one in which the Antichrist himself becomes president of the United States. Titled *The End of the Age*, this novel portrays the world on the eve of the final days.[13] Presaging the political turmoil that is to follow, natural disasters abound from the very outset. A giant meteor blasts the coast of southern California, leaving incredible tidal waves and earthquakes in its wake. Millions of people are destroyed, the West Coast is obliterated, and a residue of ash shrouds the world, threatening global winter and an all-pervasive darkness. As a result of these disasters, the president of the United States in office at the time acknowledges fault in not taking proper precautions to respond to such events, and, during a nationally televised address, he shoots himself. In his place arises one Mark Beaulieu, possessed of unearthly powers. Taking control of the Oval Office, he becomes the new president. In opposition to all who seek to undermine his power, he sends out Truth Squads. What results is Armageddon, as the Antichrist engages the Christian Resistance Movement in final combat.

Robertson's depiction of Mark Beaulieu embraces all the elements of racial stereotyping associated with notions of end time. On the surface, Beaulieu might be the "handsome" or "noble" (*beau*) "place" or "lineage" (*lieu*) suggested by his name, but underneath he is the embodiment of all that is "other," all that is suspect and ugly. Delineating how this otherness came to possess Beaulieu, the narrator returns us to the time that Beaulieu was (ironically) a Peace Corps volunteer stationed in Rajahmundry, a city in India north of Madras. There he encountered the statue of Shiva, the Hindu god of destruction, whose consort is Kali, goddess of death. Legend has it that Shiva's task is first to destroy the world and then to rebuild it to perfection. Encircling the head and shoulders of the stone god is a cobra as a sign of wisdom.[14]

Approaching this mysterious pagan statue and gazing at its face, Beaulieu first felt something "icy cold" course through his being and then heard the statue speak to him with the promise of wisdom and power. As he returned to the statue day after day, Beaulieu fell under the spell of Shiva and his cobra. He knew one day that he would "destroy the corrupt nations of the world—including America and all its greedy capitalists," after which he would gloriously rebuild what he had destroyed. Empowered by Shiva's statue, Beaulieu consulted the guru Raj Baba, who responded to his guest as one possessed: his "head jerked violently back and forth, his mouth and jaw grew slack, and spittle drooled out of the side of his mouth." From eyes no longer human, the guru looked before him with an overpowering malevolence, as a voice within him declared to Beaulieu, "You have been chosen by Shiva," who has bestowed on you "power and wisdom to destroy, then create again. You will rule your people in the name of Shiva." Taught "the secrets of the ascended masters" by the guru, Beaulieu became aware that "a powerful demonic spirit had entered the very core of his being."[15]

Here are anxieties concerning all that is deemed foreign, exotic, esoteric, and opposed to the one true belief embodied in Christianity. Infected by "foreign" influences, Mark Beaulieu becomes a prime candidate for the office of Antichrist since, by his very nature, he is opposed to Christianity and seeks its undoing. All that is not infused with the light of the one true faith is opposed to it and therefore deemed suspect. As that which is totally alien, Hinduism and the people who accept it become the site for anxieties of the "other." These anxieties are brought to the fore in Robertson's other writings as well. Discoursing on the "alien gods" in *The New Millennium*, for example, Robertson singles out Hinduism to warn the "true believer" about that which runs counter to Christianity. Hinduism, he says, has its origins in demonic power. "The product of a totally alien culture," Hinduism brings nothing but "degradation" on a nation like India. In the rigors of their worship, Hindus seek physical transcendence through "out-of-body experiences, ecstatic trances, and the visitation of alien spirits." Such a totally alien form of belief "can only lead people to poverty, anguish, and ultimately, to demon possession."[16] As the site for anxieties of the "other," Hinduism encompasses those alien beliefs manifested in all cultures that run counter to true Christianity. For Robertson, these include not just Hinduism but the Islamic religions as well as the atheistic beliefs that beset the doctrines of Communism. Such systems of belief and the cultures they infect represent a threat to all humanity. In the figure of Mark Beaulieu, that threat is fully realized.

Accordingly, in *The End of the Age*, Mark Beaulieu, as the embodiment of the "other" reflected in alien cultures, comes into his own. As one possessed, he releases his demon powers to the four corners of the globe. Now president of the United States, he is the true Antichrist. Even after an attempt on his life, he is able to perform such miracles as coming back to life. Except for the true followers of Jesus and a "small handful of Orthodox Jews," these events convince everyone (including "the leading professor of apologetics at the University of Chicago") that Beaulieu is the true Messiah. Accorded that belief, he sets about to conquer the world. This is to be accomplished through the conquest of Israel and the subjugation of the Jews. Summoning troops from Libya, Ethiopia, and the Sudan, he seeks to "envelop the entire nation of Israel and then surround the city of Jerusalem." Momentarily controlling the skies with his missiles and airpower, the Antichrist appears to prevail. "But, in the skies above the skies, another battle was being waged with much greater intensity." Under the leadership of the archangel Michael, the "principalities and powers of Satan's host" undergo a humiliating defeat. The forces of God swoop down from the skies, seize Satan and his hosts, and condemn them to the lake of fire. The Antichrist is defeated and the world saved.[17]

Whether in the form of Gog and his forces descending from the north or of the Antichrist (a.k.a. Mark Beaulieu) and his forces launching an all-out attack on Israel, the narrative of end time underscores the thinking of Pat Robertson as much as any prophecy believer we have encountered. But the pervasiveness of end-time theology in Robertson's outlook becomes all the more significant because of the political bearing it assumes in the context of what had been his aspirations to the presidency of the United States. "What difference does apocalyptic belief make?" Susan Harding asks. The answer lies in such figures as Pat Robertson through whom end-time thinking might well have become the order of the day, had he in fact become president. Conceiving such a presidency within the framework that both Colson and Robertson provide for it, one is prompted to ask who would have assumed the highest office in the land, after all: Shelby Hopkins or Mark Beaulieu? Is one, in fact, the alter ego of the other? Whereas Shelby Hopkins wreaks havoc by attempting to save the world from apocalyptic confrontation, Mark Beaulieu wreaks havoc by attempting to destroy the world through apocalyptic confrontation. Both are the culmination of all the anxieties that end-time thinking fosters. If Shelby Hopkins and Mark Beaulieu are "fictions," they are also the frightening embodiment of that "reality" created by the potential of fiction to assume the form of truth, of the desire for power to become power. Had Pat Robertson realized his presidential ambitions, end-time fiction might have become end-time fact.

Whereas Charles Colson and Pat Robertson transform end-time thinking into fictions of power, that transformation was ultimately fulfilled by one who actually did attain to the presidency of the United States, Ronald Reagan. As scholars have amply demonstrated, Reagan was an inveterate end-timer whose outlook fully reflected the impact of prophecy belief.[18] In him, the visions and oracles that distinguish the prophecy of Ezekiel flourished anew. Revealing throughout his career a deep commitment to prophecy belief in general and to end-time theology in particular, Reagan provides fascinating insight into the nature of the forces through which the Bible becomes a textbook for policy decisions. As president, he proved, however, to be neither a Shelby Hopkins nor a Mark Beaulieu. Rather, he engaged in his own form of self-fashioning, one that balanced the needs of high office with the impulse to subscribe to a particular religious ideology, an ideology that assumed fundamental importance to his frame to mind.

Inculcated in the Bible and deeply influenced by his mother, a member of the Disciples of Christ, Reagan was more than inclined to what has been defined as the evangelical point of view. Various accounts concerning his associations with such figures as Jerry Falwell and Billy Graham are well known.[19] Particularly revealing is an experience Reagan had in 1970, during his second campaign for the governorship of California. Visited by friends and associates, including Pat Boone and two evangelical-charismatic Christians, George Otis and Harald Bredesen, Reagan had a long talk with them about prophecy and end time, along with the significance of the "outpouring of the Holy Spirit." This conversation, in turn, was followed by a prayer session in which Reagan and his guests stood in a circle with heads bowed and hands clasped left and right while Otis, overcome with the Spirit as he held Reagan's hand, prophesied the governor's eventual ascendancy to the White House. As Otis's hand shook during this proclamation, Reagan's hand shook, too, in confirmation that the Spirit had also taken hold of the governor.[20] In a subsequent taped interview with Otis, Reagan discussed the Battle of Armageddon. When asked if he felt that he would be Raptured and thereby escape the Tribulation during the final battle because of the experience of being born again, Reagan responded by saying "Yes, I have had an experience that could be described as being born again."[21]

As a presidential candidate in 1980, Reagan reiterated such views, based on a firm belief in the imminence of Armageddon. Speaking to evangelist Jim Bakker of the PTL network, Reagan declared: "We may be the generation that sees Armageddon." After he became president, he continued to share these sentiments in his various addresses to the National Religious Broadcasters and the National Association of Evangelicals.[22] As a reflection of his commitment to the evangelical perspective, it appears that Reagan arranged for Jerry Falwell to attend National Security Council briefings to discuss a nuclear war with Russia. In keeping with such high-level meetings, Hal Lindsey, as noted earlier, was invited to share with Pentagon officials his views on the possibilities of an American/Russian nuclear confrontation.[23] Influenced over the years by such works as *The Late Great Planet Earth*, Reagan was clearly "hooked on Armageddon."[24] So committed was he to the idea of Armageddon that it became a well-publicized issue in his run for a second term in 1984, when the issue was raised by Marvin Kalb in the second debate between Reagan and the Democratic presidential candidate, Walter Mondale.[25] Making headline coverage in the *New York Times* and elsewhere, Reagan's obsession with the theology of Armageddon and the politics of end time assumed both a currency and an urgency consistent with his long-held belief in the "truths" augured by biblical prophecy in general and Ezekiel's oracles in particular.[26]

This fact assumes a dramatic resonance in an account provided by Senator James Mills, formerly president pro tem of the California State Senate. From the perspective of the debates concerning Armageddon that distinguished Reagan's run for office in 1984, Mills recalls an event that occurred almost twenty-five years earlier at the outset of Reagan's second term as governor of California in 1971. The event in question is a dinner conversation that Mills had with Reagan in Sacramento at an annual banquet hosted by state lobbyists. Although the importance of Mills's story has been noted elsewhere, the disturbing ironies implicit in his dramatization of his conversation with Reagan have not received the attention they deserve.[27] I emphasize this aspect to note that Mills provides just the right tone in suggesting the bizarre nature of the mise-en-scène he establishes in his account. Far from compromising what Mills calls the "serious implications" of the conversation, the incongruities that underscore it demonstrate just how profound those implications are.

As the lights dimmed to prepare the dinner guests for the "fiery entrance" of the cherries jubilee that the headwaiter brought to the table, Reagan was just completing one of his many stories, this one about the time he was a young radio announcer engaged in interviewing celebrities of various sorts. The joviality of the story, which entailed Reagan's experience of being given a hair-raising ride back to town by Barney Oldfield, the famous race car driver who had come to town for one such interview, was perfectly in keeping with the festive atmosphere. But it must have been the darkness of the room and the eerie luminescence of the cherries jubilee that abruptly altered Reagan's mood for, rather like Mona Strangely lying out under the stars on a beautiful Texas night, Reagan suddenly moved to the subject of Ezekiel. Turning to Mills, the governor asked if his colleague had ever read Ezekiel 38 and 39. A bit unsettled, Mills assured his dinner partner that, having grown up in a household of "bibliolatrous Baptists," he knew these and other oracles quite well. That was apparently not enough for Reagan: simply having read and studied the oracles did not qualify one to understand their hidden meaning, their political and apocalyptic significance as harbingers of end time. To understand what the oracles encode, one had to be a true cognoscente of the oracular, a genuine votary of prophecy belief, schooled in the art of deciphering biblical typology as a source of revelation concerning the events of our own time. Only when the "knower" is equipped with this ability will he be able to see dispensational history unfold and the ties between past (biblical time) and present (end time) fully realized. Then all will be "in place" (a recurring phrase in the parlance of end-timers) for the oracles to be fulfilled.

Imbued with prophetic fervor, Reagan as avatar began with "firelit intensity" to intone and to interpret the oracles. Ezekiel, he declared, "had foreseen the carnage that would destroy our age." In Ezekiel 38, he proceeded, "it says that the land of Israel will come under attack by the armies of the ungodly nations, and it says that Libya will be among them." Pausing to allow this aspect of the oracle to register, Reagan asked, "Do you understand the significance of that? Libya has now gone communist, and that's a sign that the day of Armageddon isn't far off." (The outlook reflected in such a statement may account in part for Reagan's authorization of an attack on the Libyan cities of Tripoli and Benghazi some fifteen years later after his conversation with Mills. Focusing on Colonel Mu'ammar Gadhafi as the source of terrorism, Reagan maintained that he was determined to undermine terrorist activities of "the crackpot in Tripoli." If the putative cause of the attack was the horror of terrorism, the politics of Armageddon no doubt underscored the president's motivations in launching the strike against Libya.)[28]

During the course of his interchange with Reagan, Mills first considered responding to the governor's Ezekiel-inspired views by blowing out the flames on his cherries jubilee but decided that doing so might violate the awesomeness of the occasion. Instead, Mills weakly protested that, although Ezekiel includes

Ethiopia (along with Libya) among the evil powers, Emperor Haile Selassie, that Lion of Judah, had not been taken into the communist fold. Reagan agreed: "The Reds have to take over Ethiopia." Nonetheless, this is inevitable, the governor assured Mills, for such an outcome must occur to fulfill Ezekiel's prophecy that "Ethiopia will be one of the ungodly nations that go against Israel." In portraying the conversation, Mills reflects that it appeared as if Reagan were actually looking forward to the downfall of Haile Selassie. (Mills notes that, when the emperor was indeed deposed by a military coup some three years after his conversation with the governor, Reagan might well have felt confirmed in his belief that the end was irrefutably at hand.)[29]

On the basis of this interchange, it appears that those committed to a particular view of dispensational history are determined to see the events of that history unfold in a manner consistent with the belief that all must be "in place." Nothing must be allowed to question the view of things falling into place for then the terms of the prophecy itself would be compromised. Even if it appears that all is not "in place" or at least in the process of falling into place, there are always explanations to justify the dispensational perspective. If the dispensationalist happens to be in a position of high political office, then perhaps events can be "influenced" in a manner to ease the course of history so that things do indeed fall into place. At the very least, the question arises whether the policy decisions of an end-timer in high political office will be founded on preconceived notions concerning the way in which history is to work itself out. For this reason, Mills expresses at the outset of his account what he calls "a degree of internal unrest." Under the circumstances, one might suspect that that unrest is a bit more than "a degree." Nor did anything else that transpired during Mills's conversation with Governor Reagan do much to alleviate the sense of foreboding that Mills felt as he learned the true meaning of Ezekiel's oracles.

Illuminated by the fiery glow of the cherries jubilee, Reagan proceeded further to demonstrate the extent to which all was "in place" for Armageddon to occur. Most important was the prophecy concerning Israel in Ezekiel 38. There the text declares that God will take the captive Israelites from among the heathen and gather them again in their homeland. Alluding to the establishment of Israel as an independent state in his own time, Reagan said that the ingathering had finally come about after two thousand years. "For the first time ever," Reagan assured his acolyte, "everything is in place for the battle of Armageddon and the second coming of Christ." Despite Mills's feeble attempts to qualify such statements, Reagan as oracle refused to be silenced. His voice rose as he recited the mantra: "Everything is falling into place. It can't be long now. Ezekiel says that fire and brimstone will be rained upon the enemies of God's people."

In the world of those who are avatars of *technē*, Reagan knew precisely what the phrase *fire and brimstone* entails. Fully prepared to undertake his own nuclearization of the Bible, he declared that Ezekiel's prophecy regarding fire and brimstone clearly means that the enemies of God will be destroyed by nuclear weapons. Taken aback, Mills felt that at this point Reagan had entirely assumed the mantle of the inspired preacher as he concluded his prophecy sermon with the assured identification of Gog with Russia: "Ezekiel tells us that Gog, the nation that will lead all of the other powers of darkness against Israel, will come out of the north. Biblical scholars have been saying for generations that Gog must be Russia. What other powerful nation is to the north of Israel? None." When Russia was a Christian country before the revolution, the identification did not make sense. "Now that Russia has become communistic and atheistic, now that Russia has set itself against God," it "fits the description of Gog perfectly." Everything is in place. Nuclear confrontation is inevitable, Armageddon inevitable, the Second Coming inevitable.

Looking back on this conversation from the perspective of Ronald Reagan's presidency, Mills reflects on the implications of what has transpired. The notion of the Soviet Union as an "evil empire" is not simply "rhetoric" but an idea rooted in the basic convictions of prophecy belief. In response to those convictions, the dispensationalist grounded in the prophecy of Ezekiel is obliged to make certain that the forces of righteousness remain powerful so that the "all-important conflict" will be won. With these thoughts in mind, Mills conjectures that, if the Reagan of the present shares the same beliefs as the Reagan of the past, then his policy decisions will no doubt be influenced by the terms of his end-time thinking. Reviewing Reagan's domestic and foreign policies, Mills concludes that such is precisely the case.[30] Certainly, it might be argued that profoundly held religious beliefs tied to notions of end time do have a bearing on decisions rendered in a secular context. For our purposes, those beliefs demonstrate the importance of the oracles of Ezekiel to both the national and the international arenas during Reagan's tenure in high public office, culminating in his assumption of the office of the presidency.

However one conceives that presidency, one point is clear: the invitation to view Reagan as president in the context of Ezekiel and his oracles gives rise to a renewed emphasis on the import of the biblical text as a source of visionary enactment. In the world of prophecy belief that Reagan inhabits, this import finds expression as much in the inaugural vision as in the oracle against Gog. As we have seen, each is in many respects a counterpart, if not a reinscription, of the other. Both emerge from the northern regions, and both share similar characteristics (whirlwinds, chariots, fire, clouds). From the perspective of prophecy belief, the visionary contexts of the chariot, on the one hand, and the oracle against Gog, on the other, invite the kinds of eschatological interpretations that are the stock in trade of end-time thinking.

In accord with that thinking, the oracle against Gog ultimately assumes a redemptive cast as it is reconfigured through the inaugural vision as a sign of God's coming forth at the end of time to redeem his faithful and to damn his enemies. Among the prophecy believers, this event draws on the apocalyptic implications associated with the defeat of the forces of Satan at the battle of Armageddon. So Merrill F. Unger speaks of God's ultimate appearance as the Son of man descending in his glory to conquer his enemies as "the outshining of God's glory emblazoning the heavens" (cf. Matt. 24:29–30). Faced with "light more resplendent than the sun," the "demon-driven armies of the world will behold the splendor of Him who is the rightful Possessor of the earth." Through "His spoken word," the "irresistible Conqueror" will "slay the huge armies" assembled at Armageddon. "In a moment," Unger says, God "will annihilate the greatest display of military might this world has ever assembled" (Rev. 19:11–16). Faced with this prospect, the satanic hordes will futilely turn their weapons on the Lord of glory. But the enemies of God will find those weapons powerless: "Not only will all munitions and weapons fail," but "the kings and their armies will be mowed down in death as the 'sharp sword' that proceeds 'out of his [God's] mouth' (Rev. 19:15) speaks death to His foes as it speaks life to His own." Confronted with "the overpowering glory of a far superior foe," the satanic hordes will be entirely thwarted in their vain attempts to repel "what they recognize as an invasion from outer space."[31] In the minds of prophecy believers, the glory of God's coming forth to overwhelm the enemy finds its aptest counterpart in an outer-space invasion. End-time thinking and UFOs go hand in hand. But this is precisely as it should be in the reconceptualization of Ezekiel's inaugural vision as end-time event. For end-timers, that vision represents nothing less than God's own celestial chariot/UFO propelled by "four angelic creatures riding in 'wheels within wheels.'"[32] In his glorious chariot/UFO, God comes forth in the final battle to destroy the demon-driven armies of Gog as a sign of his incomparable power. By means of this invasion from space, the forces of chaos will be stilled, and order will be restored in the ultimate redemption of God's faithful.

I emphasize this mode of end-time thinking because of the bearing that it assumes in Reagan's own attempt to confront the forces of chaos through all the means available to him as president. For one haunted by the prospect of Armageddon, Reagan moved toward the conviction that the most effective response to apocalyptic battle was not through an aggressive military strike against Gog and its allies (unless such a strike were absolutely warranted) but through a defensive form of self-armament that would protect God's faithful in the face of possible annihilation.[33] As important as it was to be prepared for a counterstrike should the onset of war provide no other alternative, the madness of annihilation as the result of "mutual assured destruction" (MAD) might be averted through defensive means. From the perspective of end-time thinking, this view held out the hope that the inevitability of Armageddon might possibly be compromised. Representing a new twist to the prospect of end time, such an outlook on Reagan's part emerged over a lifetime of "performance" both as an actor and as a politician.

During the period that Reagan was an actor, he performed the role of Secret Service agent Brass Bancroft in the 1940 movie *Murder in the Air.*[34] Set against the background of World War II, the film portrays a nation under siege by enemy saboteurs. The House Committee on Un-American Activities (HUAC) subpoenas Joe Garvey, the chairman of the Society of Loyal Naturalized Americans. Clearly an "alien" who speaks with a foreign accent, Garvey defends himself and his organization against the accusations of HUAC. Nonetheless, he is a saboteur who leads a ring of foreign spies. Members of the ring bear a secret tattoo mark indicating that they are "Reds." Garvey's purpose is to steal a miraculous defensive weapon that paralyzes electric currents at their source. A portable, handheld device that resembles a rocket ship, this weapon is known as the "inertia projector," a marvel of technology that emanates powerful invisible rays capable of neutralizing any target. According to the military, the inertia projector will "make America invincible in war and therefore be the greatest force for peace ever invented." To prevent Garvey from making off with the inertia projector, Brass Bancroft of the Secret Service is enlisted. The hero of the drama, Bancroft infiltrates Garvey's spy ring, undermines its plans, and saves the inertia projector. Once again, America is successful in thwarting the encroachments of its enemies. With its emphasis on Red-baiting, *Murder in the Air* presages the turn from the anti-Nazi films of World War II to the anti-Communist films of the cold war. Were it not that Brass Bancroft was acted by Ronald Reagan, the film would be forgotten. As has been noted on several occasions, the importance of the film resides in its impact on one who firmly believed that through the remarkable technology of antiweapons like the inertia projector, with its all-powerful rays, world conflict might well be resolved and the annihilation brought about by Armageddon possibly averted.[35]

On the basis of an unswerving faith in the wonders of technology that extended into the period of his political life, Ronald Reagan (now as president of the

United States, rather than as Brass Bancroft, Secret Service agent) over forty years later provided an even more compelling answer to Armageddon. That answer assumed the form of a reconceived inertia projector that would come to be known as the Strategic Defense Initiative (SDI), later designated "Star Wars."[36] No longer simply a portable handheld device resembling a rocket ship, this new laser-powered, space-based nuclear antimissile system would be Reagan's way of responding to any nuclear assault that the forces of Gog might be inclined to launch.[37] By means of the new inertia projector, God's faithful would be rendered entirely unassailable.

On 23 March 1983, Reagan introduced the idea of such an antimissile system in a widely publicized television broadcast to the nation and, indeed, the world.[38] The time was right for his speech, which was delivered during a period of heightened public debate fraught with a variety of nuclear anxieties. "From within a cauldron of anti-Soviet enthusiasm, technological exuberance, continuing attacks on the strategic status quo from both left and right, and discussion by officials in the Reagan administration about fighting and winning nuclear war," the president seemed to offer his nation renewed hope in what he called "the nuclear age." What has been termed the *grand vision* portion of his address was an add-on to a speech that defended the need for increased defense spending because of the growing Soviet threat.[39]

Beginning with a detailed account of Soviet nuclear power (the history of its development and the nature of its accomplishments), Reagan proceeded to share with the television audience detailed "aerial photographs, most of them secret until now," to demonstrate his point. Rather like Brass Bancroft in his role as Secret Service agent, the president invited the television audience to participate in the act of surveying classified documents, an act tantamount to a "debriefing" in preparation for the revelation of a tactical course of action. The audience was thereby made a party to the "secret world" of military personnel at the highest levels. Weaponry, arsenals, maps: the power of Gog was visually and verbally transmitted through the airwaves. Within this ominous context of a renewed cold war, Reagan held out hope by sharing with the world "a vision of the future" through which Armageddon might be avoided. This grand vision would entail the act of embarking on a program of defensive measures to counter "the awesome Soviet missile threat." Reagan's plan would involve harnessing the powers of a technology that could "intercept and destroy strategic ballistic missiles before they reached our own soil or that of our allies." With full faith in the resources of the scientific community, Reagan assured the world that "current technology has attained a level of sophistication where it is reasonable for us to begin this effort." He concluded his address by calling on the scientists "who gave us nuclear weapons to turn their great talents to the cause of mankind and world peace; to give us the means of rendering these nuclear weapons impotent and obsolete." To that end, he announced plans for "directing a comprehensive and intensive effort to define a long-term research and development program to achieve our ultimate goal of eliminating the threat posed by strategic nuclear missiles," an effort that would "pave the way for arms control measures to eliminate the weapons themselves."[40]

This speech was later followed by a special report from the State Department called "The Strategic Defense Initiative." The report provided a detailed rationale for the mounting of such a program and elaborated the terms on which SDI would be founded. Consistent with the "visionary" thrust of the president's speech, the report looked forward to "a future in which nations can live in peace and freedom," comforted in the awareness that "their national security does not rest upon the threat of nuclear retaliation."[41]

Although SDI had its detractors, it also had its fierce supporters, including Lieutenant General Daniel O. Graham, former head of the Defense Intelligence Agency and founder of High Frontier, a missile defense advocacy organization, and Edward Teller, a principal architect of the hydrogen bomb. Joining Graham and Teller were other influential advocates who represented Reagan's "kitchen cabinet" and advised him on pursuing a research program to develop strategies for nuclear defense.[42] At the forefront of the movement was Teller himself. He maintained that he had devised a top-secret device known as the nuclear X-ray laser, code named Excalibur, a name imbued with all the legendary implications of its Arthurian predecessor. Armed with Excalibur, he was determined that it was conceivable to "channel the energy of an exploding hydrogen bomb into beams of intense X-rays that would flash through space and destroy enemy missiles."[43] This is the thrust of Teller's essay "The Antiweapons," which addresses the nature of laser-powered devices (equipped with the ability to emit "high-intensity infrared beams") used for the purpose of disarming nuclear warheads and weapons of mass destruction. Defending the principle of SDI by means of such devices, Teller exults in the power of laser technology: "There is no such thing as an ultimate weapon," he declares, "but a laser represents the ultimate in speed. Lasers are the best antiweapons, and their development has produced the unexpected result of making defense feasible."[44]

On 23 July 1983 (four months after Reagan's "vision" speech), Teller wrote the president to urge accelerated support for research on the nuclear X-ray laser. In his letter, Teller claimed that advances in nuclear weaponry would end the era of MAD and begin a period of "assured survival" in ways "favorable to the Western alliance." The madness of MAD would be counteracted by the saneness of MAS (mutual assured survival). It is possible that this letter helped Reagan understand for the first time that the key breakthrough behind the antimissile quest was in fact to arise from nuclear weapons, those very devices he had just publicly vowed to make "impotent and obsolete." Nonetheless, Reagan pressed for funding of the antimissile defense system, and by 1986 the Lawrence Livermore National Laboratory, over which Teller exerted great influence, emerged as one of the top contractors receiving "Star Wars" funds.[45]

However the Strategic Defense Initiative is conceived, one fact remains clear: its identity as "Star Wars" reflects a convergence of high and low cultures that captured the popular imagination both in its own time and for generations to come. This convergence is largely the result of SDI's association with the immensely popular George Lucas film *Star Wars* (1977).[46] As "Star Wars," SDI is grounded not simply in that one film: the "Star Wars" epithet encompasses the entire trilogy (*Star Wars, The Empire Strikes Back,* and *Return of the Jedi*) that first appeared in the late 1970s and early 1980s. Although Reagan and his advisers were initially annoyed at the use of the "Star Wars" epithet, they not only dropped their objection to it but eventually welcomed such a usage.[47] In effect, "Star Wars" reconfigured SDI as an interplanetary (indeed, intergalactic) science-fiction phenomenon. Given the cultural implications to which the *Star Wars* trilogy gave rise, this reconfiguration is understandable. Such is made especially apparent in the remarkable events that distinguish the first film.

As the film begins, we are informed that the evil Galactic Empire with its imperial governors is in the process of dominating every planet within its reach. Having subjugated the Old Republic (once under the protection of the Jedi knights), the Empire has developed the Death Star, an immense armored space station able to annihilate entire planets. In opposition to the machinations of the Empire, the Rebel Alliance has undertaken a battle to restore the Old Republic. To be successful, the Alliance must find a way of destroying the Death Star. Enter Princess Leia Organa of Alderaan, guiding spirit of the Alliance. In possession of the stolen plans for the Death Star, she races home aboard her starship, pursued by the imperial hosts of the Empire. Attempting to regain the plans, the forces of the Empire (under the command of the dreadful Darth Vader, Dark Lord of the Sith) overtake the starship and imprison Princess Leia. Although in Darth Vader's power, she nonetheless manages to secure the plans within the robot Artoo Detoo (R2-D2), who, with the android See-Threepio (C-3PO), escapes in a pod to the planet Tatooine.

On Tatooine, a drama of recognition unfolds whereby the young, ambitious Luke Skywalker begins to discover his lineage and the nature of his mission under the tutelage of the old Jedi knight Obi-Wan ("Ben") Kenobi. To elude the forces of the Empire hot in pursuit on Tatooine, Ben obtains the services of the mercenary Han Solo, along with his traveling companion, Chewbacca, to transport him and Luke in the Corellian pirate ship the *Millennium Falcon*. Their ultimate mission is to get the plans for the Death Star into the hands of the Alliance, but first they must liberate Princess Leia, now imprisoned by Darth Vader on the Death Star. (To demonstrate the immense power of the Death Star, as well as to intimidate Princess Leia, the imperial commanders of the Empire unleash from the Death Star a huge laser beam, which obliterates Princess Leia's home planet, Alderaan.) After a series of machinations on the Death Star (including a one-on-one combat between Darth Vader and Obi-Wan Kenobi), Luke Skywalker and company manage to free Princess Leia and make their way in the *Millennium Falcon* to the secret Alliance stronghold, where they are greeted by the rebel forces. Now in possession of the plans of the Death Star, the Alliance discovers how to destroy it. In the ensuing battle that culminates the action, the destruction of the Death Star is finally achieved through the heroism and skill of Luke Skywalker.

Such in brief is the plot of *Star Wars*. For our purposes, the importance of this film (as well as the trilogy as a whole) resides in the bearing of its thematic implications not only on the culture of SDI but also on the fantasies of empowerment that underscore Reagan's end-time outlook. Underlying *Star Wars* is a belief in what is known as "the Force."[48] Articulated by Obi-Wan Kenobi, the old Jedi knight, the Force is a mysterious phenomenon through which the Jedi derives his power: "It's an energy field created by all living things. It surrounds us and penetrates us. It binds the galaxy together."[49] Although the Force has never been properly explained, scientists have attempted to analyze it. Only certain individuals have been able to recognize the significance of the Force, but even they are branded as nothing more than charlatans and mystics. Fewer still can actually make use of the Force, which is too powerful for most to master: "Knowledge of the Force and how to manipulate it was what gave the Jedi his special power."[50]

A *dynamis* that derives its potency from the *mysterium tremendum,* the Force is the most archaic of all phenomena.[51] In *Star Wars*, it is associated with what might be called the "old religion," that of "romance" and "myth," of times gone by. As such, it is spurned by those in the present, those averse to the old ways that are no longer understood. An emblem of the "romanticizing" and "mythologizing" of the Force, the weapon of choice wielded by the Jedi knights is the so-called lightsaber. Retrieving one from an old chest, Obi-Wan Kenobi introduces Luke Skywalker to the weapon. The saber that Ben gives him once belonged to Luke's late father, a Jedi knight. Taking hold of the saber, Luke

Figure 12. Luke Skywalker with his lightsaber.
Reproduced from *Star Wars, Episode IV: A New Hope,*
Lucasfilm Ltd., 1977.

pushes a button on the handle, "and a long beam shoots out about four feet and flickers there."[52] "Strangely, Luke felt no heat from it, though he was very careful not to touch it. He knew what a lightsaber could do, though he had never seen one before. It could drill a hole right through the rock wall of Kenobi's cave—or through a human being."[53] In order to assume the role of a Jedi knight, Luke must become proficient in the use of the lightsaber. By means of this weapon, the Force would find release (fig. 12).

Those divorced from the ways of the past have no time for such "nonsense." When Luke is engaged in learning how to use the lightsaber aboard the *Millennium Falcon,* Han Solo scoffs: "Hokey religions and ancient weapons are no match for a good blaster at your side, kid." Luke responds with the question, "You don't believe in the Force, do you?" Han has his answer ready: "Kid, I've flown from one side of this galaxy to the other. I've seen a lot of strange stuff, but I've never seen anything to make me believe there's one all-powerful force controlling everything. There's no mystical energy field that controls my destiny." The idea of the Force even meets with contempt among Darth Vader's own compeers. Alluding to the power of the Death Star, Vader observes: "Don't be too proud of this technological terror you've constructed. The ability to destroy a planet is insignificant next to the power of the Force." To this, another commander responds: "Don't try to frighten us with your sorcerer's ways, Lord Vader. Your sad devotion to that ancient religion has not helped you to conjure up the stolen data tapes, or given you clairvoyance enough to find the Rebels' hidden fort."[54]

Within the context of SDI, one recalls Edward Teller's top-secret X-ray laser, code named Excalibur. Channeling the energy of an exploding hydrogen bomb into energy beams, this weapon (or antiweapon) is not simply a wonder of modern technology. With its ties to ancient legend and mythopoetic belief, it is the Arthurian marvel that conjures up lost memories of courtly romance. Drawing on untold reservoirs of power, power that infuses the cosmos, it is at once the past reclaimed and the future realized. Made available through Merlins and mystics, it far transcends "mere technology": its source is the ineffable, and, properly deployed, it is used for the purpose of fostering "the Good": saving damsels in distress and regaining lost kingdoms. Equipped with his lightsaber, the Arthurian Jedi knight stands ready to defend himself against any Mordred who might oppose him. He is able to do so because his Excalibur is infused with the energy of the Force.

In the Manichean universe that George Lucas envisions, the Force, however, has a "dark side": it can be destructive as well as regenerative. Once himself a Jedi knight for the good, Darth Vader has unfortunately succumbed to the power of the dark side. Under the influence of this power, Vader becomes the "Dark Father" suggested by his name. As the Dark Father, he is intent on destruction in fulfillment of the cause of the evil Empire. A towering figure of momentous strength, Vader is described in the script as an "awesome, seven-foot-tall Dark Lord," whose "face is obscured by his flowing black robes and grotesque breath mask." Accompanying him are the imperial stormtroopers with their "fascist white armored suits." Suggested in the description is the kind of gnostic dualism through which the elemental conflict between good and evil is conceived. The politics of the conflict are equally clear. Embodied in Darth Vader and his stormtroopers are the fascistic, imperialistic forces that must be overcome if goodness is to prevail. The troops of Gog are gathered in the north, where they are prepared to destroy the true kingdom with the powerful beams from their Death Star. It is up to Luke Skywalker to save the day. As he successfully destroys the Death Star with laser torpedoes shot from his X-wing fighter, he hears the voice of Obi-Wan Kenobi resounding in the cockpit: "Remember, the Force will be with you . . . always."[55]

"May the Force be with you": this is a salutation that permeates *Star Wars*. For SDI, it is a salutation that assumes a particular resonance in the noble war of good against evil, particularly as that war assumes a place in the popular imagination. From its very inception, SDI was not only a high-tech program but also an idea that enjoyed widespread popular appeal.[56] As such, it was "a mélange of physics, psycho-politics and metaphysics." Defined by a "specialised language of evasion and euphemism," it was conceived through a discourse of "nuke-speak," equipped with its own subdialect of acronyms such as AMOS (airforce Maui optical system), TOM (threat object map), SAM (surface to air missile), and SPOCK (special purpose operating computer kernel). In the corresponding world of popular culture, the language of nukespeak was redefined through a discourse consistent with the terms of the *Star Wars* trilogy. Drawing on the conflict between good and evil dramatized in the *Star Wars* film and its sequels, the first director of the SDI Office alluded to America's cold war conflict with the Soviet Union by remarking that, "in the movie, the good guys won, because the Force was on their side." In keeping with this observation, the director then proudly declared, "I am convinced that the Force is with us."[57] Equipped with the power bestowed on it by SDI, America as Luke Skywalker will ultimately defeat the Soviet Union as Darth Vader.[58]

As Edward Reiss comments, "Star Wars is about power: technological power, economic power and military power." In their quest for power, the scientists at the Livermore National Laboratory were themselves "Star Warriors" whose ability to harness laser-equipped weapons (hypervelocity guns, space-based nuclear beam weapons, etc.) bestowed on them an almost supernatural aura characteristic of superheroes in the medium of science fiction. (The Livermore scientists were reputedly avid science-fiction readers.) The powers associated with SDI, in turn, were made available in the marketplace through an entire range of "Star Wars" video games. "To appreciate the benefit of video games," Reagan observed at the Walt Disney Epcot Center two weeks before his SDI speech, we need merely "watch a twelve year old take evasive action and score multiple hits, while playing Space Invaders."[59] (Such "Star Wars" games are now available on the Internet.)[60] Like "Star Wars" itself, these games permit us to play the hero, to become spectators as a fulfillment of empowerment fantasies.

If such empowerment fantasies reside at the core of Reagan's own belief in SDI, they must be counterbalanced by what appeared to be his professed willingness to share the workings of SDI at the appropriate time with other powers, including the Soviet Union, an idea looked on as folly by politicians of every stamp.[61] Rooted in his hope for "mutual assured survival," Reagan's offer was entirely genuine. Lou Cannon suggests that the impulse underlying such an offer can be traced back to the mid-1940s, when Reagan joined the United World Federalists, a utopian organization committed to the idea of creating a single, peaceful nation on earth. It is possible that the spirit of the United

World Federalists underscored Reagan's magnanimity. But corresponding factors might also have been involved, factors once again drawn from the cinematic world of fantasy that Reagan knew so well. These factors return us to the realm of outer space. During his career, Reagan was obsessed with the possibility of invasions from worlds beyond earth's orbit and how the nations of the world would respond to such invasions. Addressing the United Nations General Assembly on 21 September 1987, Reagan observed that, in our anxieties about international conflict, what we need is some outside threat to make us recognize our common bond. "I occasionally think how quickly our differences worldwide would vanish," he said, "if we were facing an alien threat from outside this world."[62]

In accord with such an idea, Reagan's national security adviser (then Lieutenant General Colin L. Powell) believed that the film *The Day the Earth Stood Still* (1951) was a major source of inspiration for the president's utopian gesture.[63] A science-fiction classic that portrays the alien not as destroyer but as savior, this film dramatizes the visitation of outer-space beings who come to earth on a mission of peace. What they find on arrival is a planet beset by international tensions. Having discovered the power of atomic energy, earth is on the verge of annihilating itself. The purpose of the aliens is to prevent that annihilation because of its impact not only on this planet but on the rest of the universe. Emerging from their spacecraft (which has landed on a baseball field in Washington, D.C.) are Klaatu, an alien of high intelligence, and Gort, a huge robot armed with a ray capable of neutralizing the forces of any opponent. Faced with the possibility of a deadly threat, Gort fixes his enemy in his line of vision, his visor recedes, and a blinding beam of light emanates from his eyes (fig. 13). Having enlisted the support of an eminent scientist as well as a young woman and her son, Klaatu is almost destroyed by the police and military forces on earth but is miraculously revived by the remarkable powers available to him aboard his spacecraft.

Before Klaatu and Gort return to their galaxy, Klaatu announces his intentions to those bent on destroying him (as well as each other). He declares that the planets beyond earth are aware that the inhabitants of this planet, having discovered "a rudimentary kind of atomic energy," are experimenting with rockets of potential mass destruction. "Soon," he says, "one of your nations will apply atomic energy to spaceships" that will create "a threat to the peace and security of other planets." This represents a danger that the other planets will not abide. "There must be security for all, or no one is secure." In response to this fact, Klaatu observes, "we have an organization for the mutual protection of all planets and for the complete elimination of aggression." This organization

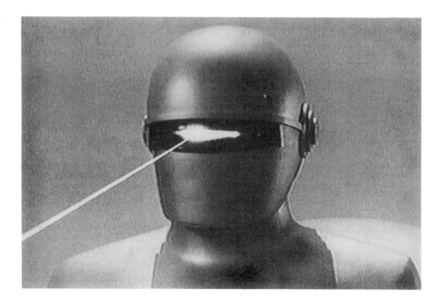

Figure 13. Gort.
Reproduced from *The Day the Earth Stood Still,*
Twentieth Century Fox, 1951.

depends on robots such as Gort, whose function is to patrol the planets in spaceships and preserve the peace. "In matters of aggression, we have given them absolute power over us. This power cannot be revoked. At the first sign of violence, they act automatically against the aggressor." As a result, "we live in peace, without arms or armies, secure in the knowledge that we are free from aggression and war." Although we may not have achieved perfection, Klaatu says, "we do have a system" that works. Earth's choice is simple: "Join us and live in peace or pursue your present course and face obliteration. We shall be waiting for your answer. The decision rests with you."[64]

With its emphasis on interplanetary space travel, the reality of UFOs, the need for international cooperation, the power of technology, and the beneficial effects of sharing remarkable discoveries for the good of all humanity, *The Day the Earth Stood Still* is precisely the kind of film that might well have appealed to the utopian visionary determined to establish the means by which genuine peace could be achieved in the present and in the centuries that lay ahead. Whether in the form of the inertia projector, the lightsaber, or Gort the indomitable, the technological instrument able to harness the powers ("the Force") that lie hidden in the inscrutable recesses of the mysterium becomes the means of ultimate salvation. For this reason, a missile defense system shared by all nations became for Reagan not simply a vehicle through which one might neutralize the enemy but a genuine force for good among all nations, a force embodying both restorative and transformative powers, the proper use of which would bring about the resurrection of a paradisiacal time of peace for generations to come.[65]

As a response to an Armageddon theology tied to the end-time attack of the forces of Gog portrayed in the oracles of Ezekiel, SDI offered Reagan a viable alternative to the threat of ultimate annihilation. Founded in fantasies of "ray technology" through which weapons are empowered to disable all that stand in their way, the fantasy of "Star Wars" became Reagan's answer to the visionary notion of the blinding light of God's glory descending from "outer space" (a "space invasion") to neutralize the satanic hordes. A veritable *ḥashmal*, this blinding light emanates from Reagan's UFO, his weapon in space. It is his Chariot of Paternal Deitie with its multitude of eyes that "glare lightning" and shoot forth "pernicious fire" among "th'accurst," whose strength is "witherd" and whose vigor is entirely "draind," as they are left "exhausted, spiritless, afflicted, fall'n" (*Paradise Lost*, 6.846–52). At the same time, it is also his fulfillment of the anxiety dream of Edwin Broome's Schreberian psychotic, whose so-called influencing machine with its potent rays is able to neutralize any enemy.

While Reagan was president, he read Robert Serling's *Air Force One Is Haunted* (1985), a potboiler inspired by the advent of "Star Wars."[66] Beset by a multitude of problems (national as well as international) that confront both the country and his administration, President Jeremy Haines is accorded the way to salvation. This salvation assumes the form of the ghost of Franklin D. Roosevelt (Ronald Reagan's own hero), who suddenly appears to Haines as he sits alone deep within the private recesses of Air Force One.[67] Like a prophet come from beyond, Roosevelt makes his presence known in order to counsel Haines on the most effective means of solving his nation's domestic problems and of staving off the menace of the Soviet Union. As the imposing vehicle through which great spirits of the past manifest themselves to their acolytes, Air Force One itself is imbued with an otherworldly aura: it becomes the "official" medium of visionary experience.[68] For Haines, such an experience ultimately resolves itself in a faith in the power of an antimissile energy beam satellite system (the heir of SDI) that is being perfected in order to guarantee the security of the United States against possible nuclear attack. A test of the antimissile system proves successful. Moving at the speed of light, "laser streaks and invisible energy beams" are projected from the "killer satellites" to neutralize their targets.[69] All those who witness the successful test respond with elation and joy. Opposing any attempts to use the antimissile system to back up a direct offensive strike against the Soviet Union, Haines takes the ultimate risk of negotiating directly with the enemy. He succeeds. Armageddon is averted, and the prospect for a lasting peace is assured.

This is a metanarrative of SDI that Reagan would no doubt have endorsed. Whether or not Haines is conceived as Reagan's alter ego, the events culminating in the deployment of the antimissile system (of which SDI is the prototype) and the successful negotiations with the enemy (later undertaken by Reagan himself in his summit meeting with Gorbachev in Reykjavík) represent the kind of "romance" to which Reagan might have responded enthusiastically.[70] With its celebration of technology as the means of salvation, its emphasis on otherworldly interventions, and its happy ending, Serling's novel represents the perfect fantasy of the thwarting of Armageddon by the very means through which the fulfillment of end time might otherwise have been brought about. As a true believer in Armageddon theology grounded in the oracles of Ezekiel, Ronald Reagan was a consummate avatar of *technē* committed to the glorious future that lay ahead.

Ideology,

Eschatology,

and

Racial

Difference

Heralding

the

Messenger

5

Whereas the first half of this study has explored a wide variety of instances in which Ezekiel's vision manifests itself in the modern world, the second half embarks on a different mode of exploration, one that focuses in depth on a particular movement, its ideology, genesis, and impact, as a means of coming to terms with the act of technologizing the visionary. In this way, the expansive nature of the first half of the study finds its counterpart in a detailed investigation of a phenomenon that has become the focus of ever-increasing attention in recent years. That phenomenon is the Nation of Islam, a fascinating, albeit controversial and politically charged, movement that has assumed genuine significance in the cultural life of America. With its all-pervasive emphasis on matters of end time, its enduring concern with the import of visionary enactment, and its determination to refashion this enactment in its own terms, the Nation of Islam offers itself as a crucial moment in the assimilation of the ineffable into the consciousness of the modern world. At the very center of this ineffability is Ezekiel's *visio Dei*, which is as crucial to the Nation, its ideology, and its theology as any other encounter with the sacred that has emerged throughout history. Particularly as a manifestation of the category of race, the vision of Ezekiel galvanizes what the Nation views as fundamental to its sense of self and its perception of that which defines it and its mission. In a sense, all that has been explored up to this point has been undertaken in preparation for the detailed investigation of the presence of the vision that inaugurates the prophecy of Ezekiel in the consciousness of the Nation of Islam.

To understand that consciousness, to appreciate what constitutes it, one must explore in depth the life and thought of those through whom the mysterious world of the Nation is made known. For that reason, the second half of this study is devoted to the Honorable Elijah Muhammad, on the one hand, and to the Honorable Louis Farrakhan, on the other. To probe the experiences of these figures, to explore their lives and their thought, is to gain a unique insight into the extent to which Ezekiel's vision is transformed in its new setting. By means of an account of the Honorable Elijah Muhammad and the Honorable Louis Farrakhan as the children of Ezekiel, all that has been associated with the biblical prophet and his message in the first half of this study will be regarded in a new light, the light of "otherness" embodied in the two leaders through whom the Nation of Islam has assumed its importance in today's world. Bestowing on these leaders their sense of mission is that most enigmatic of figures Master Fard Muhammad. In him, as shall be seen, Ezekiel's *visio Dei* comes fully to fruition. To understand the nature of this fruition, we begin with an exploration of the life and thought of the Honorable Elijah Muhammad.

According to C. Eric Lincoln, Elijah Muhammad is one of the most remarkable men of the twentieth century. His achievements include "enormous contributions to the dignity and self-esteem of the black undercaste in America, as well as the reintroduction of Islam to the United States." As the Messenger of Allah, he dedicated himself to the "restoration of the most despised and brutalized segment of American Christianity back to a level of dignity and self-appreciation." The slums of the black ghettos were his initial "parish," his potential converts the "slum-created outcasts of a developing technocratic society." His followers were those "most battered by racism and stifled by convention."[1] Scorned and repudiated by those who did not understand him, he must be considered a true prophet and liberator.[2]

Born Elijah Poole on 7 October 1897 in Sandersville, Georgia, Elijah Muhammad was the sixth of thirteen children raised by William and Mariah Poole in a sharecropper's shack on a tenant farm in a rural Southern hamlet midway between Macon and Augusta.[3] Elijah Muhammad would maintain that, because both his parents had been slaves of a white family whose surname was

Poole, this represented his "slave" name. Elijah Muhammad's father was an itinerant Baptist preacher who attempted to make a living for his family as a sharecropper. In 1900, the Pooles left Sandersville and settled in Cordele, in south central Georgia. Attending a ramshackle "colored" school until somewhere between the fourth and the eighth grades, the young boy labored at home to help his family survive. Although little is known about his early life, sources indicate that he worked as a field hand, a railroad laborer, and an employee of a sawmill and a brick manufacturing concern. During this period, he met and courted Clara Evans, and the two were married on 17 March 1919.

For the newly married couple, this was a time of intense struggle and uncertainty, a time of poverty, racism, and infestations of biblical proportions as a result of the onset of the boll weevil. It was also a time plagued by the prospect of violence against blacks, including the horror of lynchings, a commonplace occurrence. (When Elijah Poole was only a child, he witnessed such a horror in Cordele.)[4] Having undergone sufficient suffering in his native state to last him a lifetime, he joined the Great Migration and moved with his wife and two children to Detroit in April 1923.

Arriving at its destination, the family settled in a house that lacked even bare essentials, such as a toilet. A refugee from the South, Elijah Muhammad had heard extravagant praise of Detroit from the blacks of his own hometown who had moved to the city after the end of World War I. The rhapsodies, of course, proved to be entirely unfounded. Confined to a new bleak world of oppression, uncertainty, and exploitation, he supported himself and his family as well as he could. Early drawn to the notion of black separatist organizations, he joined the Marcus Garvey movement shortly after moving to Detroit.[5] From 1923 to 1929, he held various jobs, including positions with the American Nut Factory, the American Wire and Brass Company, the Detroit Copper Company, and the Briggs Body and Chevrolet Axle Company. During this period, additional children were born to the Pooles, and Clara helped the family by working as a domestic. The Great Depression set in late in 1929, after which Elijah Poole faced long periods of unemployment: "Every morning, he left home before dawn to join the lines of thousands of unemployed people that had formed in front of the gates of the local manufacturing plants. He often waited there all morning, hoping to be hired for a day's work, only to return home empty-handed to his hungry children."[6] The situation was clearly one of despair.

Certain patterns in this brief overview appear to loom large. Elijah Muhammad was shaped by a desire for liberation and the determination to forge his own identity and his own destiny. Emerging from a lineage of slavery, he sought to cast off the shackles of a culture that bound him and his people, a culture that was not his own. Although his formal education did not extend beyond the elementary grades, his range of experience was immense. His migration to Detroit represented an attempt to achieve a better world for himself and his family. With that attempt, however, came the realization that Detroit was hardly the Promised Land. Despite a struggle to support himself and his family, he faced the ravages of the Great Depression along with the ongoing indignities of racial discrimination.[7] Nevertheless, he remained undaunted and fiercely independent.

During the difficult years in Detroit, Elijah Muhammad's life underwent a transformation.[8] This event occurred as the result of the appearance in 1930 of a mysterious fair-skinned man who went from door to door selling silks and raincoats in the "Paradise Valley" neighborhood of Detroit.[9] It quickly became clear that this man was not to be regarded as simply an itinerant peddler. On the contrary, this was an individual of the greatest magnitude, one who brought and indeed embodied the message of liberation for the blacks. Maintaining that he came from the holy city of Mecca, he identified himself as Master Fard Muhammad.[10] Fostering an aura of mystery about himself both in his nomenclature and in his presence, he declared that he revealed only what was necessary concerning his person because the time had not yet come when he would be beheld in his "royal robes." Nonetheless, he assured his followers that he was their brother and that through him they would attain freedom, justice, and equality. His purpose, he said, was to awaken the black nation to its wonderful potential in a world then dominated by the white "blue-eyed devils." He taught his followers about the deceptive nature of whites and encouraged them to embrace what he portrayed as the glorious history of black Afro-Asia. As the destiny of that history came to be realized, so, too, would the ignominious history that the white race embodied. That history would culminate in the destruction of a "spook civilization" dominated by white persecution. This eschatology was fundamental to the outlook of Master Fard's converts.

Within a few years, Master Fard established an organization so effective that it featured not only its own Temple of Islam, along with its appropriate ritual and worship, but a University of Islam (actually a combined elementary and secondary school). Through Master Fard's preachings, and as part of the curriculum of his school, converts were taught that they were not Negroes but members of the lost tribe of Shabazz.[11] As a prophet, Master Fard made it clear that he came to America to restore life to his long-lost brethren, who were now able to understand that they were the original people, aptly called the Lost-Found Nation of Islam. Accordingly, they learned that to realize their destiny they had to regain their religion, Islam; their language, Arabic; and their culture, astronomy and "higher mathematics." Their foundation text was the Qur'an, along with the Bible and some of the literature of Freemasonry.[12] Inculcated in the law of Allah, they agreed to live a pure life and to abide by strict dietary and moral codes of behavior. Only then might they return to their place of origin, the holy city of Mecca, which for them was the paradise from which they had been stolen. Essential to Master Fard's teachings was the concept of end time. In these teachings, the commonplace idea that God will come on the last day to resurrect the lost sheep assumes renewed significance as a manifestation of the Mahdi or Messiah who will be revealed in his true splendor.[13]

Along with the Temple of Islam and the University of Islam, Master Fard's organization was distinguished by the Fruit of Islam (FOI), a military entity whose members were drilled by captains and charged with protecting the adherents to the Nation of Islam. Aided by a staff of assistants, a minister of Islam was appointed to run the organization. After Master Fard established permanent headquarters in the first Temple of Islam, he registered all the members and promised to restore their true names to replace the slave names imposed on them by the "Caucasian devil." Each new convert received his "original" name (Jam Sharrieff, Hazziez Allah, Anwar Pasha, and the like), which was revealed to the prophet through the Spirit of Allah. By 1933, Master Fard had organized his followers so effectively that he was able to recede into the background. Such a mysterious withdrawal served to strengthen the belief that he was indeed the "Supreme Ruler of the Universe" (i.e., the God Allah), who temporarily assumed an earthly form to fulfill his divine mission.[14]

The influence of Master Fard on the Honorable Elijah Muhammad was profound. Having met Master Fard early in the master's ministry, Elijah Muhammad underwent a spiritual transformation that determined the course of the remainder of his life. Despite the "falling out" that Malcolm X later experienced with Elijah Muhammad and the Nation of Islam, Malcolm X's own account of the extent and nature of Elijah Muhammad's conversion speaks eloquently to the issue. In his *Autobiography,* Malcolm X implicitly casts the conversion in the form of a biblical drama, one reminiscent of the interchange between Moses and the dwelling presence of Yahweh in the burning bush on Horeb (Exod. 3:1–14). There, the prophet receives his calling and, in his quest to learn the name of the deity, discovers that it is the "I AM" that addresses him. Correspondingly, Malcolm X creates a drama of conversion, one in which his own master assumes the vocation of the prophet. As an audience to this drama, Malcolm X sits galvanized as Elijah Muhammad recalls the revelation he experienced in his realization that Master W. D. Fard represented the fulfillment of all prophecy regarding the true religion of the black man. "I asked Him," says

Elijah Muhammad, "Who are you, and what is your real name?" He replies, "I am The One the world has been looking for to come for the past two thousand years." Probing yet further, Elijah Muhammad asks again, "What is your *true* name?" Only then does the master disclose that his name is Mahdi, the Messiah who has come to guide his followers onto the right path. Like the Moses of the biblical account, Elijah Muhammad heeds his calling and sets about liberating his people.[15] Such an account suggests the extent to which the circumstances surrounding Elijah Muhammad's assumption of his role are both framed and to be understood in visionary terms. From that perspective, Elijah Muhammad viewed himself in the context of his own prophetic vocation.[16]

Crediting Master Fard with removing him from "the gutters in the streets of Detroit" and teaching him the true knowledge of Islam, Elijah Muhammad devoted himself tirelessly to the master and to the movement. Indeed, he became Master Fard's most trusted lieutenant. Although Elijah Muhammad had been given the "original" surname Karriem at his initiation, Master Fard eventually acknowledged his higher status by bestowing on him the name by which he is now known, Elijah Muhammad.[17] When Master Fard dubbed him "Minister of Islam and Messenger of Allah," he was virtually designating the disciple as his successor. In that role, Elijah Muhammad was almost singlehandedly responsible for the deification of Master Fard and the perpetuation of the master's teachings. Master Fard became not only Mahdi but Allah himself. As such, Master Fard was worshiped with prayer and sacrifice. Elijah Muhammad assumed the mantle of "Prophet" that "Allah" had worn during his mission in Detroit. Eventually, Elijah Muhammad was referred to not only as the Prophet but as the Messenger of Allah. Shortly after Elijah Muhammad was named minister of Islam, Master Fard vanished in 1934 as mysteriously as he had arrived.[18]

After certain factions began to develop in the movement, Elijah Muhammad set up separate headquarters in a temple that had already been established two years earlier as the Southside Mosque in Chicago. Severing all connection with the parent group in Detroit, he bestowed the name *Temple People* on his movement and relocated his center of operations to Chicago, where the Nation of Islam thrived under his aggressive leadership. Although initially the movement grew slowly, it gained momentum as its reputation spread. The forces of oppression, however, were determined to place obstacles in Elijah Muhammad's path. Because of his opposition to the war effort, he was imprisoned from 1943 to 1946:[19] "When Elijah Muhammad entered the Federal Correctional Institution (FCI) at Milan, Michigan, to begin his prison term on July 23, the world appeared in flames, and America was at the center of the fire both domestically and abroad. The worldwide violence of 1943 seemed once again to confirm what Elijah had subconsciously believed since his boyhood in the lynch town of Cordele: at the end of the day, the white man, whether in Georgia, Detroit, Chicago, or Berlin, had nothing to offer him but slavery and death."[20]

During his imprisonment, Elijah Muhammad (admitted to the facility as Gulam Bogans) was subjected to the indignity of psychological tests. An indictment of the stereotype of the black that emerges in Western culture, the results of these tests confirm Sander L. Gilman's contention that otherness is customarily imbued with the mark of madness. So imbued, blackness is associated "not merely with pathology, but with one specific category of disease, psychopathology."[21] In the case of Elijah Muhammad, the "patient" was "diagnosed" as one afflicted by "dementia praecox" of the "paranoid type." Such a "diagnosis" was compounded by the further degradation of being classified as possessing "an IQ between 70 and 79 and a mental age between 10.6 and 11.9 years." According to the psychiatrist who subjected the "patient" to a barrage of tests, Elijah Muhammad exhibited "a marked persecutory trend both against himself and his race" and, occasionally, felt as though he were being "pursued and slandered by his enemies." Moreover, Elijah Muhammad's claim that Allah had communicated with him "in visual and auditory form" was to be seen as a symptom of schizophrenia. During the period of his imprisonment, Elijah Muhammad was repeatedly subjected to testing of one sort or another in order to demonstrate that he was "psychologically unbalanced."[22] Clearly, the determination on the part of the authorities to denigrate their ward in this manner reveals more about the misapprehensions and anxieties of the system at that crucial juncture than it does about the individual subjected to such indignities.

After Elijah Muhammad's release from imprisonment in 1946, he vigorously resumed his organizational activities. When he returned to active control, there were four temples in operation, but in the 1950s the movement underwent a period of spectacular expansion. By the end of the next decade, there were as many as sixty-nine temples distributed throughout twenty-seven states from Massachusetts south to Florida and west to California. Membership increased proportionately.[23]

When C. Eric Lincoln published his seminal *The Black Muslims in America* in 1961, he commented then that "the Black Muslims have come far under Muhammad. He has given them temples and schools, apartment houses and grocery stores, restaurants and farms. Most important of all, he has given them a new sense of dignity."[24] In his most recent estimate of the movement, Lincoln looks back on the course of Elijah Muhammad's life and achievements with pride. Leading the Nation of Islam for some four decades until the time of his death in 1975, Elijah Muhammad engendered "a pronounced American awareness of Islam, its power and its potential." As the result of Elijah Muhammad's efforts, Lincoln observes, there were temples or mosques in a hundred cities where none previously existed. A hundred thousand members of the Nation of Islam made their presence felt through their frequent rallies as well as their places of business (groceries, restaurants, bakeries, and the like). "The clean-shaven young Muslims hawking their newspapers on the street, celebrating their ritual in the prisons, debating their beliefs in the media, gave to the religion of Islam a projection and a prominence undreamed of in North America." Among those acquainted with such accomplishments, there was an awareness that members of the Nation of Islam had "done more to exemplify Black pride and Black dignity" than almost any other movement that might have been more palatable to the tastes of American society. By the end of Elijah Muhammad's seigniory, the Nation of Islam was no longer simply a community of the impoverished and downtrodden. It had attracted the allegiance of intellectuals, professionals, and celebrities in the world of sports and entertainment. Under the tutelage of Elijah Muhammad, the Nation of Islam became the most important Islamic presence in America.[25]

At the forefront of the movement stands the Honorable Elijah Muhammad. Although perhaps physically slight in appearance, he is clearly a towering figure, one whose aura of charismatic authority renders him a prophet among prophets (fig. 14). Offering an account of his impression of this prophet in his memoirs, James Baldwin records how he was drawn toward Elijah Muhammad's "peculiar authority," how his smile promised to remove the burden of Baldwin's life from his shoulders. "The central quality in Elijah's face is pain," Baldwin observes, "and his smile is a witness to it—pain so old and deep and black that it becomes personal and particular only when he smiles. One wonders what he would sound like if he could sing."[26] Figuratively, he did sing. Drawn from the lessons that he learned at the feet of the Mahdi, Elijah Muhammad's song assumed the form of his teachings disseminated both orally and in writing during the generations that the prophet fulfilled his calling as the Messenger of Allah to those who were members of the Nation of Islam.

Essential to those teachings is a new understanding of black origins and black history, one responsive to the need to forge a new sense of self-awareness and self-dignity and to the determination to see in the course of events not happenstance but the working out of a divine plan. A fundamental aspect of that plan is an all-pervasive awareness of end time toward which Elijah Muhammad's view of black history irrevocably moves. This is a history grounded in eschatology. Founded in the pain, suffering, and oppression that are the hallmarks of the black experience in this country, the song that Elijah Muham-

Figure 14. The Honorable Elijah Muhammad and His
Supreme Captain, Raymond Sharrief, St. Louis, Missouri.
Courtesy of CROE Archives, 2435 W. 71st St., Chicago, Illinois 60629.

mad sings is one of liberation from the shackles of the white race that would keep the "so-called Negro" in bondage. This song culminates in an eschatology of destruction, on the one hand, and renewal, on the other.[27] That eschatology, in turn, is centered in the vision of Ezekiel, of which Elijah Muhammad is both prime exegete and divinely inspired prophet. It is this vision that Elijah Muhammad conceives as a vehicle that signals the triumph of all that the Nation of Islam represents. To understand the nature of such a conception, one must first be aware of the terms through which Elijah Muhammad's view of black history is delineated.

These terms assume the form of a narrative of origins, an etiology that accounts for the present conditions of the black man.[28] At the core is the category of race, which provides the impetus through which the entire course of history can be explained. To gain a knowledge of that history is to acquire "the supreme wisdom."[29] According to Essien-Udom, this dimension of the teachings of Elijah Muhammad may be classified as "esoteric" (as opposed to "exoteric"). Whereas the exoteric dimension addresses the practical techniques for attaining the good life in today's world, the esoteric dimension addresses the abstruse matters of racial history centered on the essential "truths" concerning first and last things. One focuses on the "sweet here and now," the other on the "sweet bye and bye."[30] From the perspective of the "sweet bye and bye," an entire esoteric world of origins is disclosed through the authority of Allah himself.

In various forms, the narrative of origins states that, 66 trillion years ago, what is now called the earth was then called the moon, inhabited exclusively by the black man. In fact, the inhabitants were known as the "people of the moon." Among those people, a black "scientist" (the black God of the moon) became dissatisfied because he was unable to make all the people speak the same language. So he decided to destroy the people by causing a great explosion on the moon. A planetary body was blasted out from what was then the moon and traveled twelve thousand miles into space. What is now the earth, in turn, traveled thirty-six thousand miles into the atmosphere. This is the earth that we now inhabit. The part that was called the moon capsized, and its remaining life was destroyed. When it blasted off, however, it dropped water on the other planetary body now called the earth. What resulted was the ocean, which covered three-quarters of the surface of the earth.[31]

This is the cosmogony of the Nation of Islam, one that articulates its own version of the big bang theory to account for the cosmos and the creation of life on this planet. At the source of that occurrence, one finds the person of a black scientist through whom the chain of events leading to the creation of the earth itself can be traced. Such a cosmogony anchors its narrative in the language of "scientific" discourse, one distinguished by an emphasis on racial difference. At issue in the revelation of the secrets of the universe is not simply the cause-and-effect relations that underscore scientific processes. Rather, this is a distinctly black-centered cosmogony, one in which cause and effect stem from the desire for universal communication in a world populated by a black race called the "people of the moon." They may be perceived as "other" by those who do not know or understand them, but among themselves they must be made to come to terms with a unique identity that binds them in a single mission. When it is discovered that they do not all speak the same language, cosmic repercussions result. There is a disruption that ultimately occasions a new beginning, a new order. The course of this new order may be traced in the history of life on the new planetary body called the earth.[32]

In the annals of the "sweet bye and bye," the history of life on this planet assumes a form consistent with the myth of origins through which all causal relations are to be understood. Once again, the aura of "scientific precision" is brought to the service of mythic historiography. According to the "supreme wisdom" of the secret history that is unfolded, life on this planet has been in existence for trillions of years. In the beginning, Africa and East Asia were one continent, and the population was entirely black. Among this black people were twenty-four wise scientists, one of whom created a tribe that was the most powerful of all the population on the continent. This powerful tribe was the original Asian black nation known as the tribe of Shabazz. As the very well-spring of the black population, the tribe of Shabazz is the source of the so-called Negro in America. With his origin in the tribe of Shabazz, the so-called Negro is now able to understand that he is really a descendant of the powerful black people of Asia. The Original Man, then, is the black man, primogenitor of all other races. He is the alpha and the omega. All stems from him, and all is resolved in him. He is the first Adam and the last Adam as well. In fact, if there is anything of a paradisiacal dimension to this myth of origins, it may be found in the fact that, since the tribe of Shabazz was the first to occupy and explore the planet during this primordial period, it is this people who discovered the most desirable places to inhabit, including the Nile Valley and the area that was to become the holy city of Mecca in Arabia. That is at once their *locus amoenus* and their *locus sanctus*. Tracing his genealogy to this paradisiacal city, the so-called Negro becomes aware not only of his ultimate genealogy but also of his ultimate place of origin, a place that is the spiritual center of Islamic worship and identity. By virtue of this fact, the so-called Negro discovers his Muslim heritage. Once he understands and acknowledges this heritage, he will take pride in who he is. When the whole world likewise knows who the Original Man is, it, too, will acknowledge the fundamental truth of the primacy of the black race, one whose origins render it not only the most ancient but also the most holy of all the races.[33]

In his account of cosmogonic and racial origins, Elijah Muhammad as the true Prophet and Messenger of Allah may be said to have rewritten the creation account in Genesis.[34] Here is a truly black cosmogony as well as a corresponding archaeology of the formative moments in the history of the black race. That archaeology provides a local habitation and a name for the paradisiacal realm inhabited by the Original Man. By means of this reinscription of the Genesis narrative, the Prophet and Messenger of Allah transcribes his own book, his own Scripture, to account for the present condition of his people. If he offers a black cosmogony and a black archaeology, he also provides a rationale for the displacements that the so-called Negro has been made to endure over the centuries. Here, Genesis is rewritten. In that rewriting, the myth of origins on which Elijah Muhammad's concept of history is grounded finds its correspondence in a phylogeny distinguished by its own unique principles of racial evolution culminating in the travails that the black race suffers at the present time. This process involves a primal act of disruption and rebellion as momentous as any that might be associated with the biblical traditions of the Fall. Within this context of disruption and rebellion, the event that precipitates the Fall is one that C. Eric Lincoln aptly calls "the central myth of the Black Muslim movement." It is, Lincoln observes, "the fundamental premise upon which rests the whole theory of black supremacy and white degradation."[35] Anticipating and, indeed, laying the groundwork for the eschatology in which Elijah Muhammad's historical account culminates, the central myth to which Lincoln alludes provides a framework for a consideration of the apocalyptic dimensions of Elijah Muhammad's meditations on the vision of Ezekiel.

The myth in question concerns the rebellious figure of "Yakub," the antihero or devil of Elijah Muhammad's account of racial origins.[36] Yakub proved himself to be a dissident member of the original black tribe of Shabazz in the holy city of Mecca. A precocious youngster, Yakub began school when he was four and, by the time he was eighteen, had graduated from all the available colleges and universities. Exhibiting an intellect of impressive proportions, he was a figure whose very physical attributes reflected his capabilities—his head was unusually large. With a particular interest in science, Yakub was, in fact, known as "the big-head scientist."[37] A rebellious figure who did not hesitate to flaunt his accomplishments with grandiose claims, he went about preaching in the streets of Mecca to gain converts to his cause. Viewing Yakub as a potential threat to the stability of their own society, the authorities of Mecca finally exiled him, along with some sixty thousand of his followers, to the island of "Pelan" or Patmos, there to preach his incendiary "revelations" and to practice his false "science."[38]

Angered by this rebuff, and embittered toward Allah, Yakub plotted revenge. His plot harkened back to an early discovery of his when he was only six years old. Playing with two pieces of steel, he perceived in their magnetic properties the principle that unlike attracts and like repels. In this, he saw that an unlike human being, that is, the white man, could be created to rule the original black man through the "attractiveness" of his trickery and lies. As a scientist particularly skilled in the field of genetics, he accordingly conspired to create on the earth a brood of devils that would assume the form of what is now known as the white race. As the father of that race, Yakub the black scientist is thereby ultimately the figure the Bible refers to as Adam. To realize his goals, he engaged in an experiment in human hybridization through a complex process of breeding and cross-breeding. Overturning the law of nature by means of genetic engineering, Yakub established his own law of genetics, one in which the line resulting from his experiments would become not only progressively lighter but also more wicked and corrupt. He was also aware that it would take him several changes in color to get from pure black to bleached-out white. Although Yakub knew that he would never live long enough to see this white race, he left sufficient laws and instructions to those who followed after him to complete the process. In two-hundred-year intervals, the black race gave way to the brown race, the brown race to the red race, the red race to the yellow race, and the yellow race to the white race.[39] After some eight hundred years of genetic engineering, all that remained on the island of Patmos was a race of blond, bleached-out, blue-eyed devils. Naked and shameless, these hairy savages walked on all fours, lived in caves and trees, and mated not just with one another but with beasts as well.

This line of Adam continued to live on Pelan or Patmos for another six hundred years before the savages migrated to the mainland, inhabited by the natural black people. There, the white mutations transformed what had been a peaceful heaven on earth into a hell torn by fighting and disruption. As a result of its trickery and underhanded dealings, the white race was made to cover its nakedness, after which it was driven out of paradise and exiled to the caves of West Asia (Europe). After they had spent some two thousand years in this condition, Moses (Musa) was raised up to "liberate" and "civilize" them. Such an attempt was futile, however. Even though he brought them out of their caves, taught them to wear clothes and to cook food properly, and even taught them to believe in Allah, they remained savages. (In fact, they were so evil that Moses was obliged to build a ring of fire about him at night "to keep the devils from harming him.") Eventually realizing the impossibility of civilizing this recalcitrant race, Moses attempted unsuccessfully to eliminate as many of them as he could. But they continued to propagate, and, liberated from their imprisonment in caves, they succeeded finally in gaining power and have since ruled the world for six thousand years down to the present time.[40]

Their rule bears all the marks of their "civilized" nature. In our own century, their behavior includes such acts as the lynching of innocent people, the genocide of countless individuals, and the detonation of atomic bombs. Such civilized conduct also involves stealing blacks away from their homeland and transporting them to North America, where they are sold into slavery, robbed of their true names, and branded as property. As terrible as this situation might appear to be, however, it is all part of Allah's providential design. Such a turn of events, Elijah Muhammad declares, is paradoxically necessary. These savages "must conquer, and bring into subjection, all life upon the earth." They must, in fact, "master everything, until a great master of God comes." This coming will mean the end of their power and domination over the life of our earth.[41] It will also result in the ultimate liberation of those faithful who are the chosen of Allah. They shall realize their true destiny in the overcoming of Yakub's brood and the resurrection of the just. That resurrection will be accorded only to those who heed the teachings of Elijah Muhammad, the true Messenger of Allah and spiritual leader of the Lost-Found Nation of Islam.

This in brief is the central myth of the Nation of Islam, certain elements of which prove essential to an understanding of the outlook reflected in the teachings of Elijah Muhammad. Those teachings are grounded both in the mythic discourse that underlies black folklore and, as we have already seen, in Elijah Muhammad's reinscription of biblical prototypes. From the perspective of black folklore, Yakub has affinities with the figure of the trickster. Of fundamental importance to slave narratives, the trickster in his various forms is an individual who was viewed as either hero or fool, protagonist or antagonist. In either respect, the trickster attempts to "manipulate the forces in nature to rob others of their very being."[42] In his essay on the essential characteristics of mythic tricksters, William J. Hynes isolates five primary traits. These include the status of one who is viewed as an outlaw or an outcast, whose primary purpose is to deceive, who seeks to alter bodily appearance or shape, who seeks to overturn the prevailing order, and who is noted for his ingenuity as a transformer of the natural world.[43] For Laura Makarius, it is through the manifestation of this final trait in particular that the trickster becomes at once a demiurge of great power and a clown or a buffoon of laughable pretension.[44] As such, the trickster embodies the quest to become like God, to "overreach."[45] At times he is successful, at others not.

Both as demiurge and as buffoon, the "big-head scientist" Yakub shares many, if not all, of the traits of his mythic counterparts. As Theophus H. Smith observes, Yakub as trickster particularly embodies the characteristics of deceit and mischief that allowed his white offspring to rule the black man by means of lies and treachery.[46] This is what Elijah Muhammad refers to as the *tricknology* that is all pervasive in the white man's conduct as the product of Yakub's influence.[47] *Tricknology* is a potent term here: it implies all that is fallacious and delusive in Yakub as the Father of Lies whose descendants now practice what has been disseminated through their very genes. So Elijah Muhammad alerts his followers to the wiles of those who represent Yakub's brood in their dealings with the black man: "They make fools of you and then laugh at you for being dumb enough for them to trick. (The devil scientists and rulers prepare the trap for you and the others spring it on you.)"[48] In this respect, Yakub takes the form of the "signifying monkey," a trickster whose machinations cause his own people to be "signified on," that is, duped. In the genetic universe of Yakub's tricknology, such duping assumes the form of offspring whose ultimate purpose is to wreak havoc on their forebears.[49] As consummate trickster, Yakub thereby fulfills his mission triumphantly.

If Yakub as trickster has affinities with his mythic counterparts, his role as progenitor of the white race finds corresponding antecedents in the black folk culture from which he emerges. In his important book *Black Culture and Black Consciousness,* Lawrence W. Levine records several antebellum slave tales that portray God's act of beginning his creation with the black race that through the course of time degenerates into the white race. According to Levine, "the assumption of a black creation allowed slaves to stand the white creation myths on their heads." In one version of such a black creation myth, the Cain-and-Abel story features Cain as a Yakub-like figure through whom the white race comes into being. Although the first men (Adam, Cain, Abel, Seth) were all black, the treachery of Cain results in a curse that affects all his offspring. Confronted by an angry God who demands to know the whereabouts of his brother, Abel, a fearful Cain "turns white as bleech cambric" in the face, and the whole race of Cain has been white ever since. Branded with a white mark on his face as a sign of God's displeasure, Cain is then exiled to the land of Nod, there to sire the white race as the offspring of his sin. Levine records other versions of this tale, which apparently had wide currency. Such tales suggest that black slaves not only possessed their own form of ethnocentrism but also at times—as above—conceived the white race as a degenerate form of the black.[50] In some respects, Yakub and the race that he spawns are the products of such narratives. Although the concept of genetic engineering is certainly unique to Elijah Muhammad's portrayal of Yakub as a renegade black scientist, the ethnocentrism that surrounds the idea of racial origins in black culture is not.

As a counterpart to the mythic discourse that underlies the construction of Yakub, the events that distinguish biblical narrative are equally germane. Once again, Elijah Muhammad reinscribes the biblical text to forge his own version of sacred story. Such is true not only of Elijah Muhammad's cosmogony but of his demonology as well. In this respect, Yakub is Elijah Muhammad's version of Satan. In fact, in *Message to the Blackman in America*, Elijah Muhammad appropriately recounts the story of Yakub under the heading "The Devil." As the very devil incarnate, the figure of Yakub encompasses for Elijah Muhammad an entire range of biblical antecedents. In the Genesis account, these include not just the fallen Adam and the race that he begets but the serpent that comes to tempt and is punished for its sins.[51] Within the Genesis matrix, Yakub, as suggested, likewise has affinities with Cain as the first murderer.[52] Implicit in his very name, Yakub, of course, is also Jacob. In his Saviours' Day message delivered in Chicago in 1967, Elijah Muhammad focused on Jacob's wrestling with the angel in Gen. 32:24–29 as a way of suggesting that the patriarch desired to force the angel to do something he did not want to do. "This," says Elijah Muhammad, "is the history of Yakub, the father of the Caucasian people." Yakub, too, was determined to wrestle with his own kind in order to force his people to release the "whiteness" within them. "White was in black," Elijah Muhammad observes, "but no one had ever made an attempt to bring it out" before Yakub engaged in such devilish behavior. In his wrestling with the angel all night, Jacob (Yakub) was in effect "wrestling the white man out of the black man."[53]

These are only some of the biblical contexts that Elijah Muhammad draws on to fashion the figure of Yakub. The overriding theme in this depiction is a history that involves a rebellion, a fall, an exile, and the creation of despicable creatures who follow in their master's footsteps. If Yakub as devil of devils is associated with the serpent in the Genesis account, he is, moreover, associated with the beast as apocalyptic symbol. Here, Elijah Muhammad moves from the Old Testament to the New. His point of focus is, of course, the Book of Revelation, and the text that he cites has to do with the beast that rises out of the sea. Receiving its power from the dragon or serpent, the beast is worshiped by its followers: "Who is like unto the beast? Who is able to make war with him?" (Rev. 13:1–4).[54] The beast is the offspring of Yakub, the white race under which the black nation finds itself subjugated. The source of such revelation, more-over, is not Saint John the Divine but Yakub, exiled to the island of Patmos, where he grafted the white race. As the father of the white race, he looks forward to the time of triumph over the black nation when the beast will be worshiped. Turning Revelation on its head, Elijah Muhammad proclaims: "The so-called American Negroes have and still suffer under such brutish treatment from the American Christian white race, who call themselves followers of Jesus and his God."[55]

If Elijah Muhammad's reading of Revelation is of a far different sort from what the traditional interpretations of the biblical text would disclose, it is nonetheless a reading that bestows its own unique stamp on the apocalyptic dimensions of the sense of self-identity that emerges from Elijah Muhammad's teachings. As a site of visionary experience, the island of Patmos assumes nefarious implications: the exilic perspective so fundamental to the biblical outlook is displaced by a new sense of otherness and alienation. The island of Patmos is not a place where the exiled prophet experiences visions that result in the ultimate glory of those who abide by the true creed. Rather, it is a place where the exiled "big-head" trickster plies his trade as a scientist that produces a degenerate race. The sense of alienation and otherness implicit in this reading of the biblical text is true even of the prototype of captivity and liberation associated with the concept of the deliverance from bondage that the narrative of Moses entails. When the white race is exiled for some two thousand years to the caves of Europe, its release from captivity is followed not by genuine re-newal and reclamation but by recourse to further corruption. That corruption, however, becomes the source of the ultimate liberation of the black race, which, in following the teachings of the Messenger of Allah, realizes a triumphant destiny in the apocalyptic moment of end time. Standing biblical myth on its head, Elijah Muhammad generates not only his own narrative of origins but his own sacred history as well as his own prophecy of things to come. The prophet of a new order, he becomes the means through which the black man fulfills his destiny. That destiny moves inevitably toward an eschatology that finds its referent in the vision of Ezekiel, the central text in what might be called the esoteric underpinnings of Elijah Muhammad's work of the chariot.

In his recourse to Ezekiel's vision, Elijah Muhammad not only transforms the biblical text into a singular rendering of this most primal of experiences but also subsumes what had become a staple of black folk culture into his outlook as Prophet and Messenger of Allah. Whether responding directly to the biblical prototype or indirectly to the folk traditions out of which his individual experience emerged, Elijah Muhammad assumes a place as rider in the chariot. To assess the precise nature of the chariot in the thought of Elijah Muhammad, I adopt as my point of departure the traditions of black folk culture that were formative in the establishment of Elijah Muhammad's outlook. At once appropriating these traditions, Elijah Muhammad refashions them to accord with his distinctly apocalyptic point of view.

As a cultural phenomenon, the vision of Ezekiel was no stranger to the black religious experience in both the antebellum and the postbellum periods. In his study of these periods, Levine has shown how those assembled in the "prayer house" on "de Lord's day" would respond to the preacher's account of the texts of Ezekiel by exclaiming how "de Lord would come a'shinin' thoo dem pages" and inspire them to "jump up dar and den and holler and shout and sing and pat" until what emerged was "a spiritual."[56] Whatever the source and process of composition, the spiritual was a crucial means by which Ezekiel's vision assumed paramount importance to the culture of black consciousness. In song and incantation, the vision was appropriated as a form of awakening and liberation by those who sought a means of transport to the glories of the other world.[57] As a staple of the incantatory force of slave spirituals, Ezekiel's vision assumed various forms in the articulation of the world of the sacred.

In one rendering, the singer first conjures up the vision of " 'Zekus' wheel" and then exhorts his soul to "take a ride awn-uh 'Zekus' wheel." This preparatory exhortation is followed by a series of exclamations that in effect define the wheel as a source of " 'ligion," "moanin'," "prayin'," "shoutin'," "cryin'," and "laughin'." All these "acts" represent expressions of delight and confidence in the power of the wheel to transport the singer into a realm of total release from the sufferings that constrain his behavior in this world. The realm of release that represents the fulfillment of the desire for transport is confirmed as the stanzas themselves move to a point of climax through the building up of force and momentum from the first stanza to the last. The effect of this crescendo is to celebrate the vision in its very enactment. Culminating each of the seven stanzas of the spiritual, the refrain "Le's take a ride awn-uh 'Zekus' wheel" reassures the singer, along with those who participate in the visionary re-creation of the biblical event, that the experience invoked in the exhortation will ultimately be realized.[58]

In another spiritual on the same theme, the singer commences his celebration by recapitulating the cry heard in the prophecy: "Wheel, oh, wheel" (Ezek. 10:13). With that declaration, the singer proceeds to focus on the mystery of the "Wheel in de middle of a wheel" (Ezek. 1:16). Interpreting the mystery, the singer proclaims that this is "de wheel of time, / 'Way up yonder on de mountain top." Every spoke of the wheel is "human kind, / My Lord spoke an' de chariot stop." Transforming the "spokes" of the wheel into the voice of God, the singer makes clear that what Ezekiel beheld was indeed a chariot, for which the wheel itself becomes a synecdoche. As a wheel in the middle of a wheel, it is seen " 'Way up in de middle of de air." Fusing the temporal with the eternal, this object is not simply the wheel of time with its spokes as humanity; it is also a manifestation of the timeless interconnection of human faith, through which the "big wheel" revolves, and the "grace of God," through which the "little wheel" revolves. Within the framework of one spiritual, an entire theology is generated, one that emphasizes the relation of human and divine, temporal and eternal.[59] Other versions of this spiritual are extant.[60] At the center of the experience is the "wheel" both as a wheel within a wheel and as a synecdoche of the chariot through which the visionary seeks liberation.

In this respect, the chariot of Ezekiel finds its counterpart in other forms of transport, such as the chariot of Elijah (2 Kings 2:11), the subject of one of the most familiar of spirituals. There the chariot is invoked as a means of redemption and return: "Swing low sweet chariot, / Coming for to carry me home." Envisioning this vehicle as he looks over the Jordan river, the singer beholds "a band of angels" approaching to carry him home.[61] In his study of these spirituals, Theo Lehmann demonstrates the extent to which the chariot became a symbol of freedom for the black slave: this vehicle was seen as the crucial conveyance through which the slave might gain access to a world without bondage both in this life and in the next. As a means of release from bondage, the chariot became a kind of freedom train, invoked in the following manner: "Train coming, oh let me ride, / . . . Oh low down the chariot and let me ride."[62] In instances of this sort, we are presented with what might possibly be construed as a form of "coded communication" through which "some slave spirituals conveyed secret messages among conspirators engaged in underground meetings and in plots to escape from their plantations."[63]

Whether or not such "coded communication" is actually operative, the conception of the chariot as a vehicle of release in black folk culture is crucial to an understanding of the spirituals through which that culture is defined. At issue in the delineation of this culture is an ever-present sense of sacredness. Such is particularly true of the worldview of black slaves. Denied any possibility of adjusting to the external world of the antebellum South, "slaves created a new world by transcending the narrow confines of the one in which they were forced to live." In the process, they extended the boundaries of their restrictive universe until it fused with the world of the biblical text and became one with the world beyond.[64] This act of extending boundaries assumes a decidedly visionary aura in the extant transfigurational accounts of slaves converted to Christianity. In such states of transfiguration, converts commonly beheld and conversed with God or Christ: "I looked to the east and there was . . . God. He looked neither to the right nor to the left. I was afraid and fell on my face. . . . I saw God sitting in a big arm-chair. . . . I seen Christ with His hair parted in the center. . . . [The Lord] looked like he had been dipped in snow and he was talking to me." The accounts resonate with biblical nuances derived from a host of texts both prophetic and apocalyptic, including the Books of Ezekiel, Daniel, and Revelation. What is true of the spirituals, furthermore, is no less true of personal visionary accounts. Once again, the encounter with God assumes the form of a chariot vision in which the vehicle is on occasion concretized specifically as a train: "I saw in a vision a snow-white train once and it moved like lightning. Jesus was on board and He told me that He was the Conductor."[65] Whether in the spirituals or in the personal accounts, the visionary is delineated as *technē*. In this case, it is a *technē* apposite to the particular culture in which it is conceived.

Complementing the visionary dimensions of these transfigurational accounts, moreover, is the all-pervasive sense of apocalyptic warfare. Essential to the outlook that we have been exploring is the conception of the Jesus of the Gospels as the combative Lord of the Apocalypse. In their spirituals as in their personal accounts, slaves transformed the figure of Jesus into a warrior whose victories were eschatological as well as spiritual. He was both "Mass Jesus," who engaged in personal combat with the devil, and "King Jesus," who was beheld seated on a milk-white horse with sword and shield in hand (cf. Rev. 19:11: "And I saw heaven opened, and behold a white horse; and he that sat upon him *was* called Faithful and True, and in righteousness he doth judge and make war"). In their spirituals, slaves declared, "Ride on, conquering King," and, "The God I serve is a man of war." Such a transformation of Jesus is symptomatic of how slaves customarily resorted to those parts of the Bible that reinforced the apocalyptic underpinnings of their religious consciousness.[66] The language of protest among the slaves was interlaced with apocalyptic fervor.

In what Albert Raboteau calls "slave religion" one encounters a kind of black eschatology.[67] That eschatology is one in which the slaves fulfilled their desire that whites suffer just retribution for their brutality toward those they held in bondage. For this purpose, slaves found in the biblical prophets images sufficiently violent to suit the most vengeful of feelings. So it is recorded that one Mary Livermore, a New England governess on a Southern plantation, was astonished by the prophetic terms that Aggy, the normally "taciturn" housekeeper, employed to express her feelings of outrage at the beating her master had given her daughter: "Thar's a day a-comin! Thar's a day-a-comin. . . . I hear de rumblin' ob de chariots! I see de flashin' ob de guns! White folks blood is a runnin' on de ground like a riber, and de dead's heaped up dat high! . . . Oh, Lor'! hasten de day when de blows . . . shall come to de white folks, an' de buzzards shall eat 'em as dey's dead in de streets. Oh, Lor'! roll on de chariots, an' gib de black people rest an' peace. Oh, Lor'! gib me de pleasure ob livin' till that day, when I shall see white folks shot down like de wolves when they come hungry out o' de woods!"[68] The refrain of rolling on the chariots, accompanied by the flashing of guns, as part of the apocalyptic retribution that will ultimately be meted out to the "white folks" is characteristic of the desire to overwhelm the oppressor through the most devastating means possible. In this case, it is the chariot transformed into an object of warfare, equipped with firearms to annihilate the white enemy in a final confrontation. Considering the suffering that the black slave was made to endure by his white master, such retributive justice appears entirely warranted. For our purposes, the idea of conceiving this justice through the figure of the chariot as a vehicle of warfare is a particularly appropriate way of suggesting how the formulation of visionary material in black folk culture finds renewed expression in the transformation of the spiritual into the technological. No longer simply a means of transport to the other world or a form of liberation in this, the chariot becomes a vehicle of black retribution against white injustice. Rumbling on and flashing its firepower, it destroys the white foe in apocalyptic combat. At least in part, such is the milieu out of which Elijah Muhammad's concept of the chariot emerges in his vision of the apocalyptic retribution.[69]

For Elijah Muhammad, that event has its ultimate source in the vision that inaugurates the prophecy of Ezekiel. Elijah Muhammad's meditations on Ezekiel find expression in three distinct sets of writings. The first set appears in the collection of essays *Message to the Blackman in America* (1965), the second set in the corresponding collection *The Fall of America* (1973), and the third set in a series of articles in the newspaper *Muhammad Speaks* (May–September 1973). All three sets have since been conveniently gathered in a work called *The Mother Plane.*[70] Accordingly, Elijah Muhammad's meditations on Ezekiel occupied at least the decade or so before his death in 1975. Most recently, moreover, a posthumous collection, *The Theology of Time* (1992), has appeared as a transcription of some seventy-five of Elijah Muhammad's lectures delivered during the years before his death.[71] As with the works published during his lifetime, the vision of Ezekiel figures prominently in this collection as well.[72]

In *Message to the Blackman in America*, Elijah Muhammad makes it clear that his concern with Ezekiel extended back to the formative period of the 1930s when he first encountered Master Fard himself. As Elijah Muhammad describes it, his encounter with the master is framed in the dramatic context of Ezekiel's own *visio Dei*, reinforced by biblical notions of the prophet's seeking insight into the name of the deity from whom he receives his call. The scenario is one that we have already witnessed in Malcolm X's depiction of this seminal event in the *Autobiography*. Its importance is reinforced by virtue of Elijah Muhammad's recounting of the experience in *Message to the Blackman in America*. In this version, the prophetic implications associated with the figure of Moses are extended to embrace those of Ezekiel. So Elijah Muhammad avers that, on meeting the mysterious figure known as "Fard," the acolyte asks, "Who are you, and what is your real name?" As we have seen, the master responds with a kind of millenarian fervor: "I am the one that the world has been expecting for the past 2,000 years." Still yearning for the true name, Elijah Muhammad repeats the question, to which Master Fard responds: "My name is Mahdi; I am God, I came to guide you into the right path that you may be successful and see the hereafter."[73] The answer specifies the identity of the master as both God and Messiah; it also promises the bestowal on the follower of Master Fard of the ability to prophesy the course of future events. With the receipt of this ability, the prophet is made aware of end time. This is an eschatology marked by destruction, one that involves the obliteration of the world with bombs, poison gas, and fire. Although, in general, that which is to be destroyed is the "present world," the specific target of destruction is the world of the white race. Nothing will remain of that race in the final conflagration.[74]

It is at this point that the prophet is shown the means of such conflagration. According to Elijah Muhammad, the Mahdi "pointed out a destructive dreadful-looking plane" in the form of a wheel in the sky. On sighting this plane, Elijah Muhammad saw that it was a half mile by a half mile square. Beholding it then, Elijah Muhammad was given to know that it was constructed as what he calls "a humanly built planet." Those who seek the truth can see it even today. It is up there now and can actually be seen twice a week. Its presence is no secret. Although one might be astonished that such an object can be seen so often, it has been hovering in that position since biblical times. In fact, this is the same object that the prophet Ezekiel saw "a long time ago." Its purpose, then, was the same as its purpose is now: to destroy the present world, that of the white race. The fact of this destruction can already be felt by virtue of the rain, snow, hail, and earthquakes that already plague the world. These events are merely harbingers of the ultimate devastation to follow. Elijah Muhammad assures the reader that he has this knowledge on the ultimate and unquestionable authority of Master Fard. As Messenger of Allah, Elijah Muhammad has been commissioned by the master to bear the message of life (Islam) to his people here. "Islam," he declares, "is our salvation." It removes "fear, grief, and sorrow from any believer, and it brings to us peace of mind and contentment." All that stands in the way of the black race is Christianity, the religion of the slavemasters. This is a religion "organized and backed by the devils for the purpose of making slaves of black mankind." Convert to the religion of Islam, Elijah Muhammad exhorts his people, in order to be liberated. Then the plane can do its work of destroying the white race, which is also the Christian race, the race that holds the black man in bondage.[75]

What makes this account of Elijah Muhammad's calling as prophet so fascinating is the manner in which it reinscribes the circumstances of Ezekiel's vision within the context of the eschatological perspective that the Messenger of Allah is attempting to foster. As the votary of Master Fard, not only does Elijah Muhammad find himself in the position of having the vision of Ezekiel revealed to him in this century; he veritably *becomes* the prophet to whom that vision was originally revealed in biblical times. This is precisely the point that Margary Hassain, a follower of Elijah Muhammad, makes in an issue of the newspaper *Muhammad Speaks,* as a gloss that accompanies the Messenger's own commentary on Ezekiel's vision: "Messenger Muhammad," she says, "now steps forth and takes the Book of Ezekiel and shows the world that Ezekiel is no mystery to Muhammad, for Messenger Muhammad *is* the Ezekiel and the reality of Messenger Muhammad's Mission is the mission envisioned by Ezekiel the prophet in the Book of Ezekiel."[76] The source of the vision, then, is Master Fard, and its recipient is the new Ezekiel, Elijah Muhammad. In short, Elijah Muhammad as Ezekiel experiences the vision on divine authority. What he says about it is divine truth. In his words, its secrets and its mysteries are fully disclosed.

For Elijah Muhammad, the importance of the vision is further compounded by the fact that it appears to the prophet at the precise point of the founding of the new order. The vision provides the impetus for the establishment of the Nation of Islam, which owes its divinely bestowed raison d'être to the calling of the prophet at this pivotal juncture in his career. Like Ezekiel before the vision of God, Elijah Muhammad becomes aware at this juncture of what he has been commissioned to do. His commission is to communicate to his followers not just the fact of the existence of the vision but the importance of what the vision portends, that is, the inevitable destruction of the white race. As such, it assumes the form of a dangerous, indeed dreadful, object, one that may at any moment unleash its momentous forces on the unsuspecting enemy. From the psychological perspective explored in previous chapters, one is prompted to think here of the anxiety-laden dreams that Edwin Broome associates with

Freud's Daniel Paul Schreber.[77] This is not the first instance in which the vision is conceived as a manifestation of the *odium Dei*.

From the perspective that Elijah Muhammad embodies, however, the racial implications cause the vision to assume a particularly disturbing valence. According to Essien-Udom, the eschatology implicit in visions of this sort reflects the social history of the black man in America, "especially the psychological trauma and personality disturbances to which that history has given rise."[78] Considering the nature of Elijah Muhammad's own background, such an outlook is clearly anticipated in the black folk culture from which he emerges. Fundamental to that culture, the vision of Ezekiel is a case in point. There it is already transformed from a means of spiritual transport into a vehicle of destructive warfare against the white oppressor. "Thar's a day a-comin! Thar's a day-a-comin. . . . I hear de rumblin' ob de chariots! I see de flashin' ob de guns! White folks blood is a runnin' on de ground like a riber, and de dead's heaped up dat high!" Aggy cries in response to the beating of her daughter. In a very real sense, Aggy's chariots have become Elijah Muhammad's dreadful plane in the form of a wheel in the sky. Her emphasis on the "technological" capabilities of the chariots with their guns becomes his emphasis on the plane with its bombs, poison gas, and fire. In its own way, his visionary object comes "rumblin' " into view as well, but it is an object beheld in the skies.

Despite the importance of such analogues in the formation of Elijah Muhammad's point of view, one nonetheless encounters in Elijah Muhammad a radical departure from his black folk roots. It is no longer the Christian God who saves the day. Quite the opposite: for Elijah Muhammad, the Christian God is the invention of the white man against whom the black man must prevail. The white man's God is the false God that must be destroyed, along with the race he begot, a race that is keeping him alive and well while it exploits and undermines the black race with a religion of deception and hopes that can never be fulfilled. In place of the false Christian God, Elijah Muhammad gives his people the true God, who is Allah in the form of the Mahdi, Master Wallace Fard Muhammad. He has bestowed on his Messenger the truth of Ezekiel's vision and all that it represents. This radical departure from the traditional conception distinguishes Elijah Muhammad as an interpreter with his own unique sensibility, one that is distinctly anti-Christian. In his iconoclastic attitude toward Christianity as the promoter of a slave mentality, Elijah Muhammad returns to what he considers to be the roots of the original conception of godhead in Ezekiel's vision. In the process, he embraces something far more archaic, far more primal than the Christian conception allows. As a manifestation of the archaic, Ezekiel's vision for Elijah Muhammad is genuinely dreadful.

This dread is the experience to which Elijah Muhammad is attuned. It is a dread that encompasses the machine that he projects in his association of the wheel as plane with the outright retribution and terrible destruction it will most certainly wreak on its enemies. The prophet Ezekiel of the biblical account would have been in complete sympathy with such a view.[79]

As a vehicle that makes its appearance twice a week above the earth, moreover, this plane is what has been commonly designated a UFO, a phenomenon that we have already seen to be of fundamental import to the conception of Ezekiel's vision in the modern world. As a UFO that is a harbinger of the Last Day, the Day of Judgment, the plane, according to Essien-Udom, loomed large in the imaginations of those who were members of the Nation of Islam in the 1950s.[80] All Muslims believed in the imminence of the Day of Judgment. On summer evenings they were said to look for signs in the skies. "Two Muslim Sisters reported that another Muslim had told them of seeing a 'huge machine' in the skies on a Tuesday evening during the summer of 1959. When such things happen," observes Essien-Udom, "Muslims alert their neighbors and friends either by word of mouth or by telephone. Stories about 'flying saucers' are taken seriously. Recently, a Sister, hearing on the radio that scientists at an astronomical observatory had seen an 'unidentifiable object,' interpreted it as a sign of the impending destruction of the world." When confronted with the prospect of approaching doom, some felt that they had not lived long enough to enjoy the things of this world. One Sister Mildred observed, "I have mixed feelings about it. I guess I would be glad provided I was living a completely righteous life." Knowing that the day is coming, Sister Mildred felt that she would like to be granted an extension of "another day, a week or a month."[81]

Such are the sentiments that Essien-Udom recorded in his study of the Nation of Islam almost four decades ago. These feelings, moreover, prevail today. In fact, they were most recently confirmed to me by Munir Muhammad, cofounder of CROE (Coalition for the Remembrance of Elijah) in Chicago. In my discussions with Munir Muhammad, I came to appreciate the extent to which the sense of end time is an ever-present reality in the consciousness of the members of the Nation of Islam. When I inquired about the Mother Plane, I was assured that the fact of its reality is considerably more than of merely "academic" interest to those who contemplate its significance in the context of the apocalyptic milieu in which it was originally conceived. It is my distinct sense that, to this day, members of the Nation live on the edge of an awareness of the coming doom. Munir Muhammad pointed to the signs of its advent, including the diseases, earthquakes, catastrophic fires, and the like that plague us in the present world. Under these circumstances, it became abundantly clear that the subject of the Mother Plane is not to be taken lightly even now (or perhaps especially now). As the embodiment of an apocalyptic sensibility, it haunts the consciousness of those for whom its abiding presence is a constant reminder of their identity and mission in this life. In that respect, it represents the legacy that the Messenger himself has bestowed on those in the present who are the faithful stewards of CROE. It is the source of this legacy that I seek to examine here.

The

Eschatology

of the

Mother

Plane

6

In his *Autobiography*, Malcolm X recounts how voracious a reader he was in his attempt to educate himself. Reading widely in the areas of literature, philosophy, and religion, he became learned in several disciplines. Already very much under the influence of the Honorable Elijah Muhammad, he placed his learning within the context of what the Messenger taught him. One of the texts that captivated Malcolm X was *Paradise Lost*. Responding to Milton's epic from the vantage point of the narrative of origins and endings that he derived from the teachings of Elijah Muhammad, he was prompted to compare his newfound teacher with the poet of *Paradise Lost*. As a result of their respective teachings, Malcolm X observes, "Milton and Mr. Elijah Muhammad were actually saying the same thing."[1]

Although Malcolm X does not base this observation on a comparison between Milton's Chariot of Paternal Deitie and Elijah Muhammad's Mother Plane, the aspiring student might very well have concurred with the possibility of such a comparison. There is, in fact, a remarkable similarity between the

Chariot of Paternal Deitie and the Mother Plane. Both derive their primary impetus from Ezekiel's *visio Dei;* both transform that vision into a vehicle of warfare sent on a mission to destroy the satanic hosts; both are the product of the impulse to "technologize" the ineffable, to concretize it as a celestial "machine" of profound import; both place it in an eschatological context; both conceive it as the divine manifestation of the *odium Dei;* both paradoxically ascribe to it not simply destructive but generative and even regenerative dimensions. What is "paternal" in Milton becomes "maternal" in Elijah Muhammad: in both, the attribution of gender is a signature of the play of roles, one in which a dialectic of complementary forces gives rise to a grand and dynamic conception.[2]

Such correspondences, however, are as much an indication of difference as they are of similarity. That difference, of course, arises from the radically different cultures and milieus from which Milton's chariot and the vehicle of Elijah Muhammad emerge. Whereas Milton's chariot is placed in the context of the christocentric milieu that shaped the poet's vision, the vehicle of Elijah Muhammad is the product of a sensibility that finds the christocentric milieu totally alien and destructive to all that Elijah Muhammad's new Islamic perspective holds dear. At the very point that the two conceptions converge, they engage in a fierce warfare of difference. What is life for Milton is death for Elijah Muhammad. In this, paradoxically, Elijah Muhammad is at his most Miltonic for, in his reconceptualization of Ezekiel's vision, the Messenger must become an iconoclast. He must destroy the "high culture," the culture of the slavemaster, represented by all christocentric renderings, and refashion the visionary experience in his own terms. Here, one might say, high culture meets its nemesis in the culture of the street. At first glance, it would appear that the clash of cultures could not be more extreme. Nevertheless, Malcolm X is right: "Milton and Mr. Elijah Muhammad *were* actually saying the same thing." At the very point of opposition, the conceptions of the Chariot of Paternal Deitie and the Mother of All Planes converge in an act of radical destruction and renewal, as the cultures from which these two vehicles arise are transformed and reconceived. What is true for Milton is no less true for the Messenger of Allah: the vision of Ezekiel provides a foundational experience that assumes a meaning all its own.[3]

Two of Elijah Muhammad's writings, *Message to the Blackman in America* and *The Fall of America,* elaborate this meaning. Both devote pivotal chapters to a detailed explication of the vision of Ezekiel as the experience through which the Mother Plane is delineated. The essential import of both treatments is eschatological. In *Message to the Blackman in America,* the meditations on

Ezekiel are framed by a series of chapters that fall under the general heading "The Judgment." Beginning with a discourse on the history of "corruption" in the Western Hemisphere, and ending with a reminder that the "time is at hand," Elijah Muhammad encompasses the entire scope of human history through an apocalyptic account in which Ezekiel's vision plays a crucial role. It is within the context of the Judgment that Elijah Muhammad assumes the role of the prophet. As a headnote to the initial chapter, "On Universal Corruption," Elijah Muhammad cites the Qur'an: "Corruption has appeared in the land and the sea on account of that which men's hands have wrought" (30:41). This provides the impetus for the chapters that follow. Corruption begets corruption, which is visited on those who are responsible for its dissemination but which shall paradoxically resolve itself in a renewed purity once corruption has been obliterated.

Corruption, Elijah Muhammad observes, began in Europe and now covers most of the population of the earth. It is time for the true follower of Allah, the black man, to take his stand against this disease. Armed with that rallying cry, Elijah Muhammad proclaims that now is the time for "the armies of the nations of the earth" to gear themselves up for a "showdown" between the forces of good and the forces of evil. The forces of good, of course, are represented by Allah and the Nation of Islam. The forces of evil, on the other hand, are represented by the white race that would presume to do battle against the righteous. To this end, the forces of evil have called on the "mighty men of science and modern warfare" to "devise instruments and weapons against God and the armies of heaven." But, Elijah Muhammad declares, their efforts will be of no avail for "Allah hates the wicked American whites" and will certainly "remove them from the face of the earth." He will do so in the form of a final battle that will be tantamount to the War of Armageddon, one that will be a holy war to end all wars previously known on earth. The true Son of Man, the Mahdi, will then make his appearance in a form that is the consummate expression of all the hatred that the white race has brought on itself. This will be the true salvation of the black man in America. Here he will realize the kingdom of heaven on earth.[4] The manner in which this holy war is conducted represents Elijah Muhammad's version of the *odium Dei*. At the center of his conception is the vision of Ezekiel.

In *Message to the Blackman in America*, Elijah Muhammad's discussion of Ezekiel's vision constitutes three pivotal chapters, "Battle in the Sky Is Near," "The Great Decisive Battle in the Sky," and "The Battle in the Sky." As the chapter titles themselves suggest, Elijah Muhammad's meditations focus on the object that emerges from the visionary substratum as a war machine. As such, it represents Allah's response to the "instruments and weapons" that the white race has fruitlessly devised to combat God and the armies of heaven. Here, as elsewhere, the analysis is at pains to concretize and technologize the vision. Demythologizing the biblical source, Elijah Muhammad asserts that the vision of Ezekiel's wheel within a wheel is a phenomenon that can actually be seen in the sky today. It is present to all who behold and understand it as a concrete "thing." In appearance a replication of the sphere of spheres called the universe, Ezekiel's wheel is not so much a wheel as a "plane made like a wheel." As much as we might attempt to build a plane of this sort, Elijah Muhammad comments, our scientific powers are limited to our own human weapons. The plane of Ezekiel's vision is of another order. "A masterpiece of mechanics" that extends well beyond our resources, this plane is the consummate expression of a science that we cannot begin to fathom. If Elijah Muhammad demythologizes the vision, then, he bestows on it his own "construction," one consistent with the idea of Ezekiel's *visio Dei* as a wonder of technology. Through apocalyptic discourse, he generates a form of technopoetics that deconstructs the sacral implications of the vision and reconceives it as *the* war machine par excellence. In the process, he brings forth what is concealed to "the realm where revealing and unconcealment take place, where *alētheia*, truth, happens."

This technopoetics forms the basis of Elijah Muhammad's conception of Ezekiel's vision, a conception remarkable in its implications. Unlike the unsee-able, unknowable quality of the original rendering, Elijah Muhammad's rendering is not only seeable and knowable; it is almost brashly mundane. It disavows any sense of the "otherworldly," of the vision as ethereal "other." Just as Elijah Muhammad stands the biblical myth on its head in his creation of his own narrative of origins, he stands the visionary on its head in his corresponding reinscription of prophetic material. Bestowing a name on the vision, he also provides not only an account of its dimensions but a description of its operations as a war machine as well:

> The present wheel-shaped plane known as the Mother of Planes, is one-half mile by a half mile and is the largest mechanical man-made object in the sky. It is a small human planet made for the purpose of destroying the present world of the enemies of Allah. The cost to build such a plane is staggering! The finest brains were used to build it. It is capable of staying in outer space six to twelve months at a time without coming into the earth's gravity. It carried fifteen hundred bombing planes with most deadliest explosives—the type used in bringing up mountains on the earth. The very same method is to be used in the destruction of this world.

Naming the object the "Mother of Planes" is not simply a rhetorical flourish suggesting that it is the "plane of all planes": the designation bestows a gender and a function on this creation. It is seen to carry bombing planes that fly out on missions of destruction. "The small circular-made planes called flying saucers, which are so much talked of being seen," Elijah Muhammad observes, "could be from this Mother Plane," which is in effect the ultimate receptacle from which these smaller "saucers" emerge. Recalling the engineering language employed by the NASA scientist Josef Blumrich, the Mother Plane that Elijah Muhammad envisions is in keeping with the concept of the "mothership" that gives birth to objects from within her. These are the bombing planes equipped with "most deadliest explosives." Elijah Muhammad describes in detail how they operate. Propelled by motors made with the "toughest of steel," the bombs are seated at the depth of one mile within the earth's core, at which point they are timed to explode. The explosions produce mountains a mile high. Whole cities are destroyed in the process. Although the plane that sends forth these engines of destruction is visible, "do not think of trying to attack it," Elijah Muhammad warns: "That would be suicide!"[5] From this perspective, then, Ezekiel's vision is transformed into a "mother" that gives birth to engines of annihilation. As much as we might attempt to replicate this vehicle with machines manufactured to counter it, such aspirations would be fruitless. It simply cannot be replicated.

The concept of manufacture gives rise to another dimension remarkable in its implications. As a vehicle that in itself is manufactured, the Mother Plane is the product of human endeavor for, indeed, this is, as Elijah Muhammad maintains, a "man-made object." In fact, "the finest brains were used to build it" at a cost that is simply staggering. If such is the case, in what respect, then, can it be said that the Mother Plane is the vehicle of Allah as a being that is by definition divine? How, that is, can it as a man-made, manufactured object be associated with the concept of divinity? How does one reconcile the apparently contradictory aspects of human and divine in the formulation of the Mother Plane? These questions lie at the heart of Elijah Muhammad's reading of Ezekiel's vision, his reinscription of the vision as demythologized object remarkably infused with a sense of the "here and now." In order to answer the questions that his reading of the vision poses, we must briefly address the nature of Elijah Muhammad's theology.

This is a complex issue, one that Elijah Muhammad develops at length in the first part of *Message to the Blackman in America* under the general heading "Allah Is God." There, Elijah Muhammad argues that "God is a man and we cannot make Him other than man." Teaching the "reality" and "materiality" of

God in this life, Elijah Muhammad asserts that "God is in person among us today. He is a man. He is in His time, God sees, hears, knows, wills, acts and is a person (man)." Elijah Muhammad teaches not "the coming of God" as the embodiment of the invisible in the form of the visible but "the presence of God, in person," now. This does not deny the belief in the Parousia as the messianic advent in glory to judge the world at the end of time: it simply places that event in the context of a theology that insists on the perpetual presencing of God as a human being in this life.[6] Elijah Muhammad's theology is a radical and finally ideological departure from what he considers the Christian mystification of godhead that for generations has kept his people in bondage as slaves. He argues against what he feels is a false theology of God as "mystery." It is the devil Yakub (he says) who "makes the lost and found children (the American so-called Negroes) think that their real father (God) is a mystery (unknown) or is some invisible spook somewhere in space."[7] Undermining the religion of Yakub in the attempt to free his people from false conceptions, Elijah Muhammad would demystify God. In its own way, his conception of the Mother Plane is the product of this quest for demystification. As it deconstructs conventional readings of Ezekiel's vision, it generates an ideologically centered reading all its own.

Essential to that reading is the clash between good science and bad science. It is here that Yakub, the father of mystery, mystification, and trickery ("tricknology"), enters the picture once again. As the false scientist, Yakub has inspired his heirs to create weapons of destruction that have until now subjugated the human race, particularly the black man. To counter these weapons, Allah as true scientist has manufactured the true weapon of destruction. The contrast good scientist/bad scientist extends all the way back to the narrative of origins. Elijah Muhammad's cosmogony, we recall, focuses on the good scientist through whom the generation of the present world came into being as the result of a "big bang." Although we appear to learn nothing further about this primordial good scientist, by implication the Mother Plane is the product of his efforts. As the being that constitutes Elijah Muhammad's demonology, Yakub, of course, is the archetypal bad scientist by means of whose efforts the degeneration of the black man into the race of corruption is realized. To counter the forces of Yakub as false magus, degenerate trickster, then, Allah is conceived as the true magus, the sublime Mahdi, whose secret weapon is the transcendent science required to manufacture the masterpiece of mechanics, the plane of all planes, the Mother Plane.[8]

In his discourse on Ezekiel's vision as weapon of weapons, Elijah Muhammad accordingly traces the course of military history from its earliest times up to its apocalyptic present to prepare us for the warfare that awaits. Founding that history on biblical narrative, he observes that in the days of Noah God used water as a weapon, against Sodom and Gomorrah fire, and against Pharaoh fire, water, hailstones, armies of insects, droughts, and plagues. For Elijah Muhammad, both the Bible and the Qur'an prophesy that fire will be used as the final weapon. Recalling how weapons evolved from the use of the sword to the employment of firearms, Elijah Muhammad places Ezekiel's grand vision at the end point of this continuum. As a consummate means of destruction, fire is embodied in the vision that Ezekiel saw, now come to life in the present world. Wheel within wheel as universe within universe and sphere within sphere, this weapon of weapons will be brought forth to decimate the enemy. Its emergence will be announced with signs and wonders in the sun, the moon, and the stars; on the earth, its coming will be accompanied by tumultuous seas and fear in the hearts of men. The powers of heaven shall be shaken for, Elijah Muhammad declares, "they shall see the Son of Man coming in a cloud with power and great glory" (Luke 21:25–27). Who is the Son of Man? It is Allah in the person of Master W. F. Muhammad, the Great Mahdi, viewed by the Nation of Islam as a savior and by the Antichrist devils as a destroyer. The fittest designation is Fard Muhammad. Propelling the Mother Plane to its final destination, he will destroy the enemy and provide new life for the believer. Drawing on the full range of apocalyptic discourse that both the Bible and the Qur'an afford him, Elijah Muhammad constructs his own version of this event, one that he terms the "final showdown in the skies." At that showdown, "Allah will fight this war for the sake of His people (the black people), and especially for the American so-called Negroes." In his reflections on this event, Elijah Muhammad concludes that "we are Allah's choice to give life and we will be put on top of civilization."[9] His account of Ezekiel's vision as Mother Plane is finally a celebration of the ascendancy and triumph of the black race.

What is enunciated in *Message to the Blackman in America* is further delineated in *The Fall of America*, published almost a decade after the first book and two years before Elijah Muhammad's death.[10] Like the first volume, the second addresses the topic of the Mother Plane in a series of pronouncements that are tantamount to oracles through which the prophet shares his insights into this most profound of phenomena. The oracles move back and forth from one topic to the next as they illuminate various aspects of the Mother Plane from a multiplicity of perspectives. Recurrent themes emerge in the process of underscoring the nature and purpose of the Mother Plane. The first theme has to do with Elijah Muhammad as the fundamental source of revelation concerning the Mother Plane. Here, as elsewhere, Elijah Muhammad makes it clear that his authority comes directly from Allah, Master Fard Muhammad, who taught his

Messenger all that is to be known about the vehicle. Because Elijah Muhammad's pronouncements issue in such collections as *Message to the Blackman in America* and *The Fall of America,* they become "sacred texts," unimpeachable in their authority. This is how the prophet would have them viewed. Drawing on the eschatological import of these texts, Elijah Muhammad sounds a second theme that is repeated throughout. The plane is to be understood as a remarkable and invincible vehicle of retribution employed by Allah in his determination to destroy the white race in America for the evil perpetrated against its black slaves: "Allah (God) intends to repay her for what she has done."[11] Although this, too, is a theme that we have encountered earlier, it serves to remind us that, in the writings of Elijah Muhammad, the painful consciousness of the subjection to slavery endured by the black man must never be forgotten.

The Messenger's meditations on the Mother Plane in *The Fall of America* fall into two parts. Under the general heading "The Mother Plane," the first part addresses various aspects of the plane itself. Separately titled "Ezekiel's Prophecy of the Wheel," the second part places the discourse on the Mother Plane in the context of the biblical account. The first addresses the nature of the Mother Plane as both destructive and creative. Although its purpose is to destroy the old world, it is paradoxically the means by which the new world is to come into being. In this, the Mother Plane assumes a regenerative role: it is the means through which a new world of righteousness is ultimately realized. Such is only appropriate since it is this vehicle through which God created the world in its present form: "The same type of plane was used by the Original God to put mountains on His planet."[12] The source of the Mother Plane is precosmogonic: as the "mother of all planes," it was in existence "before the making of this world."[13] As the world issued from it, so will the world be annihilated and reborn through it.

Other aspects come into renewed focus as well. One of them has to do with the idea of the Mother Plane as a UFO. Elijah Muhammad observes that, in the 1930s, Canadian newspapers reported sightings of a vehicle resembling the Mother Plane. According to these accounts, the object was seen to descend out of the skies. It was alleged to have the appearance of "a great city." At the same time, the accounts maintained that something like a great tube came down from it and then went back up again. Although Elijah Muhammad does not confirm or deny such reports, he does attribute them to those who have been influenced by what he ironically calls "the devil scientists," those who attempt to explain the sightings as phenomena that exhibit certain characteristics they do not understand, including the ability of the Mother Plane to stay aloft for long periods of time. Although they might be suspicious of the sightings, the devil scientists take them seriously enough to be fearful of what they portend. As a mechanical object, the Mother Plane comes into the gravity of the earth, where it takes on the oxygen and hydrogen that permit it to stay out of the earth's gravity until it requires refueling.[14] Although Elijah Muhammad castigates the devil scientists for their suspicions, he does not discount the impulse to respond to the Mother Plane as a UFO. Rather, he welcomes that impulse as an acknowledgment of the elusive presence of the plane and the need to understand what he only is permitted to disclose.

The foregoing is preparatory to the section titled "Ezekiel's Prophecy of the Wheel." Having already indicated the precosmogonic status of the Mother Plane, Elijah Muhammad in this section makes it clear that this vehicle actually antedates the phenomenon beheld in Ezekiel's vision. If such is the case, then what Ezekiel saw was none other than "the Mother Plane." In this respect, Elijah Muhammad appropriates certain elements of the ancient astronaut hypothesis, which he refashions to accord with his own outlook. Doing so, Elijah Muhammad interprets Ezekiel's vision as a manifestation of those things that are basic to the foundational myths that underscore the teachings of Allah. For example, in his rendering, the four creatures of Ezekiel's vision become symbols of the races that reflect the effects of Yakub's genetic engineering. The four creatures represent the four colors of the original people of the earth: black, brown, yellow, and red. Each is a "power" that infuses life into the machine. Of these, the black power, of course, is the most potent. But, as the last race created before the propagation of the degenerate white race, the red (i.e., the "red Indian") is to "benefit from the judgment of the world" since it constitutes one of "the four colors of the black man."[15] The white race obviously does not share this status. From this perspective, Ezekiel's vision as the Mother Plane not only is conceived as a technological wonder that creates, destroys, and re-creates; it also becomes an emblem of racial identity and potency. The product of Elijah Muhammad's foundational myth concerning the propagation of the races, the Mother Plane is an integral part of his quest to achieve a renewed sense of self-identity and empowerment for his people.

That sense is deepened in the observations that conclude Elijah Muhammad's discourse. There the Messenger of Allah makes clear that the knowledge that he imparts is of the most profound sort. A form of black gnosis, it is a knowledge that is not shared by the white race, first because the white race is incapable of understanding or appreciating it, and second because the white race would put this knowledge to a perverse use. "If the devil would get this type of knowledge," Elijah Muhammad exclaims with an outburst of colloquial exuberance, "we could just say that we are goners." Try as he might, however, the devil is not able to attain this type of knowledge. As distinguished from that of the black race, the knowledge of the white race is distinctly limited. "The world of the white man," says Elijah Muhammad, "was made from what he found and what he has seen and learned from the work of the original black man. The white race is far from being able to equal the power and wisdom of the original black man." In fact, the knowledge that the black man possesses is such that it disables the white man even before he has the opportunity to put his plans into action. This is especially true of the black scientists on the Mother Plane. These scientists know what the white man is contemplating "even before the thought materializes."[16] Such is the extent of the black man's knowledge; such is the extent of his power. For Elijah Muhammad, knowledge (particularly occult knowledge) is to be equated with power: higher wisdom and supreme power are synonymous. This is true gnosis, a form of knowing that the Messenger would impart to those willing to understand. Disseminated directly from Master Fard, "the Wisest and Best Knower," this gnosis is the *maʿaseh merkabah* of Elijah Muhammad. In fact, Elijah Muhammad actually speaks of the Mother Plane as if it were the work of the chariot. "O mighty wheel," he exclaims; "there is plenty of significance to the make of the Mother Plane. There is much significance to the course of operation of her work." Celebrating the *merkabah* in these terms, he concludes: "let us seek refuge in Allah (God) from the destructive work to come from this Mother of planes."[17] If the work that he celebrates proves threatening to those he would deem his enemy, it is no less compelling as an instance of major import to the acculturation of Ezekiel's *visio Dei* in the modern world. In his Saviours' Day message to his people in 1967, Elijah Muhammad declared: "God has raised me in the midst of you to interpret for you."[18] This is a credo that certainly defines his role in his encounter with Ezekiel's vision as Mother Plane.

As suggested, not only do Elijah Muhammad's meditations on the vision of Ezekiel make their appearance in the texts examined thus far; they are also germane to his newspaper, *Muhammad Speaks.* Here they gain renewed impetus because of the journalistic context in which they appear. Serialized over a period of several months (May–September 1973), the meditations on Ezekiel assume an immediacy, a currency, and an urgency that only a weekly could bestow. The title of the newspaper is significant: here is the oracle consigning his message to written form. If the prophet now delivers his kerygma or proclamation as journalistic copy, he still "speaks" to those who will hear. His trumpet blast is now in the language of print. Reinforcing the printed word is the design of the copy itself. Below the title *Muhammad Speaks,* the front page of each issue displays an illustration of two black figures symbolically clasping hands across a sectioned globe that depicts the continents of Asia (on the left), America (in the center), and Africa (on the right). Behind the globe, the sun rises gloriously. In *The Theology of Time,* Elijah Muhammad interprets the symbolism of this icon. Of the gesture of handclasping, he says, "We have been lost from the path to God and the Heavens that He prepared for us. It is now coming to pass that we can shake hands with our Brothers all the way around the Earth."[19] Correspondingly important to the icon is the image of the sun, which has a pivotal position in Elijah Muhammad's outlook. The sun, he says, is a "god" because of its creative power to bring forth life and to disseminate that life-giving force throughout the planet. Embodying that power within himself, Elijah Muhammad as the Messenger of Allah is "styled as a Sun" because through Allah he empowers those he instructs with his life-giving words. He sets them on the right path and restores them to light from their realm of darkness.[20] As a kind of sun god, he assumes a cosmocratorial role in his act of overseeing and empowering his people with his words.

Beneath the iconic gesture of global unity and empowerment on the front page of *Muhammad Speaks,* the headline in large, bold typeface announces the news of the day. In the run that we are addressing, the boldface headlines include "EZEKIEL'S WHEEL," "*Messenger Muhammad's* ANALYSIS OF EZEKIEL'S WHEEL," "O WHEEL," "*O Wheel* MOTHER OF PLANES," "THE NOISE OF THE WHEEL," and "INSIDE OF EZEKIEL'S WHEEL." The sense of urgency that these headlines suggest reinforces the immediacy of Elijah Muhammad's message. What he has to say about Ezekiel's vision is truly "news." Elijah Muhammad becomes Ezekiel, his newspaper the present-day version of that scroll the biblical prophet himself was made to digest in order to deliver his message to the house of Israel: "And when I looked, behold, an hand *was* sent unto me; and, lo, a roll of a book *was* therein; And he spread it before me; and it *was* written within and without: and *there was* written therein lamentations, and mourning, and woe" (Ezek. 2:9–10).

Like *Message to the Blackman in America* and *The Fall of America,* the serialized discourses on Ezekiel in *Muhammad Speaks* extend back to the mid-1960s.[21] In fact, the article "Battle in the Sky," published in the 25 May 1973 issue of *Muhammad Speaks* (pp. 16–17), is a reprint of an earlier article that appeared on 7 January 1966. Providing a context for the series as a whole, this article draws on themes already discussed. Elijah Muhammad warns of the impending war between God and the white devil race foretold both in the Bible and in the Qur'an. To be waged in the sky, this battle will pit the so-called technological wonders of the white race against the true wonders of Allah. The 1 June 1973 issue develops the theme of celestial warfare in an article titled "The Nations Confused" (pp. 16–17), which maintains that the white race plans to poison the air in order to destroy every living thing on earth. The devil race is also attempting to establish way stations on the moon and the other planets as a way of asserting sovereignty. Clearly, Elijah Muhammad exclaims, this kind of folly must end. Its termination will occur with the final battle in the sky. The means and method of undermining the aspirations of the devil race are revealed through Elijah Muhammad's analysis of Ezekiel's vision, which follows in a series of articles addressing the issue of the biblical prophecy in *Muhammad Speaks*.

As commentaries on Ezekiel's vision, these articles are both fascinating in themselves and interesting for what they say about Elijah Muhammad's thinking during the years preceding his death. First is the extent to which they are concerned to explicate specific features of the vision. It is as if the vision has taken on greater meaning for Elijah Muhammad, who places himself in the position of illuminating what Ezekiel beholds: the enthroned figure, the wheels, the creatures, and the wings. The biblical text itself is more in evidence than in earlier commentaries. As Elijah Muhammad reflects on the meanings of the vision, his voice modulates between the prophetic and the exegetical. He is both the prophet of the vision and its interpreter. Framing each of the articles on Ezekiel's vision is a marginal commentary on the articles by Margary Hassain, referred to earlier. Her glosses represent a kind of metacommentary on the commentary of the Messenger. In fact, the reader moves from her commentary to his in the process of understanding the subject under consideration. The commentaries intersect and illuminate one another.

In the 8 June 1973 issue of *Muhammad Speaks* (pp. 16–17), Margary Hassain introduces Elijah Muhammad's discourse "Ezekiel's Wheel" through a series of admonitions concerning present corruptions in government. Her commentary is in effect a warning to those who do not heed "the fall of the prestige of the heads of this government, and the fall of the pride and the confidence of the people in the heads and the structure of the government." Railing at the same time against not only the false white leaders but their misdirected black followers as well, Margary Hassain adopts a stern and unwavering tone in her own denunciations of the present times. These prefatory remarks anticipate an entire series of such observations that frame Elijah Muhammad's meditations on Ezekiel in the issues of *Muhammad Speaks* that follow. In the final issue, that of 21 September 1973 (p. 12), Margary Hassain concludes the series with a detailed encomium of the Messenger as one who has finally provided the key to a previously sealed text. "Unfolding" for the black man "the Truth and Reality of all things," Messenger Muhammad has been able to "step forth and reveal the secrets of the Book of Ezekiel and the marvelous Ezekiel's Wheel." As one who "is the Ezekiel," the Messenger has taken on himself a mission that is precisely that of the prophet whose vision Elijah Muhammad illuminates and likewise shares. In a world faced with corruption, Elijah Muhammad's meditations on Ezekiel become the means to alleviate the travails of the present condition. If one is inclined to respond to these travails with lamentations, mourning, and woe, Elijah Muhammad will disclose the means of countering this response with the promise of hope and redemption for the true followers of Allah.

Elijah Muhammad heralds the coming of the Mother Plane in what will be the "final showdown." In that decisive battle in the sky, it is "scientifically clear" that the wheel, the Mother Plane, will be victorious in cleansing the world of white corruption and establishing the black nation as all-powerful. Even as they were constructing the Mother Ship, the black scientists knew that this engine of warfare and its crew would be called on to engage in final battle with an alien people. For that purpose, the plane was designed to be both nimble and elusive. As an immense carrier for fifteen hundred deadly planes that would be sent out on search-and-destroy missions, the Mother Plane is able at will to render itself invisible to the human eye. Addressing the issue of the Mother Plane in the first of the series in *Muhammad Speaks,* the newspaper becomes a rallying cry for the apocalyptic combat that is certain to follow.[22] The fiery zeal that infuses Elijah Muhammad's account of the Mother Plane in his opening salvo, however, is tempered by the articles that follow. Here, his voice modulates from the prophetic to the exegetical as he adopts the role of biblical interpreter, one who provides a way of "seeing" the vision from the unique perspective only he is able to bestow. This he does in a five-part series of articles that concludes his analysis of the vision and its multiple implications.

Addressing such issues as the date and authorship of the vision, Elijah Muhammad focuses on specific aspects of it that are pertinent to his cause. Crucial to his interpretive posture is the rendering of the fourfold creatures, an issue that concerns him repeatedly throughout his analysis. Focusing on Ezek. 1:7 ("And their feet were straight feet; and the sole of their feet *was* like the sole of a calf's foot"), for example, he interprets the "calf's foot" as a reference to "a people who have not been able to move fast" because of being impeded in their progress. For Elijah Muhammad the presence of such an impediment is a reflection of what he calls "the slowness of the black once-slave whose foot was shackled by his white slave-master" so that the slave would not attempt to escape from this bondage. Drawing on a vast reservoir of black slave culture, Elijah Muhammad accommodates the vision to his own singular background. Doing so, he provides essential insight into the visionary experience as a manifestation not only of the painful experience of his people but also of his desire to liberate his people from the bondage of their past lives.[23] Here, the Mother Plane becomes remarkably humanized. It is not simply a "machine" of apocalyptic import: it is also a way of talking about a people. We recall James Baldwin's observation in *The Fire Next Time* that the central quality in Elijah Muhammad's face is pain, old and deep and black. It is this pain that infuses his interpretation of the creatures of Ezekiel's vision as an emblem of struggle against oppression that the black race has endured over many generations.

Later interpretations of Ezekiel's vision focus on the fourfold creatures as symbols of the history and plight of the black race. Extolling the lion for its fearlessness, Elijah Muhammad exhorts his people to fulfill their destiny as a race determined to reflect the courage of the lion in confronting its enemies. As he moves to address each of the other creatures in the visionary configuration, Elijah Muhammad associates each with a trait in which his people can take pride. If the lion is an emblem of courage, the ox is a symbol of strength; the eagle, in turn, signifies swiftness. Finally, the man encompasses the other creatures as a unit. He is the black man as a driving force of all that the theriomorphic forms come to represent. Interpreting the directional symbolism through which the vision as a whole is articulated, Elijah Muhammad maintains that the phrase "on the left side" (as opposed to "on the right side") indicates that the black race must always be aware of its "enemy," which is said to occupy the left or evil side, an interesting speculation that suggests a sensitivity to the implications of the "sinister" elements of such a placement (cf. Ezek. 1:10). Equally fascinating are his comments on Ezek. 1:13, which describes the appearance of the living creatures as comparable to "burning coals of fire" that go up and down among the creatures. This fiery quality, Elijah Muhammad maintains, symbolizes the "anger" of the black race for the indignities that it has had to bear. At the same time, this fiery presence, especially as it flashes forth lightning (Ezek. 1:13), alludes to those bombs that the Mother Plane is prepared to send forth at the appropriate time. The living creatures become the means of swiftly moving those bombs in and out of the plane when necessary. Accordingly, for Elijah Muhammad the vision of Ezekiel alternates between the human dimension that embodies the history of black culture and the apocalyptic dimension that embodies the technological bearing of the phenomenon as machine. Recalling the discussion in *The Fall of America*, Elijah Muhammad again associates the faces of the creatures with the four major colors of the black man, black, brown, yellow, and red. As such, the creatures reflect not only the struggle for liberation from the bondage of slavery but also the racial history that underlies Elijah Muhammad's myth of origins. Out of the cultural struggles to which he was heir he forges his own unique interpretation of Ezekiel's *visio Dei* as a phenomenon of crucial import in his calling as the Messenger of Allah.[24]

The final three articles in *Muhammad Speaks* ("*O Wheel* Mother of Planes," "The Noise of the Wheel," and "Inside of Ezekiel's Wheel") bring to a close his meditations on Ezekiel's vision. Of utmost importance in these meditations is the symbol of the wheel, which for Elijah Muhammad encompasses the vision as a whole. The spirit that impels this great "mystery Wheel" is a reflection of "the Aims and Purpose of Allah," whose providence is made manifest in the motions of the wheel. To believe in the wheel is to have faith in the providence of Allah, in his desire to "carry out His Aim upon this world." As Elijah Muhammad emphasizes throughout his discourses, such providence will ultimately reveal itself in the creation of a "New World" once the old world has been destroyed. It is this promise of ultimate renewal that renders Elijah Muhammad's outlook celebrative rather than merely denunciatory.[25]

In fact, Elijah Muhammad consistently adopts Ezekiel's own apostrophic posture in the exclamation "O Wheel, O Wheel." The apostrophe alludes to Ezek. 10:13: "As for the wheels, it was cried unto them in my hearing, O Wheel." Elijah Muhammad interprets this outcry as a sign that the prophet is "admiring his vision" that he receives from "Allah (God)." This insistence on the celebrative aspects of the prophecy suggests the extent to which Elijah Muhammad is determined to transcend the immediate travails to which he feels his people have been subjected and to look forward to a new beginning. Of course, that new beginning must take into account the annihilation that precedes it, but this too is a cause for celebration because the destruction to be wrought by the wheel is a necessary part of the larger process culminating in renewal. As for the wheel itself, this is the "most miraculous mechanical" phenomenon imaginable, a fact that is also a cause for celebration: "The Wheel is capable of sitting up above earth's atmosphere for a whole year before coming down into earth's atmosphere to take on more oxygen and hydrogen for the people who are on this plane." "Oh," the Messenger exclaims, "she is a wonderful thing!" Through her, the universe came into being, through her it will be destroyed, through her it will be renewed.[26]

No white man dare affront her or call her power into question. With his own "jet planes and other military weapons," the white man might wish to destroy the Wheel, but he should not even try. In a kind of colloquial flyting or "put-down" that is so characteristic of his discourse, Elijah Muhammad admonishes the white man in a tone of playful black contempt: "You should just go home and go to sleep. No one can harm this plane, the Wheel. They are going to fix you up first, before the Wheel ever comes into sight!" Don't try to escape to other planets such as Venus and Mars, Elijah Muhammad warns "Mr. Enemy": You cannot "light" on these planets because the people there will prevent you from doing so. In his discourse on this celebrative aspect of the Wheel, then, Elijah Muhammad takes pleasure in its capabilities and moves easily between prophetic, analytic, and comic modes to suggest the all-encompassing nature of the vision he seeks to disclose.[27]

"The Noise of the Wheel" is his penultimate discourse on the vision. In it he analyzes the nature of the mysterious voice that emanates from the enthroned figure (Ezek. 1:26–28) above the firmament as well as the great noise that Ezekiel hears in the movement of the wings of the creatures: "And when they went, I heard the noise of their wings, like the noise of great waters, as the voice of the Almighty, the voice of speech, as the noise of an host" (Ezek. 1:24). Elijah Muhammad interprets this event apocalyptically as an expression of the sounds that will be made on the last day. Those sounds emanate, however, not from the race that is destroyed. Rather, they are an expression of the joy at the liberation of those downtrodden by others. Once again, the concept of the slave's breaking free of the trammels imposed by the slavemaster comes into focus. As he is lifted up from his bondage, the newly liberated black man cries out in his exuberance. The regenerative sense of experiencing a new world finds its way into every aspect of the vision. In her metacommentary on the commentary of her master, Margary Hassain expresses exactly this exuberance as she proclaims, "We the black people have no birth record and we have no beginning and no end." Renewal resolves itself into an attestation of its own timelessness.[28]

In the final discourse "Inside of Ezekiel's Wheel," the Messenger penetrates the ultimate mysteries of the Mother Plane, those "secrets" to which Margary Hassain refers in her metacommentary. Moving from the noise of the Wheel to the mysterious voice that emanates from the firmament, Elijah Muhammad maintains that this voice is that of the Master of the Plane, presumably that of Allah in the form of Fard Muhammad. He sits, as Ezekiel says, on a throne, "as having the appearance of a sapphire stone" (Ezek. 1:26). Emanating from the throne is fire "round about within it" as the appearance of amber, the *ḥashmal* that overwhelms all who come in contact with it. Depicted here is a figure who for Elijah Muhammad is the scientist of scientists: he is "*a man with all of this science on him*" (italics mine). Given the emphasis on this word *science* in Elijah Muhammad's vocabulary, the act of envisioning the supreme figure in these terms is of utmost moment. To be imbued with "all of this science" is to be possessed of consummate gnosis. This is a stunning moment in Elijah Muhammad's analysis for it accords the vision the mystery of its conception at the very point of disclosing its secrets. He remythologizes the vision in the act of de-mythologizing it. To be the scientist of scientists is to be a god indeed for that involves the power to create and to destroy and finally to re-create on a cosmic level. Such is Elijah Muhammad's *kabbalah,* and such is his celebration of this supreme of scientists. It is the "desire" of this scientist that moves the Wheel as "the most mysterious" of phenomena "that the world has ever dreamed of." As the manifestation of the power of this scientist, the Wheel becomes a token of his desire, his pleasure, to do as he will. It is with supreme faith and confidence that Elijah Muhammad accepts his role as the Messenger of one who takes plea-sure in his desire to bestow the fruits of his providence on the black nation.[29]

In his final flourish, Elijah Muhammad makes clear that he is fully aware of how provocative his interpretation will be to those who refuse to heed his call. He is unflinching in his castigation of such reprobate people. They are like scor-pions who sting and bite the very one commissioned to set them free; yet he de-clares he is not afraid. The Messenger, he says, is one who must speak the words of God to his people, "whether they will hear, or whether they will forbear (Ezek. 2:7)." Like the prophet before him, the Messenger has eaten and digested the scroll that has been handed to him by the vision. In keeping with that prophet (who himself anticipates the seer of Revelation), the Messenger knows the sweetness and the bitterness of his vision and is willing to communicate both to the house of Israel (cf. Ezek. 3:3, Rev. 10:9–10). Doing so, he will have fulfilled his divine mission as the prophet Ezekiel of the twentieth century.[30]

The same authority that imbues the writings published during Elijah Muham-mad's lifetime is present in the transcription of his lectures published in the posthumous *The Theology of Time.* Encompassing four books, this collection embraces an entire range of subjects. Titled "Knowledge of the Messenger of Allah," the first book addresses itself to the understanding of the role of the in-dividual Muslim and his relation to the Messenger. The second book, "Knowl-edge of God," concerns matters theological and epistemological. The third, "Knowledge of the Devil," focuses its attention on the figure of Yakub and his role in the generation of evil. The fourth book, "The Judgment Is Now," con-stitutes a series of lectures on eschatological matters.

Taken as a whole, *The Theology of Time* confirms his role as "knower" and as "revealer." Running throughout the book is a sense of the gnostic, and it is from such an oracular perspective that Elijah Muhammad speaks. In his lecture "The Veil Must Be Lifted," he accordingly depicts himself as the one through whom all gnosis will be attained. He will lift the veil that has occluded the supreme knowledge of secrets, the meanings of which he has been commissioned to disclose at the appropriate time. Citing Isa. 52:15, he declares in an act of self-representation that he is the "Little Fellow" who has been sent to shut the mouths of kings. Now that the time is right, he will do so through his message: "What you have been told," he declares cryptically, "was not told to you for you to understand. It was not given to you to understand before this time. This is the time that you must know the Secrets of all Truth that have been put in a symbolic manner. Now today, the veil must be lifted and you must understand."[31] *The Theology of Time* is a gnostic text issued to reveal the innermost secrets of the unknowable. Concerned with such hidden matters as the significance of secret numbers, God's self-generation, the meaning of darkness, and the meaning of light, the lectures collected in this volume constitute the Honorable Elijah Muhammad's "supreme wisdom." Within this elevated framework, his discourse on the vision of Ezekiel reemerges.[32]

Assuming the form of two lectures in the fourth book of the collection, that discourse is in keeping with the eschatological perspective explored in his other works. Present are the motifs with which we are already familiar. These include the attempts of the military to seek out and destroy the Mother Plane by launching aircraft and other missiles into outer space and by establishing observatories on other planets, the descriptions of the bombing operations of the Mother Plane, and the association of the mountains created in those operations with the mountains fashioned by God in the creation of the universe. Once again, the warfare that accompanies these final bombing missions is cast in the form of apocalyptic confrontations. Most of the bombs will be reserved for the white world of North America, but three will be reserved for England because of its limited land mass. Likewise emphasized is the notion that the Mother Plane is always present, hovering somewhere above and ready to strike. The signs of the times demonstrate that the cataclysm may occur at any moment. Elijah Muhammad exhorts his people to be prepared at all times for the final day of judgment.[33]

What is particularly interesting about the account of Ezekiel's vision in *The Theology of Time* are the political dimensions of the discourse, dimensions revealed earlier in the treatment of end time and politics. These chapters on the Mother Plane represent a kind of debriefing, a sharing of state secrets about his knowledge of the vehicle and the American government's awareness of that knowledge. In fact, he avers that he was approached by representatives of the government, arrested, and interrogated about what he knew of the Mother Plane as secret weapon. He thus depicts a scenario in which he not only describes the Mother Plane to the authorities but draws detailed pictures of it on a blackboard. After he has done so, he depicts his interrogators as taking the whole blackboard to the offices of the FBI, which keeps it in its secret files to this day.[34] As Mattias Gardell has recently demonstrated, these assertions find corroboration in documents on file with the FBI in connection with Elijah Muhammad's refusal in 1942 to register for the draft.[35] However one construes these documents, one fact remains clear: Elijah Muhammad's assertions of the Mother Plane are part of the official records that were kept on him by a government determined to view him and his calling as a threat to the nation in a time of war.

Shrouded in mystery, the secret weapon known as the Mother Plane thereby enters the world of espionage, and Elijah Muhammad becomes the source of classified information that is disclosed only under protest and coercion. As one whose occult knowledge has been forcibly extracted by the authorities, Elijah Muhammad takes solace in the fact that the efforts of the interrogator and oppressor will be undermined. The forces that constitute the Mother Plane will ultimately prevail. They will do so, moreover, in a manner entirely consistent with the stealthlike aura that surrounds the weapon itself. Specifically, they will make certain to reveal the destructive potential of the weapon at just the point when the appearance of the Mother Plane is least expected. It will come to destroy its enemies and to redeem its faithful like a thief in the night when we are all sleeping. "But the day of the Lord will come as a thief in the night; in the which the heavens will pass away with a great noise, and the elements shall melt with fervent heat, the earth also and the works that are therein shall be burned up" (2 Pet. 3:10).[36]

As the one that comes to judge in this eventful hour, the Mother Plane, Elijah Muhammad observes, is like Jesus. This christocentric allusion ironically bestows on the Mother Plane a kind of dreadful, messianic bearing that moves Elijah Muhammad to one of his most remarkable revelations about the phenomenon. Jesus, he says, was a "scientist" who had the ability to "tune in" on his enemies and tell them what they were thinking. If his enemies "said that they will go this way, he would go the other way." For that reason, his enemies could never catch him. He knew what his enemies were thinking even before they did. Although Elijah Muhammad does not develop the christocentric implications of this idea further in this context, he does elaborate on them in his teachings about Jesus. In *The True History of Jesus*, a posthumous collection of Elijah Muhammad's articles published under the auspices of CROE, the Messenger provides his own "gospel" account that seeks to "correct" those who have labored under the misapprehension of who Jesus was and what his significance is for our times.[37]

Elijah Muhammad's gospel account itself assumes the form of a "tall tale," one that has about it a homespun quality that at once domesticates the action and paradoxically attributes to it aspects of the "marvelous." We learn, for example, that Mary and Joseph were childhood sweethearts who fell in love when they were in school together but were prevented from marrying by Mary's father, a rich architect who did not want his daughter involved with a poor carpenter (and a rather middling one at that!). So Joseph then married someone else and raised his own family. Meanwhile, Mary's father tried to interest his daughter in the rich men of the village, but to no avail. Still longing for one another, Joseph and Mary were finally able to consummate their relationship when Mary's father was away on a trip. The result of their union was Jesus. Threatened by the authorities because of their illicit relationship and a knowledge (bestowed on them by an old prophetess) that their child was very special indeed, they escaped by camel to Egypt, where they raised their child, who was so bright that he finished school at twelve. He turned out to be a real scientist, trained in astronomy, geometry, and all the other sciences. But he had one talent in particular that was truly remarkable. That talent was made known to him by "an old man" who desired to teach the child how to elude those who might seek to destroy him. Having arranged to meet the child, the old man taught him that he possessed a power comparable to having a "Radio in the Head."[38] That power is compared to a radio in the head because, by means of the capability and great wisdom it imparts, it allows one to hear the thoughts of others as distinctly as one can hear messages communicated over a wire or through a radio. Having learned this lesson, Jesus was able to "tune in" on the authorities who sought to entrap him during the period that he was teaching his followers. As soon as the authorities came after him, he would dismiss his congregation and escape unharmed. Although he could draw on his radio in the head at will, he finally allowed himself to be sacrificed for the betterment of mankind.[39]

Both the use of colloquial diction and the adoption of self-conscious humor and hyperbole suggest that much of the narrative has about it almost a Chaucerian quality of high spirits. To a great extent, this is gospel as tall tale, written for the purpose of interrogating the New Testament gospel accounts that bestow on Jesus a divinity that Elijah Muhammad seeks to undermine. This fact is made clear in the probing analysis that follows the story. As Elijah Muhammad states quite directly, "These articles are written due to the mental blindness and theological knowledge of the Bible's teachings of Jesus's birth, life, teachings and death." As he goes on to declare, Jesus was a great man, even more, a prophet, but he was not the incarnation of the divine son of God. His father was human, in fact, no more than a carpenter and an adulterer to boot. Like all such prophets, Jesus fulfilled his true role, that of being a harbinger of Allah. In that role, he prefigures Allah's ultimate appearance in the form of Master Fard Muhammad, whose coming is the fulfillment of all prophecies.[40]

In Elijah Muhammad's gospel account, Jesus performs another role as well, one very much in keeping with the tall-tale quality of the narrative. That role has to do with the idea of the trickster. In this capacity, Jesus becomes a trickster par excellence. For Elijah Muhammad, his behavior is very much in keeping with his name. Engaging in his own act of creative onomastics, Elijah Muhammad observes: "The authorities chased Jesus until he became discouraging to them." They named him "Christ (which means troublemaker)" because, whenever the authorities would arrive at the place where he had been, he would already be gone.[41] As a troublemaker/trickster, Jesus offers ironic comparison with Yakub, who was likewise a troublemaker/trickster. Like Yakub, Jesus was a brilliant scientist, capable of marvelous acts of wizardry. Unlike Yakub, however, Jesus served for the betterment of his fellow men, not for their undoing. Jesus was a trickster whose acts made him a hero, not a fool. As such, he was the antithesis of Yakub. His purpose was to be an anti-Yakub. Whereas Yakub as trickster was of an evil cast of mind and determined to sow evil and discord, Jesus was of a virtuous cast of mind and sacrificed himself for others. As a brilliant scientist equipped with the truly remarkable ability to read the thoughts of others, Jesus is thereby depicted as a figure of great wisdom, of supreme knowledge and discernment. Delineating his heroic trickster in these terms, Elijah Muhammad in effect transforms his narrative of the life of Jesus into what might be called a gnostic gospel, one in which the trickster engages in the supreme act of anti-Yakubian tricknology. As trickster, Jesus is the true magus, equipped to read the minds of others.

From this perspective, Elijah Muhammad's comparison of the Mother Plane with Jesus in *The Theology of Time* becomes both clear and compelling. Embodying the gnostic qualities that Elijah Muhammad sees in Jesus as magus, the Mother Plane is grounded in a view of Jesus as a figure who knows what others are thinking and is therefore able at every turn to outwit his enemies. What is remarkable about Elijah Muhammad's rendering of the Mother Plane as a phenomenon that embodies telepathic powers, however, is the extent to which that phenomenon is conceived in racial and ideological terms. If Jesus was able to read the minds of others, Elijah Muhammad observes, "the Muslims have one out of every few hundred in the East who can tune in on us over here and tell what we are thinking about." Like the Muslims of the East, he claims, "we are pretty smart people," too, for we also have such powers if only we know how to implement them. Like the old man in the Jesus narrative, Elijah Muhammad seeks to teach his followers the supreme knowledge of telepathy. This he does from the perspective of Ezekiel's vision. Addressing his followers, he advises: "You can do it [read the thoughts of others] yourself if you will take time, clear your mind, and then go into some place where no one will disturb you, and concentrate on nothing but that Wheel or that Brother. After a while you can hear what the Brother is saying to himself. Maybe you can hear the motors going in one of the Wheels." Anyone familiar with the complex traditions of *merkabah* mysticism will see here a striking similarity between such an act of meditation as a means of achieving transcendence and the praxis that Elijah Muhammad recommends to his followers. Like the riders in the chariot of old, the Messenger internalizes the vision, casts it in a psychocentric form, and causes it to become an inner *merkabah,* a *merkabah* of the mind. In accord with the practices of the ancient riders in the chariot, this act of meditation is principally a means of self-empowerment. Through the act of meditating on the vision, one becomes capable of performing theurgic acts. He likewise has the powers of one who can read future events, Elijah Muhammad implies. These powers make him tantamount to a god for he knows what no ordinary mortal can ever know. A prophet, he knows that the end is near, that "the Time is up." Living constantly on the edge in his awareness of this fact, he is ever conscious that the time of apocalypse is upon him, that the Mother Plane is shortly to carry out its mission of terror on an unsuspecting world.[42]

The act of internalizing Ezekiel's vision in this manner represents a culminating moment in Elijah Muhammad's interpretation of this most awesome of events. From the perspectives that we have been exploring, his discourses on the vision as Mother Plane provide renewed insight into the racial and ideological foundations of visionary enactment. As the messenger of an entire movement, indeed, as the driving force behind the creation of what might be termed a nation in its own right, Elijah Muhammad appropriates the vision of Ezekiel as the foundational experience through which to articulate the message of apocalyptic proportions. A crucial part of the struggle for recognition and independence that has inspired the call for black empowerment for over six decades, that nation has evolved its own myth of origins, its own formulation of historical process, and its own articulation of eschatological events. The vision of Ezekiel is the pivotal moment through which Allah in the form of Master Fard announces his presence to the prophet who is to become the Messenger of Allah. It is that moment through which the Messenger understands the nature of his calling. It is that moment through which the entire fate of a nation coalesces into one coherent design. By its very nature, this design is alien to those it seeks to frighten, to anger, to offend. A phenomenon the originary import of which is to unleash corresponding feelings of fear and alienation among those who were made to share Ezekiel's own experience on the shores of the Chebar, the prophet's vision lends itself compellingly to its reappropriation as the "alien" event par excellence by a culture determined to declare its own "otherness" in this century. The Honorable Elijah Muhammad is the embodiment of that otherness, that alienation. He is so by design to those who do not understand his message. To his followers, the vision he recounts and the message he imparts are "as honey for sweetness" (Ezek. 3:3).

Visionary

Minister

7

On the morning of 25 February 1975, the Honorable Elijah Muhammad died of congestive heart failure.[1] Following his "departure," members of the Nation of Islam received the announcement that Elijah Muhammad's son Wallace Deen Mohammed (thereafter, Warith Deen Mohammed) was designated his father's successor.[2] As chief minister of the Nation, the new leader found himself committed to a vision that was far different from that of his father.[3] Long before he assumed leadership of the Nation, he determined that the future of his people resided in "the security of the international Muslim confraternity rather than the isolation of segments behind the uncertain palings of black nationalization." Even though he received his schooling at the University of Islam run by the Nation in Chicago, his views of Islam were consistently at odds with accepted doctrine, and he was frequently disciplined.[4] During his tenure as chief minister, he set about to reshape the Nation according to his own sense of what tenets it should embrace and what direction it should take.[5]

Under his tutelage, the ideology of the Nation was entirely reconceived.

Seeking to demythologize the narrative of origins underlying that ideology, the new leader dethroned Master Fard from his supreme station. The divine attributes associated with the Mahdi no longer obtained. Master Fard was seen to be simply mortal. Although still occupying a place of honor, even Elijah Muhammad was (to use C. Eric Lincoln's diplomatic locution) "viewed in the context of his times and limitations."[6] This is to say that Elijah Muhammad was no longer to be understood as the last Messenger of Allah. Rather, he was to be seen simply as a wise man who brought American blacks to the Qur'an. "I want to get rid of all this spiritual spookiness," the new chief minister declared.[7]

Getting rid of spiritual spookiness also involved a total reorientation in the meaning of Ezekiel's vision as Mother Plane. No longer the embodiment of a distinctly apocalyptic perspective, the vision was, according to Martha F. Lee, "de-eschatologized." In this form, the wheel within a wheel that constitutes the vision was reconfigured in communal terms. As such, the Nation of Islam became a wheel within the larger wheel of the Islamic world community.[8] In the spirit of the reformed Malcolm X become El-Hajj Malik El-Shabazz, whom Warith Deen Mohammed held in the highest regard, the Islamic world community embraced all races, including that which the ancien régime had termed the blond, blue-eyed devils.[9] Rather than being seen as intrinsically evil, the offspring of Yakub were reclaimed and acknowledged for their merits as individuals. Indeed, these offspring were even welcomed into the organization.[10]

Other reforms followed suit. Declaring himself no longer "minister" but "imam" (model), Warith Deen Mohammed, moreover, made himself one with those who were diligent in observing the requirements of the faith. The Nation of Islam itself was renamed the "American Bilalian Community," an act that honored the memory of Bilal Iban Rabah, the Ethiopian slave who was a companion of the Prophet Muhammad during his exile from Mecca. In keeping with this new designation, the newspaper *Muhammad Speaks* changed its name to the *Bilalian News*.[11] For the sake of still greater clarity and ease of recognition, the American Bilalian Community was thereafter twice renamed, first as the World Community of Al-Islam in the West, then as the American Muslim Mission.[12] These and similar reforms were the order of the day.[13] Accordingly, Lincoln observes that by 1985 "the transition from a cosmocentric, race-based community of believers in search of Armageddon to the anonymity of unchallenged inclusion in one of the world's great spiritual communions had been substantially completed."[14]

Welcomed into the community of international Islam, Imam Mohammed lectured widely at universities, churches, and synagogues. In February 1992, he was even invited to address the U.S. Senate. On the floor of the Senate, he invoked Allah on behalf of the president of the United States and each member of the Senate and the House of Representatives. In its new configuration as the World Community of Al-Islam in the West and then the American Muslim Mission, the Nation of Islam had assumed its place in the mainstream of American culture.[15] So mainstream was it, in fact, that in 1975 it hosted what the *Chicago Tribune* called an interracial "beautiful people" party, attended by dignitaries and celebrities. A gala affair, the party was "an event unequalled in the history of the Muslims." For the first time at a function sponsored by the Nation of Islam, smoking and dancing were permitted. One member was heard to remark enthusiastically, "It's boogie time in the Nation." His enthusiasm was not shared by all his brothers.[16]

In fact, it is precisely this transformation in the values and mission of the Nation of Islam that resulted in the disaffection of those who subscribed to the teachings of Elijah Muhammad in their original form. Convinced that the Nation was moving in a direction contrary to the tenets that governed its creation, the followers of the "old order" were determined that the Nation would not lose its identity. The attempt to retain such an identity in the face of Imam Warith Deen Mohammed's reforms encouraged the rise of dissenting voices that eventually regrouped into movements of their own. The situation is not unlike the various sects and schisms that populated England during the earlier decades of the seventeenth century as a reflection of the foment that was brought about with the onset of the Reformation.[17] The renewed fervor for the purity of the institution in its original, "uncorrupted" form shares the characteristics of the kind of sectarian zeal made evident in the Anabaptists, the Brownists, the Diggers, the Ranters, the Levellers, and the Fifth Monarchists, among others, who sought to cleanse what came to be viewed as a corruption of the founding discipline. As such, the reformers countered the transformation enacted by the new chief minister with a return to the ways of his father. What resulted was the creation of movements under new leaders and new chains of command.

At least a dozen of these movements are distinctive enough to be separately identified. Four of them maintain that they are either a continuation of the Nation or its only authentic resurrection, and each of these operates under the name the *Nation of Islam*. While claiming the Nation as their heritage, others have developed independent doctrines and have assumed different names. One of the four nations is under the leadership of John Muhammad of Detroit, blood brother to Elijah Muhammad. Overseen by Silis Muhammad, a second movement known as the Lost-Found Nation of Islam has a substantial following based in Atlanta. Yet another nation was established in Baltimore by Emmanual Abdullah Muhammad, who reigns as "the caliph" until the return of

Allah.[18] By far the largest and most influential share of Imam Mohammed's reformation, however, is under the leadership of the Honorable Louis Farrakhan, whose movement is based in Chicago and whose followers are dispersed throughout the United States as well as in the Caribbean, England, and Africa. This is an international movement, one customarily recognized as "*the* Nation of Islam" here and abroad.[19]

Both the emergence of this new Nation and the biographical information concerning its leader have been chronicled by Lincoln and others. For our purposes, a brief overview of Minister Farrakhan's life should suffice. Born Louis Eugene Walcott on 11 May 1933 in the Bronx, he grew up in Boston's lower Roxbury neighborhood, which was then beginning to shift from a predominantly Jewish area to a black one. A devout Episcopalian, he sang in the choir at St. Cyprian's Episcopal Church. As a child, he loved music and learned to play the violin. He was a fine student as well as an excellent athlete; he performed admirably at Boston Latin School and English High. During these years, he was imbued with a fierce pride in his heritage. This pride was fostered by his mother, a strict disciplinarian who influenced him profoundly and to whom he often refers admiringly in his speeches. A West Indian woman on the fringe of the Marcus Garvey movement, his mother inspired him with a desire to improve the lot of his people. (His father, whom Farrakhan never knew, was likewise a Garveyite.) After spending two years as a student at the Winston-Salem Teachers College in North Carolina, he pursued a career as a musician. An accomplished violinist, he also sang and played the guitar. After leaving college, he became a popular calypso singer and dancer. In fact, he performed in various venues, including nightclubs, where he appeared as Calypso Gene or the Charmer.[20]

While performing at the Blue Angel nightclub in Chicago in February 1955, he was invited to hear Elijah Muhammad proclaim his message at a Saviours' Day rally. As Farrakhan relates it, this was a defining moment for him. While sitting in the balcony of the mosque listening to the Messenger, Elijah Muhammad looked up at him as if to read his thoughts and to assure him that these were inspired teachings. After hearing Malcolm X preach a few months later, Farrakhan became a convert. On writing his letter of application for entrance into the Nation, he was granted admission in October 1955.[21] Malcolm X became Farrakhan's mentor. Realizing his follower's potential, Malcolm appointed him minister of Muhammad's Temple no. 11 in Boston. When Malcolm was suspended in 1964, Farrakhan succeeded his mentor as minister of Mosque no. 7 in Harlem.[22] Having denounced Malcolm X for his break with Elijah Muhammad, Farrakhan was subsequently recalled by Imam Mohammed from his Harlem ministry and reassigned to duties on the West Side of Chicago. Although he at first accepted his reassignment to "invisibility," he gradually came to believe that Imam Mohammed's reformation was instituted for the purpose of undermining the entire vision of Elijah Muhammad and dismantling the structure that energized it. "To be so near the scene of the destruction and yet so far from the power required to stop it eventually required a decision on Farrakhan's part that neither he nor Wallace [Imam Warith Deen Mohammed] wanted to face" for neither one could be true to his own calling and abide by the "truth" of the calling of the other. Something had to give, and so it did.[23]

In 1977, after some thirty months of Imam Mohammed's reforms, Farrakhan "stood up" and broke from the World Community of Islam in the West to reinvent Elijah Muhammad's Lost-Found Nation of Islam in the Wilderness of North America. No doubt to the surprise of many outsiders, the parting was peaceful. Farrakhan set about rebuilding the infrastructure of his "new" Nation "solidly and unequivocally on the foundation laid by the founding fathers." Concepts such as the nature of Allah, the misdeeds of Yakub, the evil of the white man, the significance of the Mother Plane, and the imminence of Armageddon were reasserted and disseminated with renewed determination and vigor. The time-honored declarations "What the Muslims Want" and "What the Muslims Believe," published in the various writings of the Nation, remained intact. The official newspaper, *Muhammad Speaks,* was resurrected as the *Final Call,* to reflect the eschatological bearing of the Nation's message. Although precise statistics concerning the size of the Nation at the present time are unavailable, Lincoln says that a figure between 70,000 and 100,000 seems consistent with what is known about other aspects of the community. Lesser numbers have been proposed by others.[24] Regardless of its size, the movement over which Minister Louis Farrakhan presides enjoys wide visibility. In its determination to restore the ideology underlying the original teachings of Elijah Muhammad, moreover, the Nation as presently constituted is a force to be reckoned with. In response to the restoration of the old order, one Johnny Lee X, a confirmed adherent to the ways of the new leader, now is able to declare with complete conviction: "We call ourselves the *new* Nation of Islam. The first Resurrection is with Elijah Muhammad; the second is with Farrakhan." The traditions of the old have remained intact with the emergence of the new. So Johnny Lee X concludes: "It's a change of command, that's all it is."[25] It is to be sure much more than that. As a result of the stamp that Minister Louis Farrakhan has placed on it, the "new Nation of Islam" comes to have a significance all its own.[26]

It would be an understatement to say that this significance derives in large part from the highly charged emotional and political climate in which Louis Farrakhan has found himself in recent years. On 11 December 1989, the *Washington Post* put the matter succinctly: Farrakhan "has gained an enduring reputation as one of the most controversial and feared U.S. public figures."[27] Such an estimate is no less true of the minister today. Branded a demagogue by the press, he has been greeted with the kind of opprobrium visited only on those viewed with utmost distrust and dismay. In 1994, *Time* magazine featured a story on Farrakhan under the heading "MINISTRY OF RAGE," accompanied by the declaration: "Louis Farrakhan spews racist venom at Jews and all white America. Why do so many blacks say he speaks for them?"[28] In 1995, the *Chicago Tribune* entered the fray with a series of investigative articles aimed at demonstrating that, for all his claims at fostering black pride and economic liberation, Farrakhan is little more than a huckster who exploits those he purports to serve.[29] The press, in short, has not viewed his cause kindly, and the fear and repugnance with which he is perceived by those responsive to the portrait of him that has been forged by the press over the years have served only to exacerbate matters.

Among the most turbulent aspects of the conflicts that underscore Farrakhan's administration are his troubled relations with the Jewish community.[30] Accused of adopting a militant attitude toward Judaism and fostering an atmosphere of fierce "anti-Semitism" and "anti-Zionism," Farrakhan has found himself in a highly charged climate of attack and counterattack.[31] This represents a departure from his early career, which was apparently free of pronounced anti-Jewish sentiment. In this respect, he adhered to the model set by Elijah Muhammad, who never made a practice of singling out Jews directly for special condemnation in his criticism of white people.[32] Like the Messenger, Farrakhan adopted a rhetoric that focused primarily on the injustices that black people had suffered under whites. Doing so, he "reviled white people, not only over slavery but also over what he [saw] as a vast white conspiracy to conceal the glorious past of blacks as the original human race and the founders of most branches of civilization and politics." In such criticism, however, he did not vilify the Jews over and above any other group of people.[33]

With the Reverend Jesse Jackson's run for the presidency in 1984, the tide turned. Initially, and quite remarkably, in fact, it moved in the direction of genuine conciliation and compromise. Farrakhan adopted a posture in keeping with the spirit and message of Jackson's campaign for a Rainbow Coalition. In an article published in *Essence*, Farrakhan looked forward to a time when all races and creeds might join hands in a mutual bond of brotherhood. Rather than being singled out for vilification in this article, Jews and Christians both were called on to play an important role in the creation of the new Rainbow Coalition that Jackson was attempting to forge. In support of that coalition, Farrakhan looked toward the realization of the dream of the Reverend Dr. Martin Luther King. "Dr. King's dream," Farrakhan declares, "was that men and women would be judged not by the color of their skin but by the content of their character. He envisioned the day when black men and white men and Christians and Jews and Protestants and Catholics would come together in the bond of brotherhood."[34] Although it might be argued that this is little more than "political rhetoric" invoked to support Jackson's campaign, the article as a whole has the genuine ring of sincerity. In any case, it is a side of Farrakhan that is rarely, if ever, noted.[35]

The year of the publication of his article in *Essence* everything changed. At that time, Fruit of Islam guards provided security for Jackson until the Secret Service was instructed to oversee that function. It was during this period that Jackson apparently ventured some unfortunate remarks about the Jews of New York. The media broadcast those comments into what came to be known as the "Hymietown" controversy.[36] This understandably provoked a strong reaction, and, according to the *Time* magazine article cited above, Farrakhan "was outraged to learn that Jewish militants were shadowing Jackson, that he had received death threats and that his family had been harassed—facts confirmed by the FBI."[37] In response to these threats, Farrakhan sought to intimidate those he deemed Jackson's harassers. In his speeches, he declared, "If you harm this brother, it'll be the last one you ever harm." Removed from its context, this warning was interpreted as an unprovoked threat directed pointedly at members of the Jewish community.[38] In his Saviours' Day rally at Madison Square Garden on 7 October 1985, Farrakhan "electrified an overflow crowd of 25,000 people" with "a message of economic and spiritual renewal coupled with a ringing denunciation of his critics," including then Mayor Edward Koch and Governor Mario Cuomo. "Mayor Koch said I should burn in hell," Farrakhan proclaimed, and then responded, "Dear Mr. Koch: The black people of New York live in hell." In protest against Farrakhan's appearance, the Jewish Defense League sponsored a "Death to Farrakhan" march.[39] The spirit of the march was thereafter reflected in the headline of the December 1985 issue of the *Final Call:* "Death Plot on Farrakhan Fails." In the article that accompanies the headlines, the following accusation is advanced: " 'The Jews tried to kill me in New York; that's the untold story,' Minister Farrakhan charged from his headquarters on October 30." According to the article, a mysterious fire broke out in a locker room not far from the auditorium where Farrakhan addressed the assembled masses. Arson was suspected, and, in connection with the attempt to put out the fire, there was fear of a possible stampede. Fortunately, chaos was averted.[40]

Whether or not the fire was the result of a plot on the life of Farrakhan and those assembled to hear his speech, one fact is certain: the inflammatory nature of the charges and countercharges clearly established a heated atmosphere of militancy that has only been exacerbated by the kind of rhetoric that has been advanced years after the incidents surrounding Jackson's candidacy occurred. Since then, accusations that Farrakhan disseminates the worst form of racist venom have become the rule rather than the exception. Farrakhan and his aides are now characteristically known as "bigots" who have labeled Jews "blood-suckers," Judaism "a gutter religion," Israel "an outlaw state," and Hitler "a very great man" (albeit "wickedly great").[41]

Such an outlook has been reinforced by the various publications that the Nation of Islam has produced. The most important of these is a book published under the auspices of the Historical Research Department of the Nation of Islam. Titled *The Secret Relationship between Blacks and Jews*, this is the document on which Farrakhan has taken his stand in his interrogation of black/Jewish relations in the twentieth century. Drawing on the testimony of "Jewish sources," the book purports to demonstrate that "Jews have been conclusively linked to the greatest criminal endeavor ever undertaken against an entire race of people—the Black African Holocaust." Jews, the book argues, "were participants in the entrapment and forcible exportation of millions of Black African citizens into the wretched and inhuman life of bondage for the financial benefit of Jews." Because of their involvement, Jews are made to bear the guilt for playing a crucial role in the implementation of the slave trade from its very inception. "The effects of this unspeakable tragedy," the book asserts, "are still being felt among the peoples of the world at this very hour." Such an argument seeks to undermine what it deems the fallacious assumption "that the relationship between Blacks and Jews has been mutually supportive, friendly, and fruitful—two suffering people bonding to overcome hatred and bigotry to achieve success." Rather, these new disclosures would establish an insidious pattern of exploitation and tyranny for which the Jews must assume ultimate responsibility.[42] Needless to say, the book has found itself at the center of a storm of controversy and refutation.[43]

However one is to view the book and its methodology, one fact remains clear: its purpose is to establish valences in the conception of black/Jewish relations that had previously not been articulated in the form that such books as *The Secret Relationship between Blacks and Jews* imply. As a result of the establishment of these valences, the tragedy of the Holocaust comes to be associated not with Jews but with blacks. In the process of this disassociation, blacks become the true Jews, that is, the true sufferers who have borne the oppression of the slavemaster in this country for over four hundred years. Those who call themselves Jews are really not so: at best, they are the "so-called Jews" whose very claims to legitimacy are open to question.[44] In the light of this disclosure, such claims must be seen as duplicitous. So the significance of Jewish cries of suffering under the Holocaust must likewise be reassessed. Despite the horrors brought about by the Holocaust, the Jewish people are no longer capable of maintaining that such an event is a tragedy uniquely their own. The black Holocaust under slavery far surpasses any Holocaust the so-called Jews might claim. When it comes to the Holocaust that the black slave was made to endure, in fact, the so-called Jews became paradoxically the very source of that suffering, indeed, the primary cause of that much greater Holocaust endured by those they have secretly exploited.[45] Those they have exploited, in turn, are the true Jews, the true "chosen people," made to undergo horrors of oppression as unbearable as those suffered by the Israelites under Pharaoh in biblical times.[46] Unmasked as the true slavemasters, the so-called Jews must assume the responsibility of their transgression: they must acknowledge their complicity in the black Holocaust that is the secret and previously unmentionable fact of their past sins. That they are unwilling to do so is one more sign of their duplicitous and intractable character as a people all too ready to claim victimization for themselves but not prepared to accept the responsibility of victimization of others. The assumptions on which this outlook is based have been termed by Henry Louis Gates Jr. *the new anti-Semitism,* one that involves both a painful dislocation in and transformation of the conception of self-identity in the ongoing relation between blacks and Jews in contemporary culture.[47] With that relation effectively undermined, the rift between the Nation of Islam and the Jewish community in the past decade has become that much greater. As the result of this rift, Farrakhan, as well as the Nation he represents, has been viewed as anathema to the main currents driving the struggle for racial and ethnic harmony in the United States at the present time.

In an assessment of the new anti-Semitism, one must be aware that, as disturbing and disruptive as the dislocations and transformations wrought by the revelation of "secret relationships" might appear to be, these reconfigurations are consistent with the originary history and, indeed, the theology on which the Nation has been founded. Moreover, whatever the immediate circumstances (charges, countercharges, and the like) that gave rise to the rhetoric of denunciation that Farrakhan has adopted in response to those he feels are a threat to his own well-being and the well-being of his people, one must look beyond "current events" in order to derive an understanding of the deep roots of the dilemma. For Farrakhan, those roots originate in a protohistory founded on the production of racial difference. As we have already seen, that protohistory is made available in the teachings of Elijah Muhammad. It is to those teachings that Farrakhan turns in his own explanation of the originary dimensions of the struggle for liberation that he feels his people are being made to undergo.

For Farrakhan, as for Elijah Muhammad before him, that struggle is centered in the figure of Yakub. In the protohistory that represents the basis of the production of racial difference, Yakub, we recall, is the ultimate source of the creation of the white race, the roots of which emanate from the Original Man, who is black. By means of a complex process of genetic engineering, Yakub was able to produce the race of devils through whose lies, deceptions, and "tricknology," the black race underwent terrible suffering from one generation to the next. In his Saviours' Day message of 27 February 1994, delivered at the University of Illinois at Chicago Pavilion, Farrakhan renegotiates the terms of this historical account by associating it with the black-Jewish struggle.[48] Resorting to the Genesis narrative of Jacob's wrestling with God (Gen. 32:24–29), Farrakhan reads the renaming of Jacob to *Israel* as a prefiguration of the Jews of today, the so-called Jews. As the true Jews, the black race must realize that the pain and suffering it has endured under the so-called Jews is like a "refiner's fire" that purifies it in its quest for union with Allah.[49] To the oppressors, Farrakhan cries out in the voice of Moses: "Let my people go!"[50]

The paradox embodied in the figure of Yakub is further intensified by the principle of attraction and repulsion. That principle is part of the very sense of self-identity through which the relation between blacks and whites, on the one hand, and blacks and Jews has been conceptualized in the teachings of the Nation. Recalling that as a six-year-old boy Yakub discovered the scientific principle that unlike attracts and like repels, Farrakhan observes that, because the Caucasian people are unlike the original people, there is a natural attraction between blacks and whites. "The problem is," he says, "that when we were attracted to them, they came among us and by the use of lies and tricknology they divided the original house one against the other until they came into power over all. This was given to them by Allah (God)."[51] The purpose of Elijah Muhammad's teachings was to reunite "the original house" by pointing out the beauty of "blackness" and the ugliness of that "whiteness" that sought to divide it with the false attractions of its lies and tricknology. The purpose of Farrakhan's teachings is to apply that idea to the new configuration reflected in the "secret relationship" between blacks and Jews. Perceiving the pitfalls of that relationship, blacks will not be seduced into believing that the so-called Jews are their friends. Rather, the so-called Jews have come as the spawn of Yakub/Jacob to seduce the original house with new lies, a new tricknology that seeks to divide the house against itself. If "blackness" can be seen for what it is, it will not be divided, it will not be repelled by itself; rather, it will reunite in a new awareness of its own grandeur, of its own selfhood. In the reversal of attraction and repulsion that this reunion accomplishes, the lies and tricknology of the bearer of the "evil name" will be fully discomfited.[52]

On the basis of the manner in which the principle of attraction and repulsion is applied to the relation between blacks and Jews in Farrakhan's teachings at the present time, it would be an all-too-easy matter to conclude that the conflict is irreconcilable: Farrakhan hates Jews just as he hates whites, and that is all there is to it.[53] Whatever else one is to conclude, the notion that "that is all there is to it" invites reconsideration. This fact is made clear in a recent interview that Farrakhan had with Henry Louis Gates Jr. While continuing to voice his certainty that "a cabal of Jews secretly controls the world," Farrakhan maintains that he reveres the Jewish people and desires to be in harmony with them. "Personally," he says, "I don't know what this argument has served. Jewish people are the world's leaders, in my opinion. They are some of the most brilliant people on the planet." Among them are the greatest scientists, thinkers, writers, theologians, and musicians. They are hated as the result of the envy of others. Sounding a point made earlier, he hails the Jews as a people whose prophets have been "the recipients of divine revelation," a gift he most admires, although he warns that this gift can be used for evil purposes as well as good.

His most fascinating reflection on this occasion came as the result of a statement about his father. His mother informed him that his father's parentage was white Portuguese. More than that, he believes that he can trace his lineage through his father to "members of the Jewish community." As one who possibly traces his own lineage to Sephardic Jewish ancestry, he recalls that, when he was a child, he loved to listen to "the Jewish cantors in Boston" singing and reciting the Torah. As a musician, he avers, all his heroes, including Jascha Heifetz and Vladimir Horowitz, were Jewish. "If in my lineage there are Jews," he says, "I would hope that in the end, before my life is over, I not only will have rendered a service to my own beloved community but will also have rendered a service to the Jewish community." Gates is convinced of the genuineness of such feelings. Farrakhan's love and loathing, Gates observes, spring from the same source: "There's a sense in which Farrakhan doesn't want his followers to battle Jews, but, rather, wants them to *be* Jews," an observation supported by the distinction between the true Jews and the so-called Jews discussed above. As a result of the conflicts embodied in this paradox, Farrakhan remains, according to his own testimony, "locked now in a struggle. It's like both of us got a hold on each other, and each of us is filled with electricity. I can't let them go, and they can't let me go."[54] That struggle, one might suggest, arises as much from the quest for individuation that rages within the very being of this charismatic leader as it does from the ongoing series of confrontations that distinguish Farrakhan's relation with the Jewish community and its relation with him.

In both respects, what is at issue is the nature of the "other," the way in which it is perceived and the way in which it is signified. Whether that "other" assumes the form of the Jew or the white race, Farrakhan confronts it, grapples with it, and seeks to overcome it at the very point of attempting to obtain reconciliation with it. His quest for harmony is at least as "strenuous" as his need to confront. At the source of both is the concept of blackness, which, in Farrakhan's thinking, assumes almost a cosmogonic significance. Within the deep structures of black identity, original blackness not only encompasses but gives rise to the "other." Accordingly, Farrakhan maintains that "Allah says in the Qur'an that he gave to man his hue and color. But when did color come? Black is not a color; Black is the essence from which all color comes. In the book of Genesis, it says that in the beginning Allah (God) created the heavens and the earth and the earth was without form and void and darkness was upon the face of the deep and Allah (God) said let there be light. And he brought light out of darkness." From this, Farrakhan concludes that "if Allah (God) could bring light out of darkness, he can bring white out of Black. And that's exactly how you (Caucasians) came; you were brought out of us. We are your father. We are your mother. And you should honor your mother and your father that your days may be long in the land which the Lord, your God, has given you."[55] Given the valences already established, the cosmogony envisioned here applies as much to Jewishness as it does to whiteness in Farrakhan's thought. In either case, the model so conceived is a far cry from the original model of genetic engineering with which the creation of white out of black is ordinarily associated in the Yakub paradigm. The underlying instinct in the Genesis model, in fact, is not only cosmogonic (light out of darkness) but nurturing (maternal and paternal) and hopeful ("that your days may be long"). It also speaks to the conviction that within the deep structures of black consciousness there resides at least the possibility of reconciliation and reunion not only with the self but also with the "other" that is, after all, buried within that self from which all else springs. Here the paradox of attraction and repulsion is reconciled not through the conflict that results from a recognition of tricknology and lies but through the reconciliation and reunion that result from a recognition of cosmic origins.

One would hope that that reconciliation and reunion model might prevail over the attraction and repulsion model. Perhaps the alternative model may eventually prevail.

However Minister Louis Farrakhan is to be seen from the vantage point of history, recent events have borne witness to the fact that he is a leader whose presence remains as paradoxical, indeed, enigmatic, as any who has appeared in modern times. As much as he might be viewed as anathema, he has also thrust himself into the mainstream of the struggle by mounting one of the most important events in the history of his ministry and certainly an occurrence of major import to the civil rights movement. Representing what has been widely perceived as a turning point in Farrakhan's administration, this event is the Million Man March to Washington, D.C., which occurred on 16 October 1995. The magnitude of the march is difficult to overstate. In sheer size alone, it drew an overwhelming number of people from throughout the United States and beyond. Estimates place that number between at least 650,000 and 1.1 million people. Perhaps there were even more; it is difficult to say. What is not difficult to say is that the march represents the largest civil rights demonstration in the history of the United States.[56] Conceived by Minister Farrakhan, who views himself as a "conduit" for the march, and implemented under the directorship of the Reverend Dr. Benjamin F. Chavis Jr., the former director of the NAACP, the march embraced a multitude whose interests and persuasions extended well beyond those of the Nation of Islam.[57] Individuals and groups who endorsed it include Mrs. Rosa Parks, the Reverend Jesse Jackson, the Southern Christian Leadership Conference, Mayor Marion Barry of Washington, D.C., Mayor Omar Bradley of Compton, California, the Detroit City Council, the National Council of Negro Women, the National Black Chamber of Commerce, the World Congress of Mayors, and a host of others.[58] Lasting the entire day, the program of the march included prayers, speeches, poems, songs, testimonials, and performances by a wide range of people, including Maya Angelou, Jesse Jackson, and Cornel West. The culminating message of the day was a speech by Farrakhan. In the words of Cornel West, "the Million Man March was an historic event—called by Minister Louis Farrakhan, claimed by black people of every sort and remembered by people around the globe as an expression of black men's humanity and decency. Never before has such black love flowed so freely and abundantly for so many in the eyes of the world."[59]

As indicated in the mission statement of the Million Man March and made evident in the speeches that accompanied it, the special purpose of the event was to proclaim a Day of Atonement, Reconciliation, and Responsibility. So the mission statement declares: "We call for a Holy Day of Atonement on this October 16, 1995, a day to meditate on and seek right relationships with the Creator, with each other and with nature." At the same time, a corresponding purpose of the Million Man March was to declare a holy Day of Absence among those who remained home. Thus, the mission statement declares that, "to observe this sacred day, we call upon all black people to stay away from work, from school, from business, and from places of entertainment and sports and to turn inward and focus on the themes of atonement, reconciliation and responsibility in our lives and struggle."[60] Judging by both the import and the outcome of the march, one can readily discern the extent to which the event represented an enterprise of the first magnitude in the quest for both betterment and empowerment not only among those who participated in the march but also among those who heeded its call in their own lives.

Farrakhan's long but impassioned speech represented a capstone to the march. Cast in the form of a sermon titled "Day of Atonement," the speech was delivered not just to the masses assembled at the march but to a national and international viewing audience that had access to the event via remote satellite. This was a remarkable performance, indeed, in its own way, a model of oratory. Although the speech is too complex to analyze in any depth, aspects of it are important to an understanding of the nature of Farrakhan's message on this occasion and the manner in which that message is to be understood.[61] One is immediately struck by what might be termed the *mystic* or *gnostic* dimensions of the speech. From the very outset, the mode of delivery is that of the hierophant or oracle disclosing the most sacrosanct of mysteries to his acolytes. The speech represents Farrakhan's "mystical theology," one very much in keeping with the "supreme wisdom" of the master who imparted the secrets of his teachings to the Messenger over sixty years ago.

Reflecting and elaborating on the import of those teachings, Farrakhan begins his sermon with a gnostic meditation on the numerological implications of his surroundings. These surroundings include the National Mall, situated on the west side of the U.S. Capitol and stretching in a long, narrow rectangular line past the Washington Monument, with the Jefferson Memorial well to the south of it, and all the way to the Lincoln Memorial at the other end, a distance of more than two miles. Facing westward at his elevated podium on the Mall, Farrakhan looks beyond the great masses gathered to hear him toward the towering obelisk of the Washington Monument and the rotunda of the Jefferson Memorial in the distance and, beyond that, toward the Lincoln Memorial, with its imposing, seated figure. In keeping with the studied grandeur of the surroundings, the space Farrakhan occupies becomes a hieroglyph, the significations of which he seeks both to capture and to decipher.

As he addresses his audience from the podium on the Mall, he assumes the role of a surveyor, one with "a golden reed to measure the city, and the gates thereof, and the wall thereof," in order to determine the shape and dimensions of the city. Like Ezekiel at the culmination of his prophecy and Saint John the Divine at the culmination of his, the new surveyor of the city of Washington become Jerusalem seeks to discover whether "the city lieth foursquare" and whether "the length [be] as large as the breadth: and he measured the city with the reed, twelve thousand furlongs" (Rev. 21:15–16; cf. Ezek. 40–48). With the exactitude of his forebears, Farrakhan seeks to record the precise measurements of the city, its temple, and its monuments. His act becomes an exercise not just in measurement, however; it is also an excursus in the art of "mystical mathematics" and numerology. So the Washington Monument, he observes, is 555 feet high. Placing the numeral 1 in front of this number, he arrives at 1555, the year, he says, that the "first fathers" of the black man set foot on the shores of Jamestown, Virginia, as slaves.[62] In the background, he notes, are the Jefferson and Lincoln Memorials, each nineteen feet high. Abraham Lincoln is the sixteenth president, Jefferson the third: 16 and 3 total 19. "What is so deep about this number 19?" Farrakhan asks. Encoded in that number is the very reason "we are standing on the Capitol steps today." So he observes that, "when you have a nine, you have a womb that is pregnant. And when you have a one that is standing by the nine, it means that there's something secret that has to be unfolded." As a hierophant in possession of the "supreme wisdom," Farrakhan is in a position to "unfold" all to those acolytes who will hear and understand. So important is the number 19 to the Nation of Islam, in fact, that it becomes a "parable," imbued with a multiplicity of meanings decipherable by the cognoscenti through an analysis of the text of the Qur'an.[63] For the hierophant, this particular occasion provides even further evidence of the fact that the signs of the times have coalesced in a manner foreordained by destiny to demonstrate the momentous nature of what is to ensue: the ultimate liberation of black people from the chains of their oppression.[64]

One need only read the mystical text encoded in the world of signs represented by the Mall with its memorials and monuments in order to decipher its meanings. As one so authorized, Farrakhan reads the text of the city laid out before him. Given his standing as a black man addressing other black men, it is appropriate that he assume this hierophantic role for, as he points out, it was under the auspices of the black scientist Benjamin Banneker that Washington,

D.C., in the form of the Federal Territory was first "laid out" or surveyed.[65] Now, Farrakhan as a black man has come to conduct a "survey" of his own. The lines of demarcation within that survey constitute a mystical map responsive to the occult resemblances of the figures there encoded. They are, Farrakhan observes, codified in a form that only those adept at the secret art of Freemasonry are able to decode. This is an art with its roots in the very founding of the Nation of Islam.[66] At the core of the secret of the rituals associated with Freemasonry is the black man. Interestingly enough, Farrakhan avers, Freemasonry is an art practiced by the first president of this land, George Washington, himself a grand master of the Masonic order.[67] When one considers the ancestry of the occult sciences embodied in this art, it is appropriate that the codes that Farrakhan discloses are particularly evident in such objects as the Washington Monument, an obelisk reminiscent of the "historical past" of the black race, centered in Egypt.[68] Ironically, Farrakhan reminds his audience, George Washington was a slaveholder. One must be reminded of his past to understand the ironies that its secrets betray.[69] This is the nature of Farrakhan's oratory on this auspicious occasion: it is an oratory that is grounded in the revelation of codes and correspondences. It is an oratory of mystery and discovery, an oratory of the hierophant come to disclose his secrets.

Such is especially true of the concept *atonement,* which lies at the heart of Farrakhan's sermon. The meaning of atonement is first placed within the larger context of achieving "perfect union" with God. For Farrakhan, it is the fifth stage in an eight-stage process through which one may be absolved of his sins. Farrakhan's discourse on this process resonates with the quality of a "penitential" or a "spiritual exercise," associated with the ecclesiastical traditions. In his account of the process, one begins with a recognition of his sins, proceeds to an acknowledgment of them, next to a confession of them, followed by the acts of repentance and atonement for them; these, in turn, culminate in the receipt of forgiveness, which is rewarded by the experiences first of reconciliation and restoration and, finally, of perfect union. Within this context, Farrakhan focuses his attention on atonement, which he calls "the satisfaction or reparation for a wrong or injury." What renders his analysis of atonement so fascinating is the delight that he takes in performing an act of creative exegesis on it. "Performing" is to the point for the exegetical "act" itself is both complemented and reinforced by the performative dimensions of the analysis. In the oral delivery of the exegesis, the voice that enacts it assumes an incantatory quality as the speaker intones the words of his sermon. This analysis, in short, must be experienced in the delivery, a feature that is true of Farrakhan's message in general.

In keeping with his occupation with occult resemblances and numerological correspondences, Farrakhan's exegesis of the word *atonement* begins by calling attention to the meanings encrypted within it. The first four letters of the word, he observes, represent the "foundation" of it: *aton*, a term in keeping with the obelisk of the Washington Monument that is beheld off in the distance. Invoking this obelisk, Farrakhan recalls the figure of Akhenaton, a pharaoh of the eighteenth dynasty who destroyed the pantheon of gods in order to bring the people to the worship of one god. This one god was symbolized by a sun disk that emanated nineteen rays, accompanied by hands holding the Egyptian ankh as "the cross of life." For Farrakhan, *atonement*, then, embodies the "parable of the 19," so important to the Nation itself as a symbol of that which becomes a source of life, of new generation, and the wellspring of meaning.[70] Deriving renewed life from the parable of the 19 encoded within *atone*, the black man will live. Moving from the historical and iconic perspective to the musicological one, Farrakhan discloses still further meanings embedded within *atone*, deciphered now as *a-tone:* "If you look at the aton, add an 'e' to it, and separate the 'a' from the next four letters," you arrive at the phrase *a-tone*. A *tone* is a sound. Since numerically and alphabetically *a* equals the numeral *1*, so the *a* sound signifies the *a* tone, the "right sound" of God's calling the faithful to the divine life, an event that Farrakhan celebrates in the rhythm of his speech as an incantation. What is the *a-tone?* he asks. Answering this question, the musicologist in him responds that *a* equals 440 vibrations. "How long have we been in America?" he asks again, responding to his own catechism with the correct answer: 440 years.

At this point in his incantatory paean to atonement, his performance builds to a remarkable coda. Underscored by the immediacy of colloquial speech patterns and turns of phrase reflected in the sermon as a whole, the rhythms of the incantation release themselves resoundingly from the text: "Well, in the 440th year, from the one God, the aton will come the a tone and all of us got to tune up our lives by the sound of the a tone. Because we've got to atone for all that we have done wrong." Intimate and inclusive in its bearing, the aura becomes characteristic of a revivalist church gathering on a Sunday morning. With the spirit that such a gathering evokes, Farrakhan moves to the high point of his coda, as he advances with alacrity to his next conclusion: "And when you atone, if you take the 't' and couple it with the 'a' and hyphenate it, you get 'at one.' So when you atone you become at one." Pause: "At one with who? The aton or the one God. Because you heard the a tone and you tuned up your life and now you're ready to make a new beginning." That new beginning, in turn, focuses on the individual soul struggling to atone: "So when you get at one, you get the next two letters. It is 'm' 'e.' Me." The sermon arrives performatively at the point of conversion as it conscripts the audience as parishioners into the space of enactment by inviting the appropriate response consistent with the crucial discovery of the *me* encoded in the word *atonement.* "Who is it that has to atone?" the speaker asks: "Me," the audience responds. Again, "Who went wrong?" Response: "Me." "Who got to fix it?" "Me." "Who should we look to?" "Me." As a sign of recognition, that *me* with an addition of an *n* becomes *men.* Thus the audience responds with *men,* to the question of adding an *n* to the *me.* The transition from *me* to *men,* then, justifies the focus specifically on men in the Million Man March. "So," the minister concludes, "Farrakhan called men."

In his final act of creative, indeed, playful, exegesis on the encoded word *atonement,* Farrakhan next addresses the suffix *ment.* This suffix, he says, implies "action," "process." "So when you say I'm atoning, you got to act on it. You got to get in the process. You got to acknowledge your wrong, confess your wrong. Repent of your wrong. Atone for your wrong." Only then, he declares, will there be forgiveness, reconciliation, and restoration. "And then you're back to the aton. Oh, Lord."[71] The performance is brought to a close with its own hallelujah. It is the triumphant performance of a minister to his flock. His church is the open space of the Mall that lies before him; his flock, the masses stretched out before him. This, of course, is only a small aspect of the sermon that Farrakhan delivered to the massive gathering on that eventful day. Whatever else one might be inclined to say about the minister and his message, the sermon and its reception became a moment on a grand scale. By means of his sermon as well as the event of which it was the culmination, Farrakhan did at least at this moment reach out and embrace the multitudes.

In its own way, that act of reaching out made a point of addressing the ongoing conflicts that the Nation has had with the Jewish community. What might well be viewed as an attempt to reconcile differences is the professed allegiance of those who formulated the march to the underlying concepts of Judaism, an allegiance that made no distinction between those deemed "true Jews" and those branded "so-called Jews." Arising elsewhere, to be sure, the distinction was all but obliterated on this occasion in what amounted not to a usurpation but to an acknowledged appropriation. For the organizers, it was accordingly no accident that the march was to be designated the Day of Atonement. Both in his speeches and in his writings, Farrakhan made clear his indebtedness to Yom Kippur in the formulation of his conception of the march. He observed in a speech delivered on Friday, 15 September 1995, in Los Angeles that the idea of atonement that he wishes to foster in the black community finds its model in the Day of Atonement that is the most sacred day in the Jewish calendar. This day, he says, is one of fasting, a day when observant Jews do not go to work, do not go to school, do not shop, and do not conduct business. It is a holy day, a day of prayer. It is a day when Jews seek reconciliation with each other and with God so that they can be assured of God's blessings throughout the following year. Having elaborated the principles of Yom Kippur, he concludes his discourse by admiring the "beauty" of this most holy day and all that the day implies.[72] This may be viewed as simply disingenuous and opportunistic by those who quite understandably react with suspicion to any gesture that incorporates the "reconciliation and reunion" model that has been delineated in at least certain aspects of Farrakhan's teachings. It is raised here only to suggest that such a gesture is not necessarily inconsistent with a certain thrust in Farrakhan's teachings. Judging by the elaborate analysis to which the concept *atonement* was subjected in his speech as the capstone of the march, one might be inclined to believe that the idea is at least of some importance to what was on that occasion a professed desire for reconciliation among those whom he has perceived as his "enemies."

In conjunction with the emphasis on atonement in his march speech, Farrakhan makes a point of appealing to the Jewish community for reconciliation. Applying the idea of atonement to the climate of conflict that continues to exist between the Nation of Islam and the Jewish community, he declares, "I don't like this squabble with the members of the Jewish community." On the authority of Elijah Muhammad, he maintains that "we would work out some kind of accord." It is "time," he says, "to sit down and talk" and to do so without any "preconditions." "You got pain," he allows; "well, we've got pain too. You hurt. Well, we hurt too." Calling for dialogue, he fervently hopes to be able to end the pain. To support this gesture of reconciliation, he recalls the situation of Israel and its peace talks with the Palestinians: "And I guess if you can sit down with Arafat where there are rivers of blood between you—why can't you sit down with us and there's no blood between us. You don't make sense not to dialogue. It doesn't make sense."[73]

As much as the march itself and such gestures of outreach moved Minister Louis Farrakhan into the mainstream, however, his multinational tour of Africa and the Middle East in the months following the march served only to thrust him once again into the maelstrom.[74] Conferring with such enemies of the U.S. government as Colonel Mu'ammar Gadhafi of Libya and Saddam Hussein of Iraq as well as with officials in Iran, Syria, Nigeria, Zaire, and Sudan, Farrakhan succeeded once again in provoking the ire of both the government and the press. Congressional hearings were threatened, to which Farrakhan defiantly responded, "Bring me before Congress!" "I was born for this moment," and "it's time for a showdown anyway."[75] These challenges were issued in his Saviours' Day message of 25 February 1996 at the University of Illinois at Chicago Pavilion. There, he made it clear that he continues to feel as if he is one pursued by forces that seek to destroy him but that the powers that are with him are far greater than those marshaled against him. Apocalyptic in tone, his speech focused on the last days with the promise of the coming of the Son of Man to exact judgment on the reprobate and finally to redeem the faithful. Although castigating those within the Jewish community who continue to seek his demise, he nonetheless hailed Judaism, Christianity, and Islam as the three great religions of the world.[76]

Judging by the Million Man March, on the one hand, and the trip to Africa and the Middle East, on the other, Farrakhan continues to confound the world at large. However he is to be understood, he emerges as a figure not only of high visibility and immense complexity but of national and, indeed, international import. Constantly in the headlines, he is at the very least perceived as a person of profound contradictions. As a result of his most recent activities, the forces of dissension appear to be even more entrenched and polarized than they ever were. Given the extent of that entrenchment and polarization, it would be an error to dismiss Farrakhan as an "opportunist," a "demagogue," a "fanatic," and a "hate monger" who deserves nothing more than contempt, denunciation, and castigation.[77] As repugnant as his message might be to those content with the delineation of him in the headlines, he is an individual who invites, perhaps even demands, analysis. This is no ordinary mortal: whether demonized or idolized, whether shunned as a pariah or worshiped as a messiah, he cannot simply be dismissed as an ephemeral, if disturbing, aberration. He is too important a figure to be consigned to the oblivion of abusive neglect. As has been noted in the past, the attempt to understand Farrakhan may be a daunting task, but it is nonetheless an urgent one.[78] It is in response to this urgency that the present undertaking sets about to examine Minister Louis Farrakhan in the context of the visionary milieu that underlies the emergence of the Nation of Islam as a pivotal event in the modern world.

8

Having explored the contexts through which the Honorable Louis Farrakhan emerged as the charismatic leader of the Nation of Islam in its present incarnation, I now proceed to an analysis of the visionary dimensions of his calling. This is a calling founded on the vision of Ezekiel as a crucial event in his ministry, a vision as fundamental to his sense of vocation as it was to that of the Honorable Elijah Muhammad. Just as the Messenger looked on himself as the heir of Ezekiel's *visio Dei* through his teacher Master Fard Muhammad, Minister Farrakhan looks on himself as the heir of this vision through his teacher the Honorable Elijah Muhammad. A phenomenon that has exerted a profound influence on the Nation of Islam and its leadership, the vision of Ezekiel represents a defining moment in the bestowal of knowledge and power, as the call of the prophet is handed down from teacher to student and from one generation to the next. Like Master Fard Muhammad and the Honorable Elijah Muhammad before him, Minister Farrakhan resorts to the vision as a means of articulating the ideology and mission of the Nation of Islam as a seminal force in contemporary society. Moreover, the visionary experience that he embodies serves as a catalyst for a renewed understanding of both his ideology and his message. Approaching Farrakhan from this perspective lends insight not simply into the present leader of the Nation of Islam but into the culture that he represents as well.

To understand the circumstances surrounding Farrakhan's receipt of his visionary calling, we might well bring to mind what for the Nation of Islam is the originary event through which Elijah Muhammad received his own calling. That event, we recall, extends back to a period over half a century ago, when Elijah Muhammad (then the unemployed laborer Elijah Poole in the city of Detroit) found himself one of many victims of the Great Depression. The time was right for visions. It was then that one who was to become Elijah Muhammad was introduced to the mysterious figure known as Master Fard Muhammad. Through him, Elijah Muhammad was shown what thereafter came to be called the Mother Plane (or Mother Wheel), a "destructive dreadful-looking plane," which appeared as an immense orb in the sky. Initiated into the meanings of this mysterious object as that which Ezekiel beheld during the period of his own calling, Elijah Muhammad in effect *became* Ezekiel and transmitted the meanings of this profound event to his followers throughout the remainder of his life. This was his message, the fulfillment of his calling. It defined for him his identity and the identity of the nation over which he was to preside.

As it occurred for the Honorable Elijah Muhammad, it likewise occurred for Minister Louis Farrakhan, who, like the Messenger before him, has felt himself called on repeatedly to bear witness to the receipt of the vision. As he does so, the experience comes to assume a life of its own. Details of his encounter with the Mother Plane emerge, as the event manifests itself both as an "actual occurrence" within the limits of time and place and as a "psychological phenomenon" within the unconscious of the visionary. In either case, the event represents a centerpiece in Farrakhan's conception of his vocation and mission.

In his articulation of the experience and its momentous impact, certain themes and characteristics of the experience loom large. First, this is clearly an event that must be told and retold. Multiple acts of bearing witness to the vision suggest that it represents a defining moment in the life of the seer, one that obviously impels him to tell it and retell it. The vision becomes for him a recurring burden that is alleviated only through the act of imparting it to others. Second, it is an experience through which the seer receives his vocation from a higher power that infuses him with its dynamism and reshapes him. He clearly views that vocation as the bestowal of prophetic and, indeed, apocalyptic knowledge that gives him the power of prescience and instills within him the need to preach and to warn. As such, the vision defines his calling as a knower, a preacher, and a warner. Third, the characteristics of the vision are very much in keeping with earlier enactments that draw on prophetic, apocalyptic, and mystical traditions of transport to higher realms and the encounter in those realms with that which is by its very nature unknowable and therefore potentially dangerous. This is surely what might be termed a *mystical* experience, but, because of its intimate ties with the fundamental enactment that is the defining moment of its own unique culture, it is also real. Its objective correlative is that "thing," that machine, that "wonder of technology" that is the driving force of the entire movement of which the seer becomes the charismatic conduit. Fourth, the terms of the vision are very much implicated in the disturbing nature of that which the visionary perceives as "enemies" that assume the form of oppressive and destructive forces (represented, e.g., by the "U.S. government," "white supremacists," or the "Jewish community"). Whatever form these enemies assume, it becomes the Yakubian "other" that must be overcome because of the extent to which it is anathema to the "cause" the visionary seeks to uphold. Clearly, the vision that the seer relates during the course of his ministry is of utmost complexity and urgency.

In the United States, Farrakhan's act of bearing witness has been given voice principally on six separate occasions, each defined by its own set of circumstances and its own milieu.[1] The six occasions constitute specific moments in the history of the Nation under Minister Louis Farrakhan. The dates and circumstances surrounding these occasions help delineate the emergence of the vision as a crucial dimension of his sense of his own calling and mission. The first act of bearing witness occurred in a speech delivered at the Final Call Administration Headquarters in Chicago on 13 July 1986. Titled "The Wheel and the Last Days," the speech marked the occasion of Farrakhan's return from a trip to Big Mountain, Arizona, where he met with members of the Navajo and Hopi nations, threatened with relocation. The second act occurred in a speech delivered once again at the Final Call Administration Headquarters in Chicago, this time on 4 January 1987. Titled "The Reality of the Mother Plane," the speech represented, in part, an opportunity for Farrakhan to enlist the support of the families of those who had recently become members of the Nation and to explain to those in attendance what the Nation conceives as its goals and aspirations. It also represented the occasion to reestablish ties with the teachings (among them, those concerning the Mother Plane) of Elijah Muhammad. The third act occurred in a speech delivered at a press conference at the J. W. Marriott Hotel, Washington, D.C., on 24 October 1989. Titled "The Great Announcement: The Final Warning," this speech provided the opportunity for

Farrakhan to broadcast his experience to the outside world, to make of it a "public event" before members of the press and the world at large. The fourth act occurred in a speech delivered at Mosque Maryam, Chicago, on 26 April 1992. Titled "The Shock of the Hour," the speech was the basis of a service at the mosque and the occasion for apocalyptic and oracular proclamations before a community of believers as well as before those on the verge of becoming believers.

The final two acts of bearing witness, in turn, occurred in the context of what was then the impending Million Man March, scheduled for 16 October 1995 in Washington, D.C., and organized by Farrakhan as a major event in his administration to bring together black men from throughout the country to proclaim a Day of Atonement, Reconciliation, and Responsibility. In preparation for this event, Farrakhan gave two speeches, both of which address the experience of his vision. Delivered at what is known as the "Vision Complex," Los Angeles, on 15 September 1995, the first speech (and, in effect, the fifth and penultimate act of bearing witness in the ongoing series of speeches) was appropriately titled "The Vision of the Million Man March."[2] Addressed both to those assembled for the occasion and to the larger public through the medium of cablevision, the speech sounded the call for the Million Man March throughout the Los Angeles metropolitan area. The second of the two speeches (and the sixth and final act of bearing witness in the ongoing series of speeches) was titled "A Special Men's Day Rally." Delivered on 17 September 1995 at the Union Temple Baptist Church in Washington, D.C., this speech addresses an audience comprising a wide variety of groups meeting in concert to hear about the Million Man March and plans for its implementation.[3]

All six speeches assume the form of "performances," the communication of visionary experience through the medium of oral delivery. These speeches reinforce Farrakhan's role as orator, preacher, and charismatic, one imbued with the sense of his calling both as prophet and as oracle to his people. It is the performative dimension of his acts of bearing witness within the space or "theater" of enactment that distinguishes his message and its mode of delivery.[4] In his press conference of 24 October 1989, for example, the vision is conceived as "enactment": the public forum becomes the "theater" of that enactment, and, in the fulfillment of this performative role, the seer is conceived as the major (indeed, the only) "player" on the stage of that theater. Both in the *Final Call* and in the subsequently published pamphlet *The Announcement*, Farrakhan appears in a photograph proclaiming his message. He is seen standing at the midpoint of a long table flanked on either side by family, friends, and colleagues, including the Reverend Al Sharpton, Kadijah Farrakhan (wife

Figure 15. "The Great Announcement: The Final Warning,"
24 October 1989, J. W. Marriott Hotel, Washington, D.C.
Reproduced from Final Call Tapes (Chicago: Final Call, Inc., 1989).

of Farrakhan), A. Akbar Muhammad, Tynnetta Muhammad, Attorney Lew
Myers, and A. Alim Muhammad. Marshaled in a row behind Farrakhan are
members of the Fruit of Islam, and on a wall behind them hangs the Nation of
Islam flag with its crescent and star. In front of the table and facing Farrakhan
are photographers and members of the press. In the videotape of the press
conference, the event is brought to life. The camera first pans the entire room to
provide a sense of the size and scope of the affair. Occupied by members of the
press with their pencils poised to write and the assembled audience waiting in
hushed expectation, the room represents the space of enactment, a receptacle
for the drama of oratory that is about to unfold. Embodied in the camera lens,
all eyes focus on the speaker, whose deportment, bodily gestures, and vocal
intonation reinforce the compelling nature of what he has to say (fig. 15). There
are moments when the speech is interrupted by applause and sounds of ap-
probation from the audience, and there is a standing ovation at the conclusion
of the speech. But questions and comments are not invited: the oracle has been
issued, the speaker and his retinue depart, and the rest is silence. This is clearly

an occasion of monumental import, one in which the vision of Ezekiel comes to be imbued with national and international significance in the modern world.

Corresponding implications are likewise discernible in the theaters of enactment that provide a setting for the other speeches as well. The speech delivered at the Union Temple Baptist Church on 17 September 1995 is a case in point. In sharp contrast to the mood of sobriety and even grimness that pervades the press conference, the aura of the church service is that of festivity, hope, and expectation. There is a carnivalesque quality about the setting. One finds himself in the midst of a revival meeting. Dancing, singing, and clapping are accompanied by the sound of organ music and the beat of drums. In its expectation of the Million Man March, this is a joyous occasion, a time of celebration. As the camera pans throughout the crowded church, one is struck in particular by the multicolored attire that graces the worshipers. Everywhere, bright colors burst on the scene. Greens, blues, oranges, reds, pinks, and yellows are the order of the day. Attire is distinctly that of embroidered dashiki and *kente* cloth indigenous to the regions of West Africa as well as traditional Muslim garb. Some members of the congregation are attired in mud cloth, some in *kufi* hats. The pulpit itself is draped in colorful *kente* cloth. Flanking Minister Farrakhan on each side of the pulpit are such dignitaries as the Reverend Dr. Benjamin F. Chavis Jr., national director of the Million Man March, and the Honorable Marion Barry, mayor of Washington, D.C., as well as the Reverend Willie F. Wilson, minister of the Union Temple Baptist Church. In attendance at this auspicious event, moreover, are a host of dignitaries, each representing his respective organization, including the Southern Christian Leadership Conference, the Caribbean Community, student associations from various colleges and universities, physicians' councils, gang members, publishers of African-American newspapers, the Universal Negro Improvement Association (founded by Marcus Garvey), the Malcolm X Education Center, members of the City Council, and the Maryland Legislative Black Caucus. On the left side of the stage, a woman signs for the deaf. In this environment, Farrakhan becomes the preacher, one who begins his "sermon" by citing and commenting on the text from 1 Cor. 1:10 ("*that* there be no division among you; but *that* ye be perfectly joined together in the same mind and in the same judgment") that is inscribed over the heads of the congregation in the back of the church. Moving from this text, Farrakhan once again bears witness to his vision of the Wheel or Mother Plane as the ultimate source of his own determination to undertake the momentous task of becoming the "conduit" for the implementation of the Million Man March (fig. 16).

Although these are only two of several of the performative contexts in which

Figure 16. "A Special Men's Day," 16 September 1995,
Union Temple Baptist Church, Washington, D.C.
Reproduced from Final Call Tapes (Chicago: Final Call, Inc., 1995).

Farrakhan bears witness to his vision during the course of a decade, they should
be sufficient to indicate the range and nature of the settings through which the
vision assumes a life of its own. These settings frame the vision, which is given
expression as the seer shares his experience with those who witness its unfolding. Because the particulars of the vision as a fully realized event are revealed
only as one comes to terms with the entire series of testimonies captured on
tape (video and audio), one must construct a composite picture of just what
transpired in the originary experience as it is portrayed in Farrakhan's various
acts of bearing witness to it throughout the years. Based on what can be gleaned
from these testimonies, a fairly consistent narrative of the details takes shape.
An account of the details will suggest what form the vision assumed in Farrakhan's experience.

Unlike the circumstances faced by Elijah Muhammad, those under which
Farrakhan received his vision were not those of a great depression, nor were
they the period before he assumed the ministry of the new Nation. Rather, the
vision occurred some eight years after the break with Imam Warith Deen

Mohammed and his World Community of Islam in the West. Significantly, the setting of the vision is not Detroit, or even Chicago, but a region in Mexico, a tiny town called Tepoztlan, to which Farrakhan had journeyed in September 1985 on one of his many pilgrimages. In this case, his visit to Tepoztlan was prompted by his need to retreat from the press of his speaking engagements. On 14 September, he had delivered an address before a very large crowd (some nineteen thousand people) at the Los Angeles Forum. During that period, he observes that it was "hot" for him in Los Angeles because he was "under attack by the Jewish community." The newspapers, he says, were filled with his "alleged anti-Semitism, bigotry, and hatred" against the Jews, and attempts on his part to rectify the situation were "too little, too late." Initially framing the onset of the visionary encounter, then, is the anxiety of being besieged by those the seer views as the "enemy." It is from this enemy that he seeks respite, if not refuge.[5]

Accordingly, Farrakhan left Los Angeles and traveled to Tepoztlan, as he had in the past, to meditate and pray. Tepoztlan, he maintains in his various accounts of his experience, is the site of the ruins of a temple dedicated to Quetzalcoatl, a temple situated on top of a mountain he has climbed several times. These references provide the topographic framework for the experience as a whole. The topography is important because it displaces the locale of the vision to a setting that is already in a sense otherworldly. Situated in central Mexico, Tepoztlan is a picturesque Aztec village some ten to twelve miles northeast of Cuernavaca. Isolated in a valley surrounded by towering cliffs, it is, according to one guide book, a favorite destination for those in search of spiritual fulfillment.[6] In the words of another guide book, it is a "magical place" frequented by soothsayers, mystics, writers, and artists, all those "in hopes of experiencing the legendary force that spawned the birth of Quetzalcoatl, the omnipotent serpent god of the Aztecs, more than 1200 years ago."[7] An editorial that accompanies Farrakhan's announcement in the *Final Call* makes a point of commenting on the significance of this locale and its people to the teachings of Elijah Muhammad, who asserts that the inhabitants of ancient Mexico were "dark-skinned beings, pyramid builders: our ancient forefathers."[8] It is only appropriate, then, that Tepoztlan be the place of Farrakhan's visionary experience, especially since Tepoztlan contains the ruins of Quetzalcoatl, whose spirit resides there.

For Farrakhan, Quetzalcoatl is a very important figure. Among the most powerful and multifaceted gods in Meso-American religions, Quetzalcoatl, the plumed or "quetzalfeathered serpent," is both a major celestial god of creation and a figure associated with the priest-king Topiltzin Ce Acatl Quetzalcoatl, the primary source of culture, political order, and religious authority in Meso-America. As a god of creation, Quetzalcoatl generates the universe, rules over various cosmogonic eras, creates fire, participates in the great sacrifice of the gods that leads to a new cosmic age, or fifth sun, and is ultimately transformed into the morning and evening star, Venus.[9] The paramount figure in the history of the New World, Quetzalcoatl is a deity around whom accrue many legends. In one of them, he is born of the virgin mother Chimalman, to whom God, the All-Father, had appeared one day under his form of Citlallatonac, "the morning." Endowed at birth with speech, great knowledge, and wisdom, Quetzalcoatl eventually becomes ruler and high priest of the City of the Gods, the Toltec city of Tula in ancient Mexico. As a priest-king, he reigns with such wisdom and purity of character that his divine realm flourishes gloriously. His cosmic temple-palace is composed of four radiant apartments that encompass the four quarters of the world. During the period of his reign, he encounters his enemy Tezcatlipoca, who attempts to undermine his reign and authority by confronting Quetzalcoatl with his "dark side," that is, his mortal or fleshly manifestation, subject to the onslaughts of those who oppose him. To overcome this aspect of himself, Quetzalcoatl embarks on a pilgrimage of sacrifice and self-awareness. This is an arduous journey that involves a period of self-burial, followed by self-immolation in a great fire from which he emerges with renewed purity as his heart ascends to the celestial spheres. In one version of the legend, his self-immolation is followed by a descent to the underworld to secure the release of mankind. Thereafter, he reascends to the celestial realms to assume his place as the Lord of the Dawn.[10]

I emphasize the topographic and mythological dimensions of Farrakhan's account of his vision, first, because he calls attention to them himself and, second, because he is at pains to define himself and his experiences in these terms. Although he certainly does not say so explicitly, he would have himself viewed as a Quetzalcoatl-like figure whose own pilgrimages represent a return to some inchoate world of beginnings, a world that draws him into its matrix, there to undergo a process of individuation and reclamation. It is this process that presumably occurs in Tepoztlan with its mountain-top temple ruins dedicated to Quetzalcoatl, the plumed serpent who ultimately takes flight to the celestial spheres. In such a guise, Minister Louis Farrakhan enacts his own visionary journey, one that represents in effect the receipt both of a new identity and of a renewed calling through his own version of the inaugural vision of Ezekiel.

So profound an impact does this experience have that Farrakhan dates it precisely to 17 September 1985.[11] One is reminded of the emphasis that the prophet Ezekiel places on the date of his own vision: "Now it came to pass in the thirtieth year, in the fourth *month*, in the fifth *day* of the month" that "the heavens were opened, and I saw visions of God" (Ezek. 1:1). In both cases, the act of bearing witness assumes a verisimilitude by virtue of an insistence on the dating of the event. Concretizing that event in time and place, Farrakhan assumes the role of delineator of his own vision. That event is portrayed as something tantamount to an ascent to the throne of God: "On the night of September 17, 1985, I was carried up on that mountain, in a vision." This is the language of apotheosis reminiscent of Pauline proclamation: "I will come to visions and revelations of the Lord. I knew a man in Christ . . . such an one caught up to the third heaven. . . . How that he was caught up into paradise, and heard unspeakable words" (2 Cor. 12:1–4). Along with canonical texts of this sort, extracanonical renderings and exegetical treatments of the ascensions of the patriarchs and prophets provide a rich context for the visionary experience delineated here. Underlying a wide range of apocalyptic and mystical texts, such ascensions represent a dominant mode from antiquity into the later Middle Ages. In the literature that draws on the vision of Ezekiel as an aspect of the so-called riders in the chariot, one thinks of many figures who have undertaken the ascent, among them, Enoch, who ascends to the heaven of heavens to behold God on his throne-chariot and to share in the secrets of the universe, and Moses, who ascends the mountain of God in his own vision as a means of attaining knowledge of the divine world. These ascensions represent the basis of the dangerous journeys on which the visionary embarks as a higher power bears him upward through the celestial halls (*hekhalot*) into the innermost chambers, where he hears the dread voice of a divine presence as it empowers him to return to the terrestrial world to warn and admonish his following of the coming doom and at times to infuse them with the promise of a glorious future.[12] If such ascents provide a context for Farrakhan's own experience, the shape of that experience is very much in keeping with the lineage of his vision founded on the conception of the Mother Plane, with its ultimate source in Ezekiel's *visio Dei.*

As Farrakhan describes his experience in his various public appearances, he finds himself and a few selected companions suddenly carried aloft to the top of the mountain he had so often climbed before. Once there, they experience a vision of a wheel that appears before them at the side of the mountain. This, it turns out, is one of the myriad of smaller wheels that reside within the Wheel of Wheels, the Mother Plane, which sends out the smaller wheels on missions of various sorts. Extending from the smaller wheel are three metal legs or spokes, which give the impression that the object is about to land. But the wheel stays aloft. Hearing a voice from the wheel call to him, the seer is afraid and looks to his companions for corroboration and support, but to no avail: "Not them; just you," a voice declares. This is a journey the seer must embark on alone. A thick and heavy beam of light emitted from the smaller wheel transports him into the vehicle itself. Characteristic of what has come to be known as *the illuminative way* in mystical theology, such a movement into the interior of the visionary object suggests the process of enlightenment that the seer undergoes in his journey.[13] As it is described, the movement is as much an internal act of being drawn into the world of the self as an experience of coming to an awareness of one's external environment. The interchange of internal and external modes of perception is held in delicate balance throughout. There is never a doubt about the presence of an actual "object," but the experience is structured to suggest that this object emerges from the depths of the seer as well as appearing to him from without. Entering the smaller wheel, the seer finds himself seated next to a "pilot," who cannot be seen: only his presence can be felt. Under the guidance of the pilot, the wheel as a spacecraft lifts off from the side of the mountain and moves at terrific speed to a final destination. As a result of this experience, the seer comes to realize that the vision in which he is enveloped is moving him ever more forcefully toward that destiny that had been delineated nearly sixty years earlier by that Messenger who was himself both the source and the recipient of visionary enactment: the Honorable Elijah Muhammad, the embodiment of Ezekiel in the modern world. In keeping with all that such a realization implies, the seer declares that he is being transported to that Wheel of Wheels, the Mother Plane.

The fact of such transport has profound significance for the seer. It means that he has been chosen, in effect, to receive the message from the source; that he has been permitted to participate in the originary experience through which that source maintains its authority; that he has been granted insight into his own calling, his own mission, and his own identity as a person and as a leader. All this is symbolized by the Mother Plane, "this great mechanical object in the sky," which for the seer exists at once within his own being and within the atmosphere that it inhabits. Farrakhan comments that such objects are what one might call "an unidentified flying object," a notion derived from Elijah Muhammad himself. If such is the case, then the seer becomes an "abductee" within his own vision. The reenactment of the vision here attests to the fact that what was a crucial and all-pervasive presence within the Nation of Islam in its original phase emerges yet once more with renewed force in the new Nation of Islam, under Minister Farrakhan. Although in his various accounts he is aware of his vision that the outside world will consider him suspect and no doubt even unstable for espousing not only a belief in UFOs but also an encounter with one, he makes it clear that, whatever the vision of the Wheel is called, it is *his* vision, *his* experience, and that, as "abductee," he has lived to tell about it. The outside world may be disdainful in its condescending references to UFOS and those who believe in them, but he is in possession of the truth bestowed by the higher powers that inspire his vision. As important as the vision was to the Nation in its formative years, it becomes even more important in its new incarnation as a phenomenon associated with the mental processes of the seer. As such, its emergence under Farrakhan's dispensation assumes perhaps a more nearly pronounced "interiority" as a manifestation of mind. In this capacity, the Mother Plane or Mother Wheel fulfills its function as a maternal source of generative enactment that produces an entirely new awareness.

In keeping with the interiority of the experience, Farrakhan relates what occurs once he is situated within the confines of the Mother Plane. He is escorted by the mysterious pilot through dark tunnels to a door that suddenly opens before him and through which he is then admitted to a room as a kind of sanctum sanctorum. This room is, as it were, a room within a room. At the center of the ceiling of this enclosure he sees a speaker that emits a voice that he immediately recognizes as that of Elijah Muhammad. If the aura initially leads one to think of Big Brother, the circumstances are much more in accord with Milton's idea of the Son's receipt of his commission from the Father in preparation for his charge of carrying forth the divine will in the Chariot of Paternal Deitie.[14] In like fashion, one might suggest, Farrakhan's commissioning within the Mother Plane represents the son's encounter with the father in the form of the Honorable Elijah Muhammad (Minister Farrakhan's true spiritual father, to be sure). Both commissionings are undertaken during a period of crisis, both authorized through the voice of "the father," and both replete with a sense of what comes to be known as the *odium Dei.*

Whether as maternal plane (gendered so in the economy of Elijah Muhammad) or as paternal chariot (gendered so in Milton's economy), the vehicle within which Farrakhan receives his commission derives its ultimate source from the biblical milieu that distinguishes Ezekiel's vision, in which the prophet hears the voice of the enthroned figure emanating from the upper reaches of the firmament before him: "And I heard the voice of one that spake. . . . And the spirit entered into me when he spake unto me" (Ezek. 1:28–2:2). Like the prophet before him, Farrakhan is the recipient of speech transformed into writing, of orality become text: "Behold, an hand *was* sent unto me; and, lo, a roll of a book *was* therein" (Ezek. 2:8). So Farrakhan relates: "He spoke in short cryptic sentences and as he spoke a scroll full of cursive writing rolled down in front of my eyes." True to the delicate balance between external and internal, thing and spirit, body and mind, Farrakhan internalizes the scroll at the very point of insisting on its reality as an object. The scroll, he says, "was a projection of what was being written in my mind." This is a nice touch for it not only authorizes the vision as an actual external enactment but also insists on its mental interiority as a processive event, an event that is in the process of being enacted as it is conceived. The idea suggests a renewed emphasis on the psychological basis of the vision as an experience of coming to awareness. That awareness is not only of the relationship between son and father but also of the true meaning of one's vocation. For Farrakhan, it is an experience that marks or inscribes the seer as prophet. To conceive the act of calling the prophet in these terms is to rewrite Ezekiel, who is made to eat and digest the scroll as a means of interiorizing the message (Ezek. 3:1–3). For the new prophet, it is not necessary to eat and digest what is already internalized, what is already part of the very being of the one who is the recipient (by voice and by letter) of that which he feels compelled to communicate to others. It is clear that the speech act embodied within the sanctum sanctorum of the Wheel is reenacted performatively in the various theaters of enactment in which the seer himself bears witness to his vision.

Farrakhan concludes his act of bearing witness, first, by providing an account of the message he receives and, second, by issuing a disclaimer that even as seer he would not be permitted to behold the being responsible for the message "face to face" (cf. 1 Cor. 13:12) until the disciple fulfills his master's calling. Dismissed, the disciple as seer returns to the pilot (whom he still cannot see), enters the small craft by means of which he was originally transported to the Mother Plane, and returns to earth. In the process of return, he is allowed to look down on the Mother Plane from a great height and to behold within it "a city in the sky." As Farrakhan attests in his accounts, this is none other than the new Jerusalem witnessed by Saint John the Divine: "And I John saw the holy city, new Jerusalem, coming down from God out of heaven" (Rev. 21:2). The idea recalls Elijah Muhammad's own quest to found a New Jerusalem for his people after the cataclysmic destruction wrought by the Mother Plane at the culmination of end time. With the conclusion of Farrakhan's vision, the seer awakens. As if to confirm the import of the vision, nature itself responds two days later with a great earthquake that strikes Mexico City and its environs. This devastating event prompts the seer, who has traveled by then to the city of Cuernavaca, to share his vision later that very morning with his wife, Khadijah Muhammad, and his associate Tynnetta Muhammad. These intimates are apparently the first to whom the seer bears witness in his mission to make public the nature of his own cataclysmic experience.[15]

From that time forth, the seer is haunted by wheels: they appear everywhere. On 7 October 1985, Farrakhan is scheduled to speak at Madison Square Garden. Mentioning that the atmosphere that greets him in New York is even more hostile than that in Los Angeles he says that the night he journeyed to New York "it came over the news that sixty of the little wheels were seen traveling from southwest to northwest, traveling at a very slow speed of 150 miles-per-hour." The army, he says, sent up its planes to track the wheels. At a press conference shortly thereafter in Washington, D.C., he once again recalls the Wheel. In the midst of his speech protesting the worsening relations between the United States and Libya, he focuses on the lights that surround him, and he recalls the vision.[16] Returning to Chicago, he learns of the explosion of the space shuttle *Challenger* on 28 January 1986: this is apparently an omen. In February 1986, he embarks on a world tour, including a trip to Africa that includes Libya. During the tour, he reads one newspaper account after another about bright, mysterious objects that are sighted in the skies on various occasions.[17] The sightings are not accidental. Responding to such sightings, Farrakhan declares: "I am telling America that wherever I am the Wheel is!" It follows him in the Middle East; it follows him in the Far East; it follows him as he flies from one country to the next; it performs for him as a way of letting him know that it is with him always. In one of his trips to Phoenix, he looks up, and there it is. As he arrives in Phoenix, he learns that "one of the little ones came out of the sky and chased a woman in a car going from Phoenix to Las Vegas, Nevada. She was terrified and called the police. It was in the front page news in Phoenix." This is a sign, he says, that "Farrakhan is in town." To anyone who might ask, "What is the Wheel?" he would respond: "Read the prophecy of Ezekiel!" As the Wheel accompanied the "Son of man" everywhere in that prophecy, so it accompanies the new seer in the modern world.[18] Like Ezekiel, Farrakhan is "connected to the Throne of God." Like Ezekiel, he has had out-of-body experiences. Lying in bed, he hears voices and beholds transcendent sights. Such is the Wheel for him. He receives his energy from it, his power from it, his authority from it, his "juice" from it. He is not a drunken man, he declares; he is not crazy. We all have had dreams, some foolish, but others real. His dream is a vision, a true vision that gives him purpose and shows him the way.[19]

It also bestows on him the powers of the mystic whose trances are able to influence the natural world. In his 1986 speech "The Wheel and the Last Days," for example, he relates his recent experience at Big Mountain, Arizona, in terms of the mystic who is able to conjure up the Wheel through fasting and praying. In its own way, the experience is reminiscent of those undertaken by the riders in the chariot who fast and pray in preparation for their own acts of summoning up the Wheel. They, too, are blessed with powers that accrue to them through the performance of precise rituals.[20] Gathering at Big Mountain with his closest associates, Farrakhan has those that accompany him form a circle. In the center of the circle, Allah's presence resides in the *kaba* as shrine or palladium. Undertaking their ritual, the acolytes engage in prayer, which involves the performance of nineteen units or *rakʿas*.[21] Consistent with his discourses on other occasions, Farrakhan makes a point of emphasizing the significance of the sacred number nineteen, through which one is able to gain access not only to the secrets of the Qur'an but also to the very workings of Providence. So, here, the number is essential in the calling up of the Wheel. Situated within the wheel of the prayer ring, and in effect making circuits around the presence of Allah, the acolytes are in harmony with the universe. They revolve around Allah even as the planets revolve around the sun. Imbued with the power of Allah, Farrakhan then asks for the Wheel to manifest itself. A wind arises, and rain comes forth, announcing the coming of the Wheel. At the time of the setting of the sun, they look up and see a bright orange object in the sky. As it grows darker, the Wheel dances for them and changes color, from red to green to yellow. In his excitement at this remarkable phenomenon, Farrakhan reassures the members of the Navajo and Hopi nations who represent the occasion for his visit that this is the power that is with them today and that, because of that power, they need not fear the confiscation of their lands. The Wheel, he says, follows him out of northern Arizona to southern Arizona. He looks up at the Wheel and takes comfort in the knowledge that through it he is "the cornerstone of a kingdom that will never be removed from the earth."[22]

Such are the circumstances that both frame Farrakhan's experience at the time he receives his vision and define his outlook thereafter. As important as these circumstances are in determining the visionary setting and content of what he beholds, equally significant is the actual message or commissioning that he receives at the climactic point of his vision. As we recall, after he is transported aboard the Mother Plane, he is ushered into the sanctum sanctorum, the room within the room at the utmost interior of the vehicle, where he hears the voice emitted from the speaker at the center of the ceiling of the enclosure. The voice, we are told, issues its message in the form of short cryptic sentences as if a scroll with cursive writing were unrolled before his eyes. This message represents the substance of the vision. Proclaiming that message in the various accounts of his vision recorded above, Farrakhan enunciates it with dramatic force in his press conference of 24 October 1989 and records it in print in the *Final Call* (30 November 1989) and in *The Announcement*. Heard through the voice of Elijah

Muhammad as the ultimate source of gnosis, that message bestows on the vision an immediacy and an urgency that catapult the seer's experience onto the stage of world politics: "President Reagan," the voice declares, "has met with the Joint Chiefs of Staff to plan a war. I want you to hold a press conference in Washington, D.C., and announce their plan and say to the world that you got the information from me on the Wheel."[23]

What makes this message so stunning is not only the fact of its topicality but the extent of its departure from the kind of visionary experience that had earlier defined the mission of Elijah Muhammad. The two aspects are intertwined. As the defining experience of his role as Messenger, Elijah Muhammad's vision remained throughout his career an externalized manifestation of what was to become an apocalyptic warning of approaching doom for those who did not heed its call. As a visionary object, the Mother Plane was always "out there" ready to appear at any moment in order to fulfill its dreadful mission. Although the Mother Plane possessed the uncanny power to read one's thoughts, the psychocentrism implicit in such an idea only begins to suggest the internalization of the vision that one discovers in the experience of Louis Farrakhan. As "abductee" of that vision, he both "enters into" its inner core and in a sense becomes the source of that which generates the abduction he envisions. What reinforces this departure even further, moreover, is the issue of topicality implicit in the message proclaimed by the respective visions of the two seers. In its message, Elijah Muhammad's vision does not move beyond a generalized conception of universal destruction that reinforces distinctions of racial identity and adherence to a set of beliefs and doctrines. Elijah Muhammad never names names, specifies dates, or concretizes immediate locales or settings. In its message, Farrakhan's vision does precisely these things and in so doing causes his message to assume an immediacy and an urgency it would otherwise lack. Authorizing that message through the visionary presence of Elijah Muhammad allows his oracle to move beyond the generalized apocalyptic outlook he has formerly embraced and to sanction specific pronouncements on what shall occur in the immediate future and therefore must be heeded posthaste. Armageddon exists not simply at some indefinable time in the future: it is now.

It is this sense of urgency that distinguishes Minister Farrakhan's message. In the process, it transforms Ezekiel's vision from a staple of biblical prophecy into a highly charged message of warning concerning the machinations of officials at the highest reaches of political power and authority, actors on the international stage of assault and counterassault, surveillance and countersurveillance, conspiracy and counterconspiracy. Infused with the prophetic fervor of the visionary, precisely this kind of discourse informs the speech delivered at the press conference and transcribed in the *Final Call* and in *The Announcement*. As a public message issued to an assembled press, the speech is both formal and declamatory. Observing all the niceties of deliberative oratory, it is conscious of its purpose to persuade to action. But the tone is decidedly prophetic and denunciatory as well. Farrakhan begins by invoking Ezek. 3:18–19: "When I say unto the wicked, You shall surely die; and you give him not warning, nor speak to warn the wicked from his wicked way, to save his life; the same wicked man shall die in his iniquity; but his blood will I require at your hand. . . ." The context of the citation is that which occurs after the prophet in his bitterness and in the heat of his spirit has been transported by his vision to the captives at Tel-abib. There he resides until the word of the Lord returns to inform him that he has been "made a watchman unto the house of Israel." To this house he is to proclaim his prophetic warning of the doom to come (Ezek. 3:12–17). Having beheld his own vision, Farrakhan proclaims his own warning to the assembled press and thence to the world.[24]

Recalling the effects of the vision on him and the message that the oracle issued within it, he slowly realizes what the words of the oracle mean. Between the time of the vision in September 1985 and January 1986, it begins to dawn on him that the oracle's statement that "President Reagan has met with the Joint Chiefs of Staff to plan a war" signifies the impending bombing of Libya on 14 April 1986. It is this event that Farrakhan is determined to make known to the world. Scheduled to embark on his world tour in February 1986, Farrakhan makes certain to arrange his itinerary so that it includes a trip to Tripoli, where he alerts Colonel Gadhafi of the impending attack.[25] Speaking before the representatives of some eighty nations at the Second Mathaba Conference in Tripoli on 16 March 1986, he delineates his vision and all that it implies, as a warning to President Ronald Reagan and Secretary of State George Schultz not to attack Libya.[26] Appearing in the *Final Call* on 27 April 1986, the full text of that speech places Farrakhan's vision in the context of his address to the Libyan parliament (the Jamahiriyat), with Colonel Gadhafi in attendance. In the *Final Call* article, photographs of both Minister Farrakhan and Colonel Gadhafi accompany the text as visual evidence of the transmission of the vision and the message it bears.[27]

Not heeding the warning contained in the message, the United States launched its attack on 14 April 1986.[28] As he relates these incidents in *The Announcement*, Farrakhan observes with satisfaction that "during the confrontation in the Gulf of Sidra, between the United States Air Force and the Libyan Air Force, it was reported in the press that a bright orange object was seen over the Mediterranean. The Wheel was, in fact, present and interfered with the highly sensitive electronic equipment of the aircraft carrier, forcing it to return to Florida for repairs."[29] Transplanted to the shores of Tripoli, Ezekiel does, indeed, make the international scene as a prophet of superpower intrigue and high-tech combat. At strategic moments, the Wheel of the prophet mysteriously appears to interfere with the bombing raids launched by the satanic enemy. In his account of the attack and its aftermath, Farrakhan declares with the satisfaction of a prophet whose message has been not only fulfilled but acknowledged (albeit unknowingly) by the world at large that the *very words* of Elijah Muhammad's message to him on the Wheel thereafter appear in newspaper articles covering the bombing of Libya.[30] The prophetic is coupled with the uncanny.

It is at this point in *The Announcement* that Farrakhan brings to bear the full weight of his visionary experience and its relation to the attack on Libya. The agenda has to do only in part with the fact of the attack. More precisely, the attack provides a context for Farrakhan's announcement of what he envisions as a new conspiracy, one that associates the clandestine meetings of the former administration (that of President Ronald Reagan) with corresponding meetings of the present administration (that of President George Bush). "I am here to announce today," Farrakhan declares, with the full force of his oratorical powers, "that President Bush has met with his Joint Chiefs of Staff, under the direction of General Colin Powell, to plan a war against the black people of America, the Nation of Islam and Louis Farrakhan, with particular emphasis on our black youth, under the guise of a war against drug sellers, drug users, gangs and violence—all under the heading of extremely urgent national security."[31] Greeted with resounding applause, the announcement prepares the way for Farrakhan to argue that the Bush administration shares the same responsibility as the Reagan administration for engaging in a conspiracy against those it looks on as the enemy, whether at home or abroad. The particular object of his denunciation is the FBI, "which has been working," he declares, "to destroy the Nation of Islam since 1940." With its so-called antiterrorist task force and its crime racketeering task force, the FBI seeks to cripple and ultimately destroy the Nation of Islam, an event culminating in an "attack on Louis Farrakhan with the purpose of discrediting, embarrassing and ultimately causing the death of Louis Farrakhan."[32]

Such acts of discomfiture are coupled with and reinforced by the attempt of President Bush to elicit the support of the Jewish community in the portrayal of Farrakhan as an "anti-Semite." This is a portrayal that Farrakhan declares to be particularly offensive to him, so much so that he warns the president and the government of the United States in a declaration the syntax of which is as oblique and complex as it is inward turning: "If I in your mind am before you of myself; if I in your mind am a hater, an anti-Semite, a wild-eyed radical; then you have nothing to fear from my presence; for I, like those who are actually like that will go the way history has decreed for that kind of person." This is a fascinating declaration: at the very point of acknowledging with pain the deleterious effects of a reputation from which Farrakhan desperately seeks to distance himself, he vehemently disclaims what he considers to be the misconceptions on which that reputation is founded. These misconceptions have arisen as a result of conspiracies he does his utmost to explode. If I am truly a hater, an anti-Semite, and a wild-eyed radical, Farrakhan declares, then I deserve the opprobrium that such individuals have justly suffered and that history has decreed for these scoundrels. But I am decidedly not these things that you have made of me "in your minds." Seeking to dispel the world of appearances that becloud the minds of his detractors, Farrakhan interposes himself, his own "reality." What results is the interposition of the "I" of the speaker between the illusion of his presence in the "eyes" of those who behold him and the reality of his being as it is projected in his speech. The speaker is authorized to engage in such an act of self-fashioning as a result of the visionary empowerment that has been bestowed on him by the experience of the Wheel. Thus, he declares, "I am really in front of you by Allah's Divine Will, as an extension of the Divine Warnings given to you from Elijah Muhammad." The Wheel authorizes his presence, confirms the fact of his "reality," and undermines the creation of the "appearances" that have been heaped on him through the conspiracies of others. "You would do well," he concludes, "to leave me alone; to leave the Nation of Islam alone; and to cease and desist from your evil planning against the future of the black people of America and the world."[33] The "I" thereby comes to represent not only the person of the speaker but the larger communities included within it: the Nation of Islam most immediately, but ultimately black people throughout the world.

At the center of all resides the Wheel as a source of authority and empowerment. It is by means of the Wheel that Farrakhan is able to fashion himself and to issue his warning at every stage of his career from the prophecy of the bombing of Libya to the revelation of the government conspiracies against all black people. "I warn you," he proclaims, "that I am backed by the Power of Allah (God) and His Christ and the Power of that Wheel in which I received This Announcement." Prepared to be mocked for what has been revealed to him in the Wheel as a manifestation of Ezekiel's *visio Dei* in the modern world, he prophesies that his mockers will one day see a multitude of wheels over the major cities of America, at which time "the calamities that America is presently experiencing will increase in number and in intensity." This event will cause those who mock him to humble themselves "to the Warning contained in This Announcement."[34] One might note in passing that, in the Qur'an, the Surah titled "An-Naba" (The announcement) speaks of a time that is fixed for the Day of Judgment, the day when the trumpet blast is sounded and the heavens are opened in preparation for the final cataclysm (Sur. 78.17–30).[35] Such is the fervor that underlies Farrakhan's own "announcement," which represents a warning of the apocalypse culminating in an Armageddon that has in effect already begun in our own times.

This is precisely the message that Farrakhan communicates in "The Shock of the Hour," one of his most apocalyptic and disturbing speeches, delivered at Mosque Maryam, Chicago, on 26 April 1992. Presented in the wake of the appearance of *The Secret Relationship between Blacks and Jews* (1991), the speech reflects the climate of hostility that prevailed during this period and, for the most part, continues to prevail. Holding forth on the theme that "the United States of America is about to come under the divine chastisement of God" in response to America's own putative plans to "attack" the Nation of Islam, Farrakhan adopts this occasion to engage in a severe and unrelenting chastisement of the Jews. It is as if some deeply embedded animus suddenly rises to the fore in the process of marshaling all of one's power for the purpose of confronting the enemy in combat. The mode of the speech is invective (at times, amounting to derision), and, by its very nature, the purpose of invective as the expression of verbal warfare is to wound. Giving no ground, it identifies, pursues, and seeks to discomfit the enemy. However one may wish to respond to the arguments on which this invective is founded, it becomes particularly important for our purposes because of the extent to which it is infused with an apocalyptic fervor that culminates in a discourse on the Mother Plane. Farrakhan's castigation of the Jews provides the modus operandi for that discourse.

Allegation after allegation is discharged in a series of salvos against the "enemy." After claiming that the Jews rule the currency and control the banks, including the Federal Reserve Bank, Farrakhan maintains that it is even possible to see the Star of David embedded as a secret sign in the depiction of the eagle on the dollar bill. Disclosing the "conspiracy" between the United States and Israel, he observes that "the real Israel is over here." It is called "the United States of America, run by Jews." From these allegations, he proceeds to controvert any claim that Jews were ever the supporters of blacks. "The same Jewish people that say they are our friends were," he asserts, "the principal architects of the slave trade." Indeed, "they were the ones that helped to rob us of our minds, our means, our culture, our religion, our God." It was the Jews, he says, who fought against Nat Turner, who worked against Frederick Douglass, Marcus Garvey, Malcolm X, Dr. Martin Luther King, "Jews in the government, in the Justice department, Jews that worked against the prophets of God," and Jews who are now seeking to undermine Farrakhan as one who is in the line of these prophets. "I have greater enemies against me in this movement," he says, "than Moses had against him, than Jesus had against him, than Prophet Muhammad had against him. My enemies are greater than all their enemies combined."[36] It is against these enemies that he takes up arms. Besieged on all sides, he warns his enemies that the powers that back him are far greater than those that are against him. It is clear to him that those who are against him warrant the full force of his powers to call his enemies to account. Those powers are embodied in the Mother Plane. In doing battle against his enemies in a final confrontation, the Mother Plane will reveal itself in its full authority and force.

Farrakhan's apocalyptic sermon, then, is about power, the conflict of power, and the means by which power reveals itself. If his enemies have a secret force, a secret power, that which protects him is of far greater magnitude than that which protects them. In keeping with its emphasis on power and the revelation of the source of power, much of the discourse concerns itself with the notion of the *Deus absconditus* and the manner in which the "hidden God" is made known. If the enemy claims its own hidden God that seeks to control and overpower, such hiddenness will be exposed for what it is. The hidden God who protects the true servant, in turn, will be revealed in all his grandeur. That hidden God is one who reveals his glory through Farrakhan's own source of power, that is, the Honorable Elijah Muhammad, who resides at the right hand of ultimate power, Master Fard Muhammad (cf. Matt. 26:64). Drawing on the representation of the ancient of days in Daniel 7, Farrakhan reassures his followers through the idiom of colloquial immediacy: "He ain't no cracker. He's a Black man that look like you. Got hair like lamb's wool . . . feet like burnished brass. . . . God raised Him, God taught Him, God empowered Him, and the two of them [Master Fard Muhammad and the Honorable Elijah Muhammad] are together like two arrows shot from the same bow, and the two of them is backin' me." Embodied in the ancient of days, this is a terrifying figure that resides on his throne-chariot. He is one whose throne "*was like* the fiery flame, *and* his wheels *as* burning fire" (Dan. 7:9–10). Such a depiction serves to underscore the emphasis that Farrakhan places on the Mother Plane as a manifestation of the power that drives him.

Recalling the Honorable Elijah Muhammad's teachings about the Mother Plane, Minister Farrakhan describes its military functions, its aircraft, its bombs, its explosives, its destructive capabilities, its speed and aerodynamics, and its ability to create as well as to destroy. It is a vehicle that runs by the grace of God. The white man may call this phenomenon a "UFO," Farrakhan says, but it is truly the vehicle of God. First beheld by the prophet Ezekiel, it is a vision the presence of which is to be felt everywhere. With its multitude of eyes, it is all-knowing. With its revolving wheels, it manifests itself in the world about us. Indeed, it can be seen even in the dome of the mosque in which Farrakhan is delivering his sermon for the dome is made like a wheel and almost appears to be turning. Its windows are the eyes through which we see and by means of which we are beheld. So the dome itself attests in its own way to the omnipresence of the Wheel. All it takes is the ability to read the signs of the time. As if swept up in the spirit of his utterance, Farrakhan is moved to intone the words of the well-known spiritual: "Swing low, sweet chariot, comin' for to carry me home. A band of angels is comin' after me, comin' to carry me home." "Swing low, sweet chariot," he continues, "the chariot of fire! Swing low, sweet chariot! Let 'em know there's another power beside white power." That power will make itself known at the appropriate time and at the appropriate occasion. Portents of its coming are already felt. "The ants are stirred up; the bees are stirred up; the flies are stirred up. . . . We've got a God that's going to kill white folks today," the seer assures his following. In the words of the Qur'an, this is "the Shock of the Hour" for "the great upheaval of the Hour will indeed be terrible" (Sur. 22.2). It is when all will see the Wheel come down, when God declares "open war" on the devil. Then all will see terror as they have never seen it before. Although the enemy will be coming soon with tanks, machine guns, and helicopters, his power will be totally eclipsed by the Wheel in a final confrontation.

Meditating for a moment on the charged climate that he has created through his discourse, Farrakhan himself marvels at the force of his oracles: "I've never talked like this before," he exclaims. But he feels that he has an obligation to do so. Just as these are not ordinary times, he maintains that he is not an ordinary person, nor are those who follow him ordinary people. Quite the contrary: they are a people defined by their difference, by their "otherness." Nothing will change that: as much as the world at large may seek to assimilate these people into the "mainstream," they will continue to maintain their identity through difference. "We are a strange people in your midst," he declares to the world at large. "You look at us, but you do not know us." We are the people of the Wheel: it protects us, it guides us, it stands as a constant reminder of our difference, our otherness. As a manifestation of the "Shock of the Hour," it betokens the final end.[37]

If such is the outlook that Farrakhan adopts in "The Shock of the Hour" (1992), the message that he declares in the two speeches preparatory to the

Million Man March is no less urgent. "The Vision of the Million Man March" (1995) and "A Special Men's Day Rally" (1995) both emphasize the importance of continuing to be imbued with the apocalyptic fervor that pervades all Farrakhan's speeches. At the core of that mentality is the presence of the Mother Plane as the vehicle of divine wrath and judgment. In both speeches as well, Farrakhan reiterates the events surrounding his ascent to the Wheel and the message that is communicated to him through the voice of Elijah Muhammad. Both speeches view the vision as the ultimate source of inspiration for undertaking the march in the first place. Moreover, because of the constant presence of the Wheel wherever Farrakhan travels, he is certain that it will be with him on the occasion of the march, just as it accompanied him on all his pilgrimages. Its presence at the march will make certain that the marchers are safe no matter what weaponry those who might distrust them have in reserve. All the artillery stored in the armories of Washington, D.C., will be as nothing compared with the force of the Wheel. This visionary phenomenon will be as a "cloud by day and a pillar of fire by night" (cf. Exod. 13:21–22) to deliver the marchers, just as it did the children of Israel, out of the bondage imposed by Pharaoh.[38]

Despite the apocalyptic overtones present in the two speeches that preface the Million Man March, this dimension of Farrakhan's thought must be seen in the larger context of what amounts to the tone of hope, even festivity, that underscores his message on the occasions of his talks. Here the outlook is finally not one of confrontation, castigation, and conflict. Rather it is one of celebration and the desire for conciliation. At the center of this new emphasis on coming together, of atoning for one's sins, of seeking peace with others and with God are once again the Jews, their religion, and their practices. As indicated in the previous chapter, Farrakhan does not hesitate to acknowledge that the event that represents a high point in his ministry is founded squarely on the most sacred day in the Jewish calendar, Yom Kippur. Calling attention to this fact in "The Vision of the Million Man March," he emphasizes it once more in "A Special Men's Day Rally." Admiring in the first speech the way in which observant Jews fast and absent themselves from work, school, business, and any other activities that might distract them in pursuit of seeking reconciliation with each other and with God, he praises the "beauty" of Yom Kippur as a most holy day. In "A Special Men's Day Rally," he goes even further in his pursuit of this line of thinking. He not only extols the Jews as the people of God from the time of the Exodus on but also views them as a people "receptive to new visions" after they have undertaken reconciliation with God during Yom Kippur. Given Farrakhan's own view of visions and the visionary, one suspects that he could conceive of no greater form of attestation to the worthiness of a people. It is difficult to know what to make of such a gesture of conciliation, except to observe that it was reiterated before an audience of millions on the occasion of the speech that Farrakhan delivered at the Million Man March.

In the context of the present discussion, it is indeed appropriate to emphasize the importance of realizing the extent to which Farrakhan's relation with the Jewish community is implicated in the vision of the Wheel. For all practical purposes the two are almost inseparable. They engage with one another; they tug at one another; they frame one another in the course of the narrative of the process of self-realization and individuation. The space they inhabit is one of paradox, if not contradiction. One might suggest yet once more that the principle of attraction and repulsion embedded in the Yakubian paradigm is as operative in the visionary realm as it is in the social, political, and religious realms. The interaction (in Farrakhan's terms, the "relationship") that emerges from the visionary realm simply heightens and intensifies the dilemma. It is ironic that at the visionary level, in particular, the fascination with and indebtedness to the experiences to which Ezekiel's *visio Dei* gives rise assume such profound significance for those whose points of view are grounded in contexts and ideologies so apparently at odds. The ancient *merkabah* mystics with their throne-chariots and their celestial ascents and Minister Louis Farrakhan with his wheels and his own experience of transport would no doubt have a good deal to talk about (if not to argue about) if they could only communicate. For both, the vision of Ezekiel provides an experience through which opposing contexts and ideologies might well converge. One thing is certain: for both cultures, Ezekiel's *visio Dei* is an event of utmost sacrality, one that represents a defining moment in the history of visionary enactment.

If the foregoing provides some insight into the way in which Farrakhan's acts of self-testimony concerning his experience of the Wheel have a profound bearing on the nature of his calling and mission, the various meanings that accrue to the Wheel as visionary phenomenon are made clear in the interpretations that appear in such publications as the *Final Call*. These interpretations suggest even more compellingly the extent to which the Wheel is as pervasive in the Nation of Islam under Minister Louis Farrakhan as it was in the Nation under the Honorable Elijah Muhammad. All one need do is scan the issues of the *Final Call* to corroborate this impression. As with the Honorable Elijah Muhammad, so with Minister Louis Farrakhan: the Wheel is consistently conceived as an object that the modern world customarily terms a UFO. Under the present dispensation, this phenomenon is part of the very fabric of the visionary imagination of the members of the Nation of Islam and a crucial dimension of their spiritual life.

In the issue of the *Final Call* (30 November 1989) that carries *The Announcement* under the title "A Final Warning," practically the entire paper is devoted to the subject of UFOs. Accompanied by photos of UFOs, articles such as "UFOS: 'Above Top Secret'" and "UFOS: Fantasy or Reality" address corresponding aspects of the concept. These articles, in turn, are highlighted by editorial commentaries on the subject of UFOs. Finally, Farrakhan's own press conference is followed by the republication of the chapter "Battle in the Sky is Near" from Elijah Muhammad's *Message to the Blackman*. To reinforce the apocalyptic sense of the whole, the *Final Call* cites a passage from the Qur'an: "And if they were to see a portion of the heaven coming down, they would say: Piled-up clouds. Leave them then till they meet that day of theirs wherein they are smitten with punishment. The day when their struggle will avail them naught, nor will they be helped" (Sur. 52.44–46). In short, these are UFOs with a message: as harbingers of the coming doom, they are (as they have always been for the Nation of Islam) "signs of the time" that indicate an all-pervasive awareness that the presence of the Wheel cannot, indeed, must not, be forgotten. It is always in the sky to remind those who inhabit the realm of end time of who and what they are, what constitutes their identity and mission in life.

In the article "UFOS: 'Above Top Secret,'" for example, Abdul Wali Muhammad provides coverage (pp. 1, 22–23) of an extensive international UFO conference ("Dialog mit dem Universum") held in Frankfurt, Germany, 26–29 October 1989.[39] In his coverage of the conference, Abdul Muhammad tellingly observes that it occurred "only two days after Minister Louis Farrakhan appeared before the national and international press detailing a personal experience that he had with one of these planes" and warning that they would "soon be seen over the major capitals of the world in response to a U.S. government plot to assault the black community." Attended by official representatives from the Nation of Islam, the conference hosted participants from twenty-five countries, including a delegation of scientists and journalists from the Soviet Union as well as American, European, Asian, and Central and South American participants. The purpose of the conference was in part to prepare people for contact with UFOs, which many participants believed would be seen throughout the world in the 1990s. Describing the site of the conference (the Congress Hall in Frankfurt), Abdul Muhammad suggests that its interior had a carnivalesque quality. Complete with a "booming high-fidelity ceiling-hung sound system," the rectangular auditorium was ringed on three sides by a second-floor balcony on which were arranged tables "teeming with books of the presenters, crystals, copper wire pyramid hats, audio and video tapes, herbal teas and medicines, Indian art work, alpha-wave eye glasses and a variety of other 'new age' staples." Such trappings, however, in no way compromised the highly serious nature of the gathering itself. (In fact, the "high-fidelity ceiling-hung sound system" recalls in its own way the speaker that emits the voice of Elijah Muhammad on the ceiling of the Mother Plane described by Farrakhan in *The Announcement*.) The high seriousness of the enterprise as a whole, Abdul Muhammad suggests, was meant to undermine what in effect amounted to a government conspiracy of maintaining silence about UFO sightings throughout the world. Those assembled within the auditorium enclosure of the Frankfurt conference were determined to break that silence. As strenuous as the U.S. government had been to classify UFOs as "above top secret," the most confidential of information had come to light through files released by the CIA, the National Security Agency, and the FBI, among other agencies, in adherence to the Freedom of Information Act (FOIA). According to Abdul Muhammad, such files conclusively demonstrated that even enterprises like the Strategic Defense Initiative (SDI or "Star Wars") were being launched to do battle with UFOs.

When it comes to international events of this sort, the Nation of Islam is quick to seek confirmation of its outlook within the largest spheres possible. Receiving extensive news coverage in the Nation's official organ, these events provide additional evidence of the far-reaching implications of convictions that carry the weight of religious belief. This is theology of the highest order. For the Nation, the key to that theology is to be found, as suggested, in the UFO par excellence, the Mother Wheel or Mother Plane as a concretization of Ezekiel's *visio Dei* in the modern world.

The coverage of the Frankfurt conference is only one example of many in which Ezekiel's vision as UFO finds expression in the news accounts of the Nation of Islam.[40] As early as the *Final Call* for 31 January 1987, the idea received detailed examination. "UFOS KNOWN!!" the headline of that issue declares in anticipation of a story by Abdul Wali Muhammad (pp. 1–2, 5) concerning a recent sighting by the Japanese pilot of a commercial aircraft. On a routine flight from Iceland to Anchorage on 17 November, Japanese pilot Kenju Terauchi and two crewmen sighted three strange craft that first appeared as three lights following their Japan Airlines jumbo jet, flight 1628. According to Captain Terauchi, the lights were emitted from two "walnut-shaped" aircraft that instantly disappeared and reappeared. Captain Terauchi also saw the silhouette of a much larger ship, which, along with the two walnut-shaped craft, followed his plane a long distance. (Diagrams of both the walnut-shaped and the larger craft accompany the story.) This incident, states Abdul Muhammad confirms the fact that "Ezekiel's Wheel is Real." It exists as the Mother Plane, designed as the ultimate weapon to protect black people. In the same issue, Farrakhan has an article ("Warning, Mr. Reagan: God's Power Is Real," pp. 20, 29) that places this story in the context not only of the Messenger's discussions of the Mother Plane but also of the visionary experience that Farrakhan had at Tepoztlan. Bearing witness here as so many places elsewhere to this experience, he discloses yet again just how important the Wheel has been for him: "It was," he says, "my constant companion as I traveled throughout the world, and when I returned to the United States, it remained with me." The Wheel, of course, is the very object "that Ezekiel saw in a vision 595 years before the birth of Jesus." As much as President Reagan might seek to attack the Wheel with his Star Wars program, his effort will prove futile, Farrakhan declares, for through the Wheel black people will ultimately be liberated in a star war to end all star wars.

Along with the apocalyptic dimension of the Wheel in the thinking of Minister Louis Farrakhan, the "inward-looking" dimension is made the subject of commentary and speculation by those commissioned to interpret what he has experienced. Two such figures are Jabril Muhammad and Tynnetta Muhammad, both of whom offer their own perspectives on the significance of Farrakhan's visionary encounters with the Wheel. Each is a kind of "theologian" for the movement, and each has a visible presence in the issues of the *Final Call* as well as in individually authored works (including books and tapes).[41] In the *Final Call*, Jabril Muhammad's reflections appear under the heading "Farrakhan the Traveler," an ascription that is adopted as the title of his collected essays.[42] As Jabril Muhammad indicates, the ascription speaks at once to the many journeys or pilgrimages (national and international) that Farrakhan has undertaken during his career and to the spiritual movements of his mind. "The core of what this column is about," Jabril Muhammad observes in the *Final Call*, "is to indicate that as Minister Farrakhan travels outwardly, he is traveling inwardly towards his Lord."[43] Jabril Muhammad assumes the responsibility of mapping Minister Farrakhan's travels, to provide a kind of vade mecum that seeks to illuminate what those travels signify.

Accompanying Jabril Muhammad's columns in the *Final Call*, Tynnetta Muhammad's columns likewise interpret the significance of Farrakhan's journeys, but her outlook is essentially metaphysical or "mystical." Her concern is very much with the numerological implications of the events surrounding Farrakhan and what these events portend. Tynnetta Muhammad's reflections have appeared in book form as *The Comer by Night*, which in its own way is a commentary on Ezekiel's vision from the perspective of the Nation of Islam in its present form. A composer as well as an author, Tynnetta Muhammad has published books concerning the place of women in Islam and a "mathematical theology" of Islam.[44] Although space does not permit a detailed discussion of the reflections of Jabril Muhammad and Tynnetta Muhammad, a glance at their writings should suggest the extent to which Farrakhan's experiences have generated an interpretive culture all their own.

In a series of articles in the *Final Call,* Jabril Muhammad undertakes an analysis of Ezekiel's vision in the context of Farrakhan's 1986 visit to Libya. Jabril Muhammad recounts how he first came to know of the vision of Ezekiel through the teachings of Elijah Muhammad, how his conversations with Tynnetta Muhammad enlightened him further, and how all the pieces seemed to fall into place with Farrakhan's own vision. Determined to establish a kind of historical continuity between the Nation of Islam under the Honorable Elijah Muhammad and Minister Louis Farrakhan, Jabril Muhammad moves as an exegete from verse to verse in Ezekiel 1 and following, as he considers such matters as chronology, geography, and environment. Through an analysis of each of these factors, Jabril Muhammad painstakingly attempts to demonstrate those all-important connections that confirm and legitimate the global centrality of the Nation he represents, to provide it with a providential center anchored in the biblical visionary milieu.

For example, the reference to the thirtieth year in Ezekiel 1 finds its counterpart in the appearance of Master Fard in 1930. In keeping with this dating, the fourth month refers to 4 July, the putative time of Master Fard's appearance. Correspondences of this sort are insisted on as evidence of the grand design. These factors, in turn, are viewed in the context of recent events, specifically, the bombing of Libya as an indication of the provoking of the Mother Plane to retaliate, an event foretold in Matt. 24:36–44 and Luke 17:26–36, having to do with the coming of the Son of Man "as the lightning cometh out of the east" and the fall of the stars from heaven (Matt. 24:27). The apocalyptic is never far from Jabril Muhammad's thinking. Nor are UFOs: time and again, the two phenomena are coordinated so that one is seen as an expression of the other.[45] Citing newspaper accounts of alien craft that appear at various times throughout the world, especially during times of great crisis like the bombing of Libya, Jabril Muhammad invokes one Dr. Axel Hertlein, a German UFO expert, as saying that "it is as if the world is at the brink of Armageddon and somebody or something has taken an interest," so much so, in fact, that "there is nothing in the history of UFO research to compare with the number of sightings that are being reported from troubled places." Responding to such an observation, Jabril Muhammad asks why Dr. Hertlein used the word *Armageddon* to describe the hot spots that are accompanied by UFO sightings. The answer is to be found in the vision of Ezekiel, which portrays the prophet spiritually transported to the future, there to behold "the end of the time of the wicked ruling powers and the end of the time of the captivity of God's people." That time, Jabril Muhammad declares, is now, and the prophet of these times is Elijah Muhammad. In him is embodied the Son of Man as well as he who is called faithful and true in Rev. 19:11. He may be beheld coming from heaven with an army of angels. This is the very being who spoke with Farrakhan in the Wheel. Farrakhan entered the Wheel even as the prophet went in between the wheels, "even under the cherub," and filled his hands with "coals of fire from between the cherubims" (Ezek. 10:6–9). These were Farrakhan's "inner travels," his movement within himself into the center of his own being, there to converse with Elijah Muhammad in the receipt of his calling.[46]

In another article on UFOs in the *Final Call*, Jabril Muhammad first warns the world that Farrakhan is under the protection of the Mother Ship during his various travels. That warning is next followed by a candid acknowledgment that all this talk of UFOs will no doubt appear to be bizarre and even mad to the outside world. Jabril Muhammad realizes that the very notion that there is above the atmosphere a huge plane with people who are guarding Farrakhan "may seem crazy to most people," but this is one of the guiding principles of those who subscribe to the tenets of the Nation of Islam. They know that in biblical times "there was an object above the heads of the exslaves, as they struggled for freedom in Egypt." From this object in the skies, God defeated the armies of Pharaoh.[47] The Nation and its people, then, are not insensitive to how they will be looked on by those who do not understand, but the vision and its implications are phenomena that they willingly accept, indeed, to which they adhere with conviction as manifestations of an essentially apocalyptic mentality. Everywhere they look, they see portents of the fact that they are living in end time. "We are living at the end of the world that we have known," Jabril Muhammad declares. UFOs are seen everywhere: these are signs of the time. That time marks the certain arrival of Armageddon.[48]

Complementing these reflections of Jabril Muhammad's are those of Tynnetta Muhammad. Contributing her own columns to the *Final Call*, Tynnetta Muhammad is often in dialogue with Jabril Muhammad in the interpretation of the significance of Farrakhan's journeys and in the recognition of the grand purpose behind his every move. Recalling the Nation of Islam's own fascination with number mysticism dating back to Master Fard Muhammad's *Teaching for the Lost Found Nation of Islam in a Mathematical Way*, Tynnetta Muhammad repeatedly explores the numerological significance of names, dates, and events.

Her writings focus on the "parabolic" nature of certain key numbers (principally the "Parable of 19") as they unlock the profound secrets of the universe and reveal "the signature of God."[49] For Tynnetta Muhammad, "Islam is Mathematics and Mathematics is Islam."[50] Her foundational text is the Qur'an, in conjunction with the Bible, and, as she would argue, her perspective is uncompromisingly Islamic in its commitment to her understanding of the teachings of the Prophet Muhammad. As a close reader of the qur'anic and biblical texts, Tynnetta Muhammad is determined to perceive the grand providential design that codifies the working out of history, particularly as that history culminates in the line of vision extending from Master Fard Muhammad through the Honorable Elijah Muhammad to Minister Louis Farrakhan. In her universe, everywhere is correspondence; everywhere is meaning; nothing occurs by happenstance. To know history is to decipher its code, to understand its signatures as they are revealed numerologically.

As the most complete statement of what might be called Tynnetta Muhammad's "mystical theology," her collection of essays *The Comer by Night* provides insight into her approach to the vision of Ezekiel and all that it represents in the context of the thought and ideology of the Nation of Islam under Minister Louis Farrakhan. Her stated purpose in the expression of the Nation's thought and ideology is to disclose "the esoteric or inner meanings of the Holy Qur'an." The key text to which the title of her book refers is Surah 86 ("At-Tariq" [The night star]) of the Qur'an. This surah declares: "I call to witness the heavens and the night star—How will you comprehend what the night star is?" (Sur. 86.1–2). For Tynnetta Muhammad, the Night Star (or, in her terms, "the Comer by Night") is the revelation of truth that emerges from the profound darkness. A phenomenon of "piercing brightness," the Comer by Night discloses in its brilliance all that can be known of the mysteries in which the world is veiled. "Under the cover of darkness," she observes, "a great sign appears in the star of piercing brightness, as God is about to bring to birth a new Nation out of the old world of chaos and deception."[51] At the spiritual core of this vision of the Comer by Night is Elijah Muhammad, whose features are depicted on the right-hand side of the cover of *The Comer by Night* in astral outline, including a fez adorned by the familiar crescent moon and star. (From the left-hand corner of the cover is seen the trumpet blasting out the message of "the final call.")[52]

If the Night Star (the Comer by Night) is the key qur'anic text of Tynnetta Muhammad's book, the key biblical text is that which portrays the vision that inaugurates the prophecy of Ezekiel. For Tynnetta Muhammad, this vision in its own way is a Comer by Night that emerges from a profound and enveloping darkness to reveal to the cognoscenti the deep mysteries that it encodes. Those mysteries are manifested in the form of concrete incidents, such as the periodic appearance of Halley's comet in the heavens as a sign of the working out of providential history. Foremost among all such events, however, is that which Tynnetta Muhammad suggests is the underlying theme of her book. This theme is one in which the Comer by Night is conceptualized specifically as the Mother Plane, which appeared to Minister Louis Farrakhan as the maternal source of his visionary ascent on 17 September 1985. All springs from that event, and all future events are to be understood in reference to it. Having discovered the "real" meaning of Ezekiel's vision as manifested in Farrakhan's experience, Tynnetta Muhammad says that "the words of Ezekiel were so explicit, and they matched so perfectly with the details of the Minister's Vision, that I wanted to record them in this book."[53] Such is precisely what she purports to do as she goes about to apply the specifics of Ezekiel's vision to Farrakhan's experience.

In the process of venturing correspondences between the biblical text and the experience of Farrakhan, Tynnetta Muhammad bears witness to her own personal experience in the form of a dream that serves to confirm for her the reality of the minister's visionary experience that she is explicating. She maintains that, while she was in the company of Farrakhan in the west African country of Benin, and shortly before their departure, she dreamed that Elijah Muhammad "entered a landing field where a plane was waiting." He directly faced Minister Farrakhan, on whom the Messenger focused his eyes intensely. "Shortly thereafter," she relates, "they both moved swiftly towards the front cockpit of the plane until they were separated from [her] sight." The dream becomes a foundational moment for Tynnetta Muhammad, who relates it not simply to confirm the "truth" of what she is expounding but also to suggest that she, too, is a participant in the visionary community whose experiences she seeks to illuminate. As covisionary, she is thereby authorized to speak with a kind of oracular assurance about the significance of matters that would otherwise be hidden from sight. These involve the unprecedented sightings of ufos in the world, including the Mediterranean, North Africa, the Persian Gulf, and North, Central, and South America. Such sightings are harbingers of the Armageddon that the world itself inevitably faces. Underlying these events is the symbolism implicit in mathematical codes such as that associated with the number nineteen, among other holy numbers. As mystical integer, the number nineteen in particular is seen to "revolve around its link to the Mother Wheel, a celestial phenomenon that relates to God's Throne of Power."[54] These and other relations underscore Tynnetta Muhammad's act of disclosing the significance of the visionary implications of the Mother Plane. As she develops them in *The Comer by Night*, these implications assume an entire range of meanings that extend from the esoteric to the political. The point of focus is the Wheel as a concretization of the vision that the prophet Ezekiel delineates in his account of the *visio Dei*.[55]

What began over half a century ago as the vision of a mysterious figure who went from door to door selling silks and raincoats in the Paradise Valley neighborhood of Depression-wracked Detroit has since evolved into a multifaceted experience that represents the foundational moment of an entire movement. A reflection of the religious, ideological, and racial dimensions of the movement as a whole, this vision came to assume an essentially apocalyptic bearing, one that is as pervasive in the movement as it is presently constituted as it was in the movement from its very point of origin. Centering on the *visio Dei* that inaugurates the prophecy of Ezekiel, that vision traces its lineage directly from Master Fard Muhammad through the Honorable Elijah Muhammad to the Honorable Louis Farrakhan. This lineage constitutes a visionary milieu that binds a community with its own distinctive characteristics and practices. As disturbing and, indeed, as threatening as this movement might appear to the world at large, it remains a phenomenon to be reckoned with, to be analyzed, and, as much as possible, to be understood. This movement, this Nation, is as visionary now as it has been in the past. That is to say that it is as revolutionary as it ever was. In the words of Tynnetta Muhammad, "all revolutionaries must be visionaries for there can be no revolution without vision."[56] Such would seem to be a guiding principle of the Nation of Islam from its point of conception until its present incarnation.

Conclusion

One of the most remarkable features of the vision that inaugurates the prophecy of Ezekiel is the presence of all those eyes: "As for their rings [*vegabeyhen*], they were so high that they were dreadful [*veyir'ah*]; and their rings [*vegabotam*] *were* full of eyes round about [*mel'et 'eynaim sabib*] them four" (Ezek. 1:18). Within the wheels (*'ophannim*), the rings are imbued throughout with eyes. Specifically, the rings are "full" of eyes, "full to overflowing." So replete are the rings with eyes that there is no containing them. Translated in later versions as *rims,* the term *rings* (from *gab*) moves the whole into the realm of trope: it suggests that which is elevated, curved, perhaps the brow of the eye.[1] These are the rings of the wheels within the wheels (*'ophannim* within *'ophannim*): inwrought with rings, the wheels revolve within and without, moving in all directions. Rings within rings, wheels within wheels, eyes within eyes, full to overflowing, within and without, eyes inset within their dreadful hollows, their own dreadful rings, within and without, moving in all directions: this is the consummate vision of eyes. The rings with their eyes provoke dread. Indeed, the very core of the vision is that which is "dreadful" (*yir'ah*), that which causes not just fear but reverence and awe.

Inspired both by dread and by reverence and awe, the prophet must have been taken with the multiplicity of the eyes that permeate the rings for, when he beholds the vision at a later point in his prophecy, he multiplies the eyes well beyond their immediate surroundings. As such, they appear not only within the rings of the wheels but within the wheels themselves and the very fourfold creatures (*ḥayyot*) that accompany the wheels: "And their whole body, and their backs, and their hands, and their wings, and the wheels, *were* full of eyes round about, *even* the wheels that they four had" (Ezek. 10:12). Bodies, backs, hands, wings, wheels, rims: all are replete with eyes. This is a remarkable, indeed, frightening, transformation. Unstable at its core, the vision is constantly changing, moving, revolving, assuming new forms. Even within the biblical text, its features are always reinscribing themselves in new ways. The basis of the reinscription is the presence of the eyes: they keep popping up in new places.

To behold the vision with one's eyes is to behold eyes. To see is to be seen. To be the subject (the one who sees) of the vision is to become the object (the one who is seen). For not only is this a vision that is seen; it is a vision that *sees back.* The more the seer beholds, the more he is beheld; the more he penetrates, the more he is penetrated; the more he seeks to know, the more he is known. Other circumstances compound the dilemma. This is not simply *a* vision of God: it is many visions. At the very outset of his prophecy, the seer declares that "the heavens were opened, and I saw the visions of God [*mar'ot 'elohim*]" (Ezek. 1:1). All those wheels, all those rings, all those creatures, all those eyes, are replicated in a multitude of visions that open on a multitude of perspectives. If there is one seer with one set of eyes, what his eyes perceive are visions on visions, each replete with multiple eyes that in turn perceive him, scrutinize him, penetrate him. The "*mar'ot 'elohim*" are the appearances *of* God, the seeings *of* God, the appearances *by* God, the seeings *by* God: subject and object, they are the mirror of the seer, who is also subject and object. What he sees is himself, yet not himself either. What he sees is "other," wholly other. Even then, he is at a loss for the seer does not behold God directly in these visions. He does not even behold the glory of God, not even the likeness of the glory of God. Only at several removes he beholds "the appearance of the likeness of the glory of God" (Ezek. 1:28). So removed, the seer is still enveloped with dread. He is still in danger of annihilation. The *hashmal* forever threatens to flame out and destroy him. He is haunted by those eyes. Encompassed by fire, the enthroned figure above the firmament stares down at him. It is little wonder the prophet falls on his face at the culmination of the vision.

Mindful of what the vision entails, I have undertaken to explore the many perspectives, the many *mar'ot*, the many "sightings" to which Ezekiel's *visio Dei* has given rise in modern times. These *mar'ot* are themselves reinscriptions of that which is already in the process of being reinscribed as the vision unfolds. Reinscribing the vision in this manner, the new visions, the new *mar'ot*, have bestowed new names on it. Naming the unnamable, they have concretized it and technologized it in new ways. Newly envisioned, the *mar'ot 'elohim*, conceived as a *merkabah* or chariot, reemerges in a multitude of forms, of constructs that constitute the rebirth of the *ma'aseh merkabah* or work of the chariot in the modern world. As such, the *merkabah* becomes a locomotive, a tank, a flying machine, an unidentified flying object, an inertia projector. The embodiment of power in technological form, the new *merkabah* is able to overwhelm any enemy. Armies fall at its presence. Glaring lightning from its fierce eyes, its laser-equipped beams render all opposition futile. It is the *odium Dei* become machine.

Those who construct and reconstruct Ezekiel's *visio Dei* as machine are the avatars of *technē*, the new riders in the chariot. They assume the role of prophet, of seer. Tracing their lineage to Ezekiel, they are his children: Melchior Bauer, Erich von Däniken, George Adamski, Josef F. Blumrich, prophets all. In them, high culture and low coalesce. Real space or cyberspace is their domain: Mona Strangely and Drunvalo Melchizedek cavort in the empyrean. Pastor Charles Taze Russell and Judge Joseph Franklin Rutherford watch for the end. Hal Lindsey, Jerry Falwell, and Pat Robertson do battle against Gog and company in preparation for the final battle, but that true child of Ezekiel, Ronald Reagan, saves the day.

If the zeal that empowers such prophets inspires the work of John Milton with his Chariot of Paternal Deitie, it is no less compelling in the work of the Honorable Elijah Muhammad and the Honorable Louis Farrakhan, each with his Mother Plane. Reformation zeal and racial confrontation go hand in hand. So Malcolm X: "Milton and Mr. Elijah Muhammad were actually saying the same thing." With this observation, the frontiers of difference are conjoined, only to be exploded. Comparison provokes contrast: the antinomies of crisis remain intact. Confrontation becomes the defining feature in the full realization of the *visio Dei* as a category of race. With that realization, the discourse of conflict resolves itself in the annihilation of all who stand opposed to the true path. Be they whites, Christians, or Jews, Allah will prevail.

Here, in particular, is where the *merkabah* assumes so frightening an aspect. This, once again, is a rearticulation of what David J. Halperin terms "the dark side of the *merkabah*."[2] Recalling the rhetoric of *Star Wars*, the phrase conjures up images of Darth Vader, powerful lord of darkness and dread. An encounter with Ezekiel's *visio Dei* comes at a price. One constantly risks madness or annihilation in a confrontation with the "other." It is for this reason that the Schreberian psychotic of Edwin Broome continues to haunt those who have come in contact with the "influencing machine," the mechanism that Ezekiel's vision becomes in the world of the present.[3] If one is determined to perform a "technopoiesis" by which the visionary reveals itself as machine, one must be willing to face the consequences of his actions. Those actions may result in the revelation or unconcealment of that which is hidden, but such a disclosure of *alētheia* is not without its dangers. The eyes of the vision gaze out at the seer, enveloping him, consuming him, destroying him.

One brings to mind yet once more the narratives (*'aggadot*) swirling around the *ma'aseh merkabah* in rabbinic Judaism. To violate the interdictions surrounding the *merkabah* is to tempt the most horrible of fates. For anyone foolish enough to disregard these interdictions, it were a mercy if such a person had not come into the world (*Ḥagigah* 11b). Ignorance of the law is no excuse. The fate of the poor child who encounters the text of Ezekiel at the house of his teacher can never be erased from memory: apprehending the meaning of *ḥashmal*, the child is consumed by fires that flash forth from the mysterious electrum (*Ḥagigah* 13a).

In the spirit of Broome's psychotic, I conclude with a narrative of my own. Holding a fellowship at the Newberry Library in 1981, I was responsible for making a seminar presentation based on my research concerning Ezekiel. The seminar was scheduled Friday, 7 November. To introduce my presentation, I had planned to warn those assembled that we were all in mortal danger in broaching a subject as sacrosanct as the *merkabah* and that, like the child referred to in the rabbinic story, we all ran the risk of being consumed by the fires from the *ḥashmal*. This sort of opening to my presentation was certain to be an attention getter, and I looked forward to using it. Having placed the finishing touches on my talk Monday, 3 November, I was satisfied that I had done all that I could to prepare early for the impending seminar, four days later. Between Monday and Friday, I anticipated a week of scholarly serenity. This was not to be. On Tuesday, 4 November at 6:20 P.M., a bolt of lightning from an unexpected thunderstorm struck my house, and, although the storm passed as quickly as it arrived, the fires caused by the lightning consumed much of the third floor—but fortunately left intact my study (with all my notes on Ezekiel).

When I broached the subject of the *merkabah* on the Friday of my presentation, the tone of jocularity that I had planned to adopt was tinged by the fact of my recent personal trauma. I was thereafter informed by a number of people who attended the seminar that, during the weekend following my presentation, they, too, experienced disasters of one sort or another. I wish to issue a disclaimer here that I refuse to bear any responsibility for those disasters or any future ones that might ensue. I have enough problems of my own.

Notes

Introduction

1   The idea of the "will to power" (*der Wille zur Macht*) is fully developed by Friedrich Nietzsche in *The Will to Power*, trans. Walter Kaufmann and R. J. Hollingdale, ed. Walter Kaufmann (New York: Random House, 1967). According to Nietzsche, "the victorious concept 'force,' by means of which our physicists have created God and the world, still needs to be completed: an inner will must be ascribed to it, which I designate as 'will to power,' i.e., as an insatiable desire to manifest power; or as the employment and exercise of power, as a creative drive" (pp. 332–33).

2   Unless otherwise indicated, biblical references are to *The Bible: Authorized King James Version, with Apocrypha*, ed. Robert Carroll and Stephen Prickett (Oxford: Oxford University Press, 1997). The Hebrew text is quoted from the *Biblia hebraica stuttgartensia*, ed. R. Kittel (Stuttgart: Deutsche Bibelstiftung, 1968). For ease of recognition, transliterations of the Hebrew are keyed to the table of foreign alphabets in *The Random House Dictionary of the English Language*, gen. ed. Laurence Urdang (New York: Random House, 1968). Vocalizations are rendered phonetically.

3   Although matters of authorship, dating, and text are extremely vexed, it is possible to provide

some information about the prophet and his prophecy here. According to the superscription that precedes the vision (1:1–3), Ezekiel, the son of Buzi, was of priestly lineage. Accompanied by other Judeans, he suffered deportation to Babylon after the surrender of Jehoiachin in 598/597 B.C.E. (Ezek. 1:1; cf. 2 Kings 24:12–16). Ezekiel received his calling on 31 (?) July 593 B.C.E. The site of his calling, the river Chebar, is mentioned in the Babylonian records as a canal that flows southeast from its fork above Babylon, through Nippur, and rejoins the Euphrates near Erech. The prophet continued to receive divine communications until at least 571 B.C.E. (Ezek. 29:17). His age at the time that his inaugural vision (as well as his subsequent visions) appeared to him is not recorded, nor are the circumstances of his death specified. It is conjectured that he (or perhaps a "school" of the prophet) composed the prophecy in 563 B.C.E. For a brief background summary, see the entry on the Book of Ezekiel in *The Oxford Companion to the Bible*, ed. Bruce M. Metzger and Michael D. Coogan (New York: Oxford University Press, 1993), 217–19.

4 I allude to Herbert J. Muller's important book *The Children of Frankenstein: A Primer on Modern Technology and Human Values* (Bloomington: Indiana University Press, 1970). Muller mounts a powerful critique of those who have written about the history of technology and its impact on modern culture. The "secular" text in question is, of course, Mary Wollstonecraft Shelley's classic novel *Frankenstein: or, the Modern Prometheus* (London, 1818).

5 The oracle against Gog extends to the thirty-ninth chapter as well. There, it is rearticulated in a form that reinforces the notion of prophecy belief and the emphasis on end time culminating in graphic portrayals of devastation. As I shall discuss, the site of interpretation represented by Gog provides a new way of understanding the complex nature of the vision that inaugurates the prophecy of Ezekiel.

6 Michael Fishbane, *The Garments of Torah: Essays in Biblical Hermeneutics* (Bloomington: Indiana University Press, 1989), 60–61.

7 Among other studies of the pictorial renderings, see Wilhelm Neuss, *Das Buch Ezechiel in Theologie und Kunst bis zum Ende des XII. Jahrhunderts* (Münster: Aschendorffsche, 1912).

8 For a full discussion of the elusive dimensions of the vision, see my *The Visionary Mode: Biblical Prophecy, Hermeneutics, and Cultural Change* (Ithaca, N.Y.: Cornell University Press, 1994), 15–41.

9 This is a point that I develop in ibid., 26–34. The technological dimensions already implicit in the vision have been well documented. See, e.g., Ralph W. Klein, *Ezekiel: The Prophet and His Message* (Columbia: University of South Carolina Press, 1988), 16–24. Addressing the "wheels within wheels," Klein observes that they "may represent an interpretation of the concentric circles formed by a heavy rim and the hub and other parts of ancient wheels. The eyes on the rims presumably connote the all-seeing character of the omnipresent deity, though their location on the rims is not readily explained. Perhaps the very notion of eyes represents a misunderstanding of nails that were driven into the rims of ancient chariot wheels to give them added strength" (pp. 23–24).

10 Lieb, *The Visionary Mode*, 32–34. I use *mysterium* here in its etymological sense as that which is transcendent as well as that which is both secret and mysterious. See the entry on *mysterium* in *A Latin Dictionary*, ed. Charlton T. Lewis and Charles Short (Oxford: Clarendon, 1879).

11 In keeping with prevailing usage, I adopt the letter *b* to indicate the *vet* in *merkabah*. It should be noted that the term is occasionally transliterated *merkavah*, a form that reflects its pronunciation.

12 The term *chariot,* however, *is* present in the rendering of the Septuagint for Ezek. 43:3, which has "the appearance of the *chariot [harmatos]* which I saw" (italics mine). References are to *The Septuagint Bible with Apocrypha: Greek and English* (1851; reprint, Grand Rapids, Mich.: Zondervan, 1972).

13 In the King James Version of both 1 Chron. 28:18 and Ecclus. 49:8, the phrase used is *the char-iot of the cherubims.* For both texts I silently amend *cherubims* to *cherubim,* the correct designation.

14 See André Parrot, *Babylon and the Old Testament,* trans. B. E. Hooke (New York: Philosophi-cal Library, 1958), 128–36; P. S. Landersdorfer, *Baal Tetramorphos und die Kerube des Ezechiel* (Paderborn: Druck/Ferdinand Schöningh, 1958); and Lorenz Dürr, *Ezechiels Vision von der Erscheinung Gotes (Ez. c. I u. 10) im Licht der vorderasiatischen Altertumskunde* (Münster: Aschendorffsche, 1917).

15 Hans Peter L'Orange, *Studies on the Iconography of Cosmic Kingship in the Ancient World* (Oslo: H. Aschehong, 1953), esp. 48–63; and James B. Pritchard, *The Ancient Near East in Pictures Relating to the Old Testament,* 2d ed. (Princeton, N.J.: Princeton University Press, 1969), pls. 644–51. The archaeological evidence for the emergence of the chariot in ancient Near Eastern culture has been ably surveyed in M. A. Littauer and J. H. Crouwel, *Wheeled Vehicles and Ridden Animals in the Ancient Near East* (Leiden: E. J. Brill, 1979). For a corre-sponding history of Europe and western Russia, see Stuart Piggott, *The Earliest Wheeled Transport from the Atlantic Coast to the Caspian Sea* (Ithaca, N.Y.: Cornell University Press, 1983).

16 Amber is "a pale yellow, sometimes reddish or brownish, fossil resin, translucent, brittle, and capable of gaining a negative electrical charge by friction" (*The Random House Dictionary of the English Language,* s.v.).

17 *The New Brown-Driver-Briggs-Gesenius Hebrew and English Lexicon,* trans. Edward Robin-son, ed. Francis Brown (Peabody, Mass.: Hendrickson, 1979), s.v. For additional speculation, see, among other studies, G. R. Driver, "Ezekiel's Inaugural Vision," *Vetus testamentum* 1 (1951): 60–63; William A. Irwin, "HASHMAL," *Vetus testamentum* 2 (1952): 169–70; and Wal-ther Zimmerli, *Ezekiel,* 2 vols. (Philadelphia: Fortress, 1979–83), 1:122–23.

18 For the Authorized Version, see *The Bible: Authorized King James Version, with Apocrypha.* For the New Revised Standard Version, see the *New Oxford Annotated Bible with Apocrypha,* ed. Bruce M. Metzger and Roland E. Murphy (New York: Oxford University Press, 1991), a revision of the Revised Standard Version. For the New International Version, see *The NIV Triglot Old Testament* (Grand Rapids, Mich.: Zondervan, 1981). For the Revised Standard Version, see *The New Oxford Annotated Bible with the Apocrypha,* ed. Herbert G. May and Bruce M. Metzger (New York: Oxford University Press, 1973). For the Septuagint, see *The Septuagint Bible with Apocrypha.* For the Vulgate, see *Biblia sacra iuxta vulgatam versionem* (Stuttgart: Deutsche Bibelgesellschaft, 1983).

19 References to talmudic material can be found in *The Babylonian Talmud* (hereafter *BT*), trans. Isidore Epstein, 8 vols. (London: Soncino, 1938), 8:59–91. See Lieb, *The Visionary Mode,* 85–103.

20 Despite the idea that the rabbis sought to suppress the Book of Ezekiel, the text was widely interpreted in rabbinic circles. See David J. Halperin, *The Merkabah in Rabbinic Literature* (New Haven, Conn.: American Oriental Society, 1980), 182–83, and *The Faces of the Chariot: Early Jewish Responses to Ezekiel's Vision* (Tübingen: J. C. B. Mohr, 1988).

21 This notion is explored in the classic work of Rudolf Otto, *Das Heilige: Über das Irrationale un*

der Idee des Göttlichen und Sein Verhältnis zum Rationalen (Breslau: Trewendt and Granier, 1917), translated by John W. Harvey as *The Idea of the Holy: An Inquiry into the Non-Rational Factor in the Idea of the Divine and Its Relation to the Rational* (London: Oxford University Press, 1928).

22 On the subject of how old one must be before he undertakes speculation on the *merkabah*, see, among other studies, Lieb, *The Visionary Mode*, 92–93, 252.

23 See the definitions under *electricity* in its various forms and combinations in *The Complete English-Hebrew/Hebrew-English Dictionary*, comp. Reuben Alcalay, 3 vols. (Tel-Aviv: Massadah, n.d.).

24 In Martin Heidegger, *Basic Writings: From "Being and Time" (1927) to "The Task of Thinking" (1964)*, ed. David Farrell Krell (New York: Harper and Row, 1977), 294.

25 Michael E. Zimmerman, *Heidegger's Confrontation with Modernity: Technology, Politics, Art* (Bloomington: Indiana University Press, 1990), 225. In this connection, Heidegger engaged in a controversial etymology of the term *physis*, to which he ascribed two meanings. Linking *physis* to *phyēin*, he defined *physis* as "the event of self-emergence, as when a bud bursts forth into a flower." Linking *physis* to *phainesthai*, moreover, he defined *physis* as "appearance, shining, showing forth." For Heidegger, then, *physis* implied both self-emergence and appearing or disclosure (p. 224). The revelatory was integral to his outlook as a philosopher.

26 Heidegger, "The Origin of the Work of Art," in *Basic Writings*, 173–78.

27 Zimmerman, *Heidegger's Confrontation*, xx, 96–97. As Zimmerman makes clear throughout his study, one must keep in mind Heidegger's own ambivalence toward technology as well as the development of his views concerning the function of technology in society. Zimmerman's study is valuable for examining Heidegger's philosophy of technology within the context of his engagement with the National Socialist "revolution."

28 Jacques Ellul, *The Technological Society*, trans. John Wilkinson (New York: Alfred A. Knopf, 1970), 403–5. Ellul makes a distinction between what he calls "technique" and "technology" as such. By *technique*, Ellul means not *technology* in its strictest sense but rather "the *totality of methods rationally arrived at and having absolute efficiency* . . . in *every* field of human activity" ("Note to the Reader"). For our purposes, technology and technique will be viewed synonymously since *technology* is one category of *technique*. Among Ellul's other books in the same vein, see *The Technological System*, trans. Joachim Neugroschel (New York: Continuum, 1980), and *The Technological Bluff*, trans. Geoffrey W. Bromiley (Grand Rapids, Mich.: W. B. Eerdman, 1990).

29 In the May and Metzger edition of the Revised Standard Version (1973), see the editorial note to Ezek. 14:1–11, which glosses Ezekiel's use of the term *gillulim* (literally, "dung balls"). This term is found thirty-nine times in Ezekiel, compared with nine times in the rest of the Hebrew Bible.

30 Among other studies, see George Widengren, *Literary and Psychological Aspects of the Hebrew Prophets* (Uppsala: A.-B. Lundequistska, 1948); Max Weber, *Ancient Judaism*, trans. Hans H. Gerth and Don Martindale (Glencoe, Ill.: Free Press, 1952), esp. 286–87; Kelvin van Nuys, "Evaluating the Pathological in Prophetic Experience (particularly Ezekiel)," *Journal of Bible and Religion* 21 (1953): 244–51.

31 Edwin C. Broome, "Ezekiel's Abnormal Personality," *Journal of Biblical Literature* 65 (1946): 277–92, 286. See Daniel Paul Schreber, *Denkwurdigkeiten eines Nervenkranken* (1903), trans. Ida Macalpine and Richard A. Hunter as *Memoirs of My Nervous Illness* (London: Wm. Dawson and Sons, 1955). Influenced by Zoroastrianism, Schreber depicted a bipartite deity

conceived as a "lower God," Ariman, and an "upper God," Ormuzd. For Schreber's detailed accounts of these figures, see esp. pp. 124–25, 228–29. Schreber speaks of the "mechanical" nature of the rays and compares the course they take to what he calls the circular or parabolic patterns of Roman chariots (p. 228). He also has a conception of what he calls the "forecourts of heaven" (p. 49), an idea that has some affinities with the so-called *hekhalot* (celestial halls) of *merkabah* mysticism, discussed below. For Freud's discussion of Schreber, see "The Case of Schreber," in *The Standard Edition of the Complete Psychological Works of Sigmund Freud,* trans. James Strachey, 24 vols. (London: Hogarth, 1958), 12:12–82. In a postscript to his study of Schreber, Freud acknowledges the wealth of mythological material implicit in Schreber's visions (pp. 80–82). See also Victor Tausk, "On the Origin of the 'Influencing Machine' in Schizophrenia," in *The Psychoanalytical Reader: An Anthology of Essential Papers with Critical Introductions,* ed. Robert Fliess (New York: International Universities Press, 1948).

32 See David J. Halperin, *Seeking Ezekiel: Text and Psychology* (University Park: Pennsylvania State University Press, 1993), esp. 12–38, 219–20.

33 As scholars have pointed out, the phrase *yordei merkabah* is most accurately translated "descenders to the chariot" (see Gershom Scholem, *Major Trends in Jewish Mysticism* [1941; New York: Schocken, 1954], 47, 85). Although the phrase "riders in the chariot" has been challenged, it is widely enough known to justify its metaphoric use here. For precedent, see texts ranging from the intertestamental Similitudes of Enoch (1 Enoch 70:1–4) to the Hekhalot Rabbati (chap. 21) in the fifth–sixth centuries. For a contemporary, fictionalized version of the idea, see Patrick White, *Riders in the Chariot* (New York: Viking, 1961).

34 See Lieb, *The Visionary Mode,* 42–84. The foundational works on the subject include Gershom Scholem, *Jewish Gnosticism, Merkabah Mysticism, and Talmudic Tradition* (New York: Jewish Theological Seminary of America, 1960), and *Major Trends;* Ihamar Gruenwald, *Apocalyptic and Merkavah Mysticism* (Leiden: E. J. Brill, 1980), and *From Apocalypticism to Gnosticism: Studies in Apocalypticism, Merkavah Mysticism, and Gnosticism* (Frankfurt: P. Lang, 1988); Ira Chernus, *Mysticism in Rabbinic Judaism* (Berlin: Walter de Gruyter, 1982); Halperin, *The Merkabah in Rabbinic Literature,* and *Faces of the Chariot;* Moshe Idel, *Kabbalah: New Perspectives* (New Haven, Conn.: Yale University Press, 1988); and Elliot R. Wolfson, *Through a Speculum That Shines: Vision and Imagination in Medieval Jewish Mysticism* (Princeton, N.J.: Princeton University Press, 1994).

35 See Gruenwald, *Apocalyptic and Merkavah Mysticism,* 119–21. Compare Ezekiel's own experience of abduction: "And it came to pass in the sixth year, in the sixth *month,* in the fifth *day* of the month, as I sat in mine house, and the elders of Judah sat before me, that the hand of the Lord God fell there upon me. Then I beheld, and lo, a likeness as the appearance of fire: from the appearance of his loins even downward, fire; and from his loins even upward, as the appearance of brightness, as the colour of amber. And he put forth the form of an hand, and took me by a lock of mine head; and the spirit lifted me up between the earth and the heaven" (Ezek. 8:1–3).

36 Lieb, *The Visionary Mode,* 62–63 (quotation, 76–78).

37 I use the term *avatar* here as that which embodies the very essence of the idea its presence conveys. In this case, the avatar is the incarnation of the idea of technology. Later in the study, I appropriate the term in the context of its cyberspatial implications.

38 The term *technopoetics* is derived from Strother B. Purdy's "Technopoetics: Seeing What Literature Has to Do with the Machine," *Critical Inquiry* 11 (1984): 130–40. The phrase *machina ex deo* was appropriated by Lynn Townsend White Jr. to serve as the title of his book

on technology and society (*Machina ex Deo: Essays in the Dynamism of Western Culture* [Cambridge, Mass.: MIT Press, 1968]).

I Technology of the Ineffable

1 All references to Milton's poetry are from John T. Shawcross, ed., *The Complete Poetry of John Milton,* 2d rev. ed. (Garden City, N.Y.: Doubleday & Co., 1971). References to Milton's prose are to *The Complete Prose Works of John Milton,* 8 vols. in 10, gen. ed. Don M. Wolfe et al. (New Haven, Conn.: Yale University Press, 1953–82), hereafter designated *YP.* Corresponding references to the original Latin (and on occasion to the English translations) are to *The Works of John Milton,* 18 vols. in 21, ed. Frank Allen Patterson et al. (New York: Columbia University Press, 1931–38), hereafter designated *CM.*

2 As is well known, the Son's act is numerically, as well as thematically, central. As Shawcross (*Complete Poetry,* 383n) points out, in the 1667 edition of *Paradise Lost,* the word *ascended* (6.762) occurs at the precise numerical midpoint of the epic "since 5275 lines precede it and follow it." For an analysis of structural and numerological placements, see Gunnar Qvarnstrom, *The Enchanted Palace: Some Structural Aspects of "Paradise Lost"* (Stockholm: Almquist & Wiksell, 1967), 55–85; Claes Schaar, *The Full Voic'd Quire Below: Vertical Context Systems in "Paradise Lost"* (Lund: C. W. K. Gleerup, 1982), 310–32; and John T. Shawcross, *With Mortal Voice: The Creation of "Paradise Lost"* (Lexington: University of Kentucky Press, 1982), 50–51, 73–74. See also Gunnar Qvarnstrom, *Dikten och den Nya Vetenskapen: Det astronautiska motivet* (Lund: C. W. K. Gleerup, 1961). In Milton studies, the mystical background of the *merkabah* is treated by J. H. Adamson, "The War in Heaven: Milton's Version of the *Merkabah," Journal of English and Germanic Philology* 57 (1958): 690–703 (reprinted in *Bright Essence: Studies in Milton's Theology,* ed. William B. Hunter et al. [Salt Lake City: University of Utah Press, 1971], 103–14); with a rejoinder by Jason P. Rosenblatt, "Structural Unity and Temporal Concordance: The War in Heaven in *Paradise Lost," PMLA* 87 (1972): 31–41. For discussions of Milton and Ezekiel in general, see E. C. Baldwin, "Milton and Ezekiel," *Modern Language Notes* 33 (1918): 211–15; and Mary Jane Doherty, "Ezekiel's Voice: Milton's Prophetic Exile and the *Merkavah* in *Lycidas," Milton Quarterly* 23 (1989): 89–121.

3 For comprehensive studies, see, among others, my *Poetics of the Holy: A Reading of "Paradise Lost"* (Chapel Hill, N.C.: University of North Carolina Press, 1981), 246–312; Stella Revard, *The War in Heaven in "Paradise Lost"* (Ithaca, N.Y.: Cornell University Press, 1980), 123–28, 256–60; Boyd Berry, *Process of Speech: Puritan Religious Writings and "Paradise Lost"* (Baltimore: Johns Hopkins University Press, 1976), passim; Roland M. Frye, *Milton's Imagery and the Visual Arts: Iconographic Tradition and the Epic Poems* (Princeton, N.J.: Princeton University Press, 1978), 156–57; John G. Demaray, *Milton's Theatrical Epic: The Invention and Design of "Paradise Lost"* (Cambridge, Mass.: Harvard University Press, 1980), 97–100; Georgia Christopher, *Milton and the Science of the Saints* (Princeton, N.J.: Princeton University Press, 1982), passim; and Robert Thomas Fallon, *Captain or Colonel: The Soldier in Milton's Life and Art* (Columbia: University of Missouri Press, 1984), passim. Laura Lunger Knoppers has examined the Chariot of Paternal Deitie from the illuminating perspective of Royalist triumphs based on Roman models (*Historicizing Milton: Spectacle, Power, and Poetry in Restoration England* [Athens: University of Georgia Press, 1994], 96–122).

4 For an interesting discussion of Milton's figure of Zeal as a "metonymy" for Milton himself, see John Guillory, *Poetic Authority: Spenser, Milton, and Literary History* (New York: Colum-

bia University Press, 1983), 101–3. According to Guillory, "Zeal is the most authoritative trope in the *Apology* and the final vindication of the private name" for Milton as polemicist and as poet. Milton projects himself onto this trope of unleashed power as "a phase preparatory to the self-aggrandizement of the ego, like the shrinking of a star before its explosion as a nova" (pp. 101–2).

5   Among many other commentaries on the subject, see Austin Farrer, *A Rebirth of Images: The Making of St. John's Apocalypse* (Boston: Beacon, 1963), 130–31. In Revelation 4, Saint John the Divine beholds an enthroned figure in heaven. Surrounding the throne are twenty-four elders, and in the midst of the throne are "four beasts full of eyes before and behind." As in Ezekiel's vision, the first beast is like a lion, the second like a calf, the third like a man, and the fourth like an eagle. Out of the throne "proceeded lightnings and thunderings and voices." Ezekiel 1 (cf. also chaps. 2, 3, 8–11, 43) and Revelation 4, in turn, were traditionally viewed in conjunction with Isaiah 6 and Daniel 7.

6   Henry Bullinger, *A Hundred Sermons upon the Apocalypse of Iesu Christ, reveiled by the angell of the Lord But seene or received and written by the holy Apostle and Evangelist St Iohn*, trans. John Daus, 2 vols. (London, 1573), 1:68–69. Although Bullinger's principal focus here is Ezekiel 10, he has the first chapter in mind as well. The remainder of Bullinger's discourse is of interest to the parodic role assumed by Satan in *Paradise Lost:* "There is in poets much mention of the chariots of the Goddes, taken haply by the first writers out of the holy Scriptures. For Sathan, the Ape of God, goeth about alwayes to diffame the worde of trueth. But we omitting the triflinges of poetes, wyll consider the sober description of thys caryage of God, or rather of Gods throne."

7   Katharine Firth, *The Apocalyptic Tradition in Reformation Britain, 1530–1645* (Oxford: Oxford University Press, 1979), 6 n. 10, 269–73.

8   The description of this vehicle is entirely in keeping with one who is earlier described as exalted "High on a Throne of Royal State, which far / Outshon the wealth of *Ormus* and of *Ind,* / Or where the gorgeous East with richest hand / Showrs on her Kings *Barbaric* Pearl and Gold" (2.1–4). Such, of course, is the posture of the enthroned Satan in hell, defeated in his efforts to supplant God in the War in Heaven, but nonetheless determined in his hopeless insatiability to pursue "Vain War with Heav'n" yet once more (2.5–9). Both before and after his fall, Satan remains a figure of blasphemy.

9   In the *Reason of Church-Government,* Milton portrays the blasphemous equipage of the prelates as a kind of tottering chariot, unworthy to bear the ark of the reformed church in the realization of its destiny: "But no profane insolence can paralell that which our Prelates dare avouch, to drive outragiously, and shatter the holy arke of the Church, not born upon their shoulders with pains and labour in the word, but drawn with rude oxen their officials, and their owne brute inventions." As putative stewards of the church, the prelates can do nothing but profane it. "Let them make shewes of reforming while they will," Milton observes, "so long as the Church is mounted upon the Prelaticall Cart, and not as it ought betweene the hands of the Ministers, it will but shake and totter." What results in the attempt of the prelates to subject the church to their profane and blasphemous ways is the invention of an "unlawfull waggonry" (*YP,* 1:754–55). Milton compares the futile efforts of the prelates with the fate of Uzza (Num. 3:31; cf. 1 Sam. 6:14). In *The Readie and Easie Way to Establish a Free Commonwealth,* Milton associates the backsliders in the cause of Reformation with the "tigers of Bacchus" or Dionysus, yoked (along with the lynx, the panther, and the lion) to his chariot, which they draw in procession (*YP,* 7:452–53).

10 Michael Lieb, " 'Hate in Heav'n': Milton and the *Odium Dei*," *ELH* 53 (1986): 519–39.

11 For treatments of the *odium Dei* in Reformation theology, see John Calvin, *Commentaries on the Prophet Malachi, Commentaries on the Twelve Minor Prophets*, trans. John Owen, 5 vols. (Edinburgh: Calvin Translation Society, 1849), 5:467–68. See also John Calvin, *Commentaries on the Epistle of Paul the Apostle to the Romans*, trans. John Owen (Edinburgh: Calvin Translation Society, 1849), 353–62, and *Commentary on the Book of Psalms*, trans. Rev. James Anderson, 5 vols. (Edinburgh: T. Constable, 1849), 5:221–23. Also of interest are William Perkins, *Revelation: Preached in Cambridge Anno Dom. 1596* (London, 1604), 152–53; and Thomas Wilson, *A Commentary on the Most Divine Epistle of S. Paul to the Romans*, 2d ed. (London, 1627), 355. In *The Marrow of Theology* (1628), trans. John D. Eusden (Boston: Pilgrim, 1968), William Ames likewise acknowledges the significance of divine hatred as theological enterprise. For Ames, hatred represents the most extreme form that the wrath of God is liable to assume. In this respect, God is said to "hate" the reprobate (Rom. 9:13). Those who follow God manifest a "zeal of the will" that reflects a comparable hatred of evil, just as it reflects a comparable love of good (pp. 87, 156, 223, 315). Concerning the psychologizing of hatred as an attribute of deity, Edward Reynolds takes full account of how the *odium Dei* manifests itself as psychological phenomenon (*Treatise of the Passions and Faculties of the Soule of Man* [1640], ed. Margaret Lee Wiley [Gainesville, Fla.: Scholars' Facsimiles and Reprints, 1971], 111–30). From the perspective of the early church, the idea of the *odium Dei* is fully developed by Lactantius (ca. 250–ca. 330), who argues on behalf not only of a God of wrath but also of a God of hate (see *De ira dei* [On the anger of God], in *The Works of Lactantius*, trans. William Fletcher, 2 vols. [Edinburgh: T. & T. Clark, 1871], 1:5–8, 32). For the influence of Lactantius on Milton, see Kathleen E. Hartwell, *Lactantius and Milton* (Cambridge, Mass.: Harvard University Press, 1929).

12 Although recent discussion has raised questions concerning the authorship of this work, I here assume its place in Milton's canon. At present, it is unclear whether the matter of authorship will ever be settled conclusively. Until it is, I shall continue to believe that the work is Milton's. The debate was set in motion by William B. Hunter and continued by several scholars, including John T. Shawcross, Barbara K. Lewalski, Christopher Hill, and Maurice Kelley. See Hunter's *Visitation Unimplor'd: Milton and the Authorship of "De Doctrina Christiana"* (Pittsburgh: Duquesne University Press, 1998).

13 On the invention of firearms, see the explanatory notes in John Milton, *The Poems of John Milton*, ed. John Carey and Alastair Fowler (London: Longman, 1968), esp. 751. Following earlier editors, Carey and Fowler point out that the invention of gunpowder and the use of firearms is mentioned in Lodovico Ariosto (*Orlando Furioso*, 9.28, 91), Edmund Spenser [*Faerie Queene*, 1.7.13), and Samuel Daniel (*Civil Wars*, 6.26). Erasmo di Valvasone, like Spenser, Milton, and others, associates the invention of gunpowder and firearms, including cannons, with the machinations of the devils (*L'Angeleida*, 2.20). For the scatological implications of these inventions in *Paradise Lost*, see my *The Dialectics of Creation: Patterns of Birth and Regeneration in "Paradise Lost"* (Amherst: University of Massachusetts Press, 1970), esp. 117–24.

14 Such is Milton's epic rendering of a concept that he had formulated years earlier in his poems on the Gunpowder Plot and on the inventor of gunpowder. So, in the poem "On the Gunpowder Plot" ("In proditionem Bombardicam"), Milton ironically conceives the conspiracy of 5 November 1605 to blow up Parliament as an act that would send the king and the British lords "to the courts of high heaven / in a sulphurous chariot with flaming wheels

[*Sulphureo curro flammivolisque rotis*]," an event, he quips, reminiscent of Elijah's own experience of being borne aloft by a fiery chariot in a whirlwind (lines 4–6). A corresponding notion is evident in the poem "On the Inventor of Gunpowder" ("In inventorem Bombardae"). There, Milton playfully compares the inventor of gunpowder with Prometheus, the son of Iapetus. Whereas Prometheus is praised for having brought down celestial fire from the chariot or axle of the sun, the inventor of gunpowder might be said to have stolen from Jove his "ghastly weapons and threeforked thunderbolt" (lines 2–4). If Milton's disposition toward king and Parliament changed between the period of the Gunpowder poems and the period of his epic, the inclination to conceptualize the devilish technology surrounding the invention of gunpowder and its use in cannon warfare through the conceit of the chariot remains intact. Judging by the earlier poems, one even discovers a certain admiration for the technological wizardry implicit in the ability to invent so potent a substance as gunpowder, that combustible substance of utmost force.

15   The history of the invention of gunpowder and its use in weaponry is treated in a wide variety of sources. Introduced into Europe in the fourteenth century, it was quickly adapted to military use. After 1325, cannons of various sorts existed throughout western Europe and were used by Edward III at Crécy (1346). The science of ballistics assumed scientific precision in sixteenth-century Italy. In the seventeenth century, that science was perfected even further ("Gunnery," *The New Encyclopaedia Britannica*, 30 vols., 15th ed. [Chicago: Encyclopaedia Britannica, 1982], 8:488–98). Among other works, see A. V. B. Norman and Don Pottinger, *English Weapons and Warfare, 449–1660* (New York: Dorset, 1979); and Richard A. Gabriel, *The Culture of War: Invention and Early Development* (New York: Greenwood, 1990). In *History and Warfare in the Renaissance Epic* (Chicago: University of Chicago Press, 1994), Michael Murrin has an illuminating discussion of the use of gunpowder and related weaponry in a wide range of epics, including *Paradise Lost*.

16   Arnold Stein, *Answerable Style: Essays on "Paradise Lost"* (Seattle: University of Washington Press, 1953), 25.

17   In *De doctrina christiana*, Milton defines *zeal* as "an eager desire to sanctify the name of God, together with a feeling of indignation against things which tend to the violation or contempt of religion" (*YP*, 6:697; *CM*, 17:153). As biblical exemplars of zeal, Milton offers such figures as Lot, Moses, Phinehas, Elijah, Jeremiah, Stephen, Paul, and, of course, Christ himself. In *Paradise Lost*, the spirit of zeal is dramatically manifested in the servant of God, Abdiel, "then whom none with more zeal ador'd / The Deitie" (5.805–6). Opposed to zeal is the act of "talk[ing] about God in an impious and shameful way, which is commonly called blasphemy, from the Greek *blasphēmia*," and from the Hebrew *gedupha* and *kelalah* (*YP*, 6:698; *CM*, 17:154–55).

18   See the discussions in William B. Hunter Jr., "Milton on the Exaltation of the Son: The War in Heaven in *Paradise Lost*," *ELH* 36 (1969): 215–31 (reprinted in *Bright Essence*, ed. Hunter et al., 115–30); Joseph H. Summers, *The Muse's Method: An Introduction to "Paradise Lost"* (Cambridge, Mass.: Harvard University Press, 1962), 135; and Michael Fixler, *Milton and the Kingdom of God* (London: Faber & Faber, 1965), 227–28.

19   The biographical summary that follows is derived from Clive Hart and Helen Hart, "Melchior Bauer's Cherub Wagon," *Aeronautical Journal of the Royal Aeronautical Society* Paper no. 758 (January 1980), 22–27, reprinted in Clive Hart, *The Prehistory of Flight* (Berkeley and Los Angeles: University of California Press, 1985), 164–76. What is known about Bauer is incorporated in a novel by Peter Supf, *Der Himmelswagen: Das Schicksal des*

Melchior Bauer (Stuttgart: Drei Brunnen, 1953). See also Clive Hart, *The Dream of Flight: Aeronautics from Classical Times to the Renaissance* (London: Faber & Faber, 1972). Important as well is Marjorie Hope Nicolson, *Voyages to the Moon* (New York: Macmillan, 1960). The literature that constitutes the prehistory of flight is immense. Both Nicolson and Hart explore the writings and accounts of a long line of enthusiasts convinced of their ability to construct flying machines of various sorts. As Hart's directory of flying machines (*Prehistory of Flight,* 195–208) in western Europe (850 B.C.–A.D. 1783) makes clear, those intent on constructing a means of aeronautical transport despite insuperable odds were both inventive and determined. Thus, we find that in 850 B.C. King Bladud of Troja Nova (London) attached wings to his arms in his attempt to fly but fell onto the Temple of Apollo and was killed. In the fourth century B.C., Archytas of Tarentum constructed the model of a small flying object as a prototype for flight. In A.D. 875, Abu'l-Kasim of Andalusia, Spain, constructed feather-covered wings that transported him a considerable distance before he alighted at his staring point. These were followed by such figures as Eilmer (980–1066), a monk of Malmesbury, England, who attached wings to his hands and feet and flew more than a "stadium" (606.75 feet) from the top of a tower before falling and breaking his legs. The list continues through the eighteenth century and beyond. For our purposes, the figure of Eilmer the Flying Monk is of more than passing interest because Milton addresses him in *The History of Britain* (*YP,* 5:394). For an account of Milton's depiction in the context of his antimonasticism, see my "Milton among the Monks," in *Milton and the Middle Ages,* ed. John Mulryan (Lewisburg, Pa.: Bucknell University Press, 1982), 103–14.

20 References to the original are from the facsimile *Die Flugzeughandschrift des Melchior Bauer von 1765,* introduced and transcribed by Werner Querfeld (Gütersloh: Greifenverlag, 1982). The Harts date the composition of the manuscript to 1764. Translations are drawn from Hart and Hart, "Melchior Bauer's Cherub Wagon." Writings concerning the possibilities and theories of flight abound from the earliest times. The period extending from the early modern era through the eighteenth century is a case in point. In 1505, Leonardo da Vinci summarized his ideas concerning flight in a notebook known as the *Codice sul volo degli uccelli.* There, he proposed not only a theory of bird flight but also the construction of what Hart calls *ornithopters* as devices for conquering the air. In the following centuries, such works as Friedrich Hermann Flayder's *De arte volandi* (1627), John Wilkins's *The Discovery of a World in the Moon* (1638) and *Mathematicall Magick* (1648), Johann Caramuel Lobkovitz's *Mathesis biceps, vetus et nova* (1670), Francesco Lana Terzi's *Prodromo overo saggio di alcune inventione nuove premesso all'arte maestra* (1670), Emanuel Swedenborg's *Suggestions for a Flying Machine* (1714), and Carl Friedrich Meerwein's *Der Mensch: Sollte der nicht auch zum Flügen gebohren seyn?* (1783) provide ample evidence of the extent to which the subject of the invention of flying machines proved fascinating to theorists and practitioners alike. The subject of flight as the product of one's belief in one's own ability to transcend the physical proved no less fascinating to Samuel Johnson, who devoted a chapter to it, "A Dissertation on the Art of Flying," in *The History of Rasselas, Prince of Abissinia* (1759).

21 *Die Flugzeughandschrift,* 12–18 (facsimile), 1v–2r (manuscript); Hart and Hart, "Melchior Bauer's Cherub Wagon," 22–23.

22 *Die Flugzeughandschrift,* 20–24 (facsimile), 2v–3r (manuscript); Hart and Hart, "Melchior Bauer's Cherub Wagon," 23–24.

23 Hart and Hart, "Melchior Bauer's Cherub Wagon," 23.

24 Ibid., 26.

25 *Die Flugzeughandschrift*, 24 (facsimile); 3r (manuscript); Hart and Hart, "Melchior Bauer's Cherub Wagon," 23.

26 *Die Flugzeughandschrift*, 14 (facsimile); 1v (manuscript); Hart and Hart, "Melchior Bauer's Cherub Wagon," 22.

27 Hart and Hart, "Melchior Bauer's Cherub Wagon," 27.

28 Desiderius Erasmus, *The Praise of Folly*, in *Prose and Poetry of the Continental Renaissance*, ed. Harold Hooper Blanchard (New York: Longmans, Green, 1949), 572.

29 The impulse to technologize the vision finds its counterpart in the impulse to cosmicize it. From this perspective, one might consider the tractate *Ezechiel's Vision Explained* (London, 1778), attributed to one Charles Walmesley, Bishop. Responding to the vision, the author views it as a divine manifestation of the "worldly system," composed of the natural world with its elements and the divine world with its planetary bodies and stars. The world as a whole is embodied in the Chariot of God. Thus, the depiction of the wheels within wheels "represents the Globe of our Earth, encompassed with the three spheres of air, water, and fire." All these "bear a round figure like a wheel." Adopting the new cosmology, the author observes that the four spheres that compose each wheel turn all around a common center, the center of the earth, "by that motion which is called the diurnal motion, and is well known to be performed in twenty-four hours." This return makes the course of day and night. The spirit that moves the wheels, in turn, infuses those forces that produce and regulate earth's motion around the sun. As part of the system that the author establishes, God is represented as "residing above the highest part of the sky; from whence he extends his dominion over all the spheres beneath him." From there, God "directs the System of nature, and governs our world." This act of systematizing and cosmicizing Ezekiel's vision is the author's method of seeing in the vision "the general plan of Divine Administration over the Universe." In this way, the vision provides "a noble view of the Dispensations of Divine Providence towards mankind" (pp. 24–51).

30 T. S. Ashton, *The Industrial Revolution, 1760–1830* (London: Oxford University Press, 1948), 61.

31 Among other studies, see C. Hamilton Ellis, *British Railway History: An Outline from the Accession of William IV to the Nationalisation of Railways, 1830–1876*, 2 vols. (London: George Allen & Unwin, 1954).

32 Leo Marx, *The Machine in the Garden: Technology and the Pastoral Ideal in America* (London: Oxford University Press, 1964), 27, 190–95.

33 Ibid., 198–203. As Marx makes clear, this view was not universally endorsed. In "Signs of the Times" (1829), Thomas Carlyle, e.g., speaks with some animus of the times in which he lives as the Mechanical Age, a period of "industrialism" in which the impulse to technologize can lead to that which is destructive and alienating. Left unchecked, the mechanistic is capable of stifling the "forces and energies of man." What results is the possibility of psychological deterioration. Marx observes that this crisis becomes most obvious in the figure of Professor Diogenes Teufelsdröckh in *Sartor Resartus* (1833–34). Admirably living up to what is implied by his name (Diogenes as "born of God," as well as the Cynic, and Teufelsdröckh as "devil's dung"), Professor Teufelsdröckh undergoes a depression tantamount to total despair (Marx, *The Machine in the Garden*, 169–79). The language of this despair is formulated in technological terms: "To me," says Professor Teufelsdröckh, "the Universe was all void of Life, of Purpose, of Volition, even of Hostility: it was one huge, dead, immeasurable Steam-engine, rolling on, in its dead indifference, to grind me limb from limb" (quoted in ibid., 179).

34   Ashton, *The Industrial Revolution*, 61.

35   John Francis, *A History of the English Railway, Its Social Relations & Revelations, 1820–1845* (1851; reprint, New York: Augustus M. Kelley, 1968), 140–63. See also Muller, *The Children of Frankenstein*, 45–46. According to Muller, the importance of the locomotive to the Industrial Revolution prompts one to designate this period "the Railway Age" (p. 51).

36   *An Illustration of Ezekiel's Vision of the "Chariot"* (London: J. Hatchard, 1843), 14–17, 23.

37   Ibid., 16–17.

38   G. Wilson Knight, *Chariot of Wrath: The Message of John Milton to Democracy at War* (London: Faber & Faber, 1942), 158–59. For the idea that *Paradise Lost* is to be looked on as a source of the spirit of technology in modern times, see also Humphrey Jennings, *Pandae-monium, 1660–1886: The Coming of the Machine as Seen by Contemporary Observers*, ed. Mary-Lou Jennings and Charles Madge (New York: Macmillan, 1985).

39   Specific information on the tanks is from *Jane's Armour and Artillery, 1981–82* (London: Jane's, 1981–82), 23–25, 797. See also Gunther E. Rothenberg, *The Anatomy of the Israeli Army: The Israeli Defence Force, 1948–78* (New York: Hippocrene, 1979).

40   All references are to Ze'ev Klein (21 Hapninim St., Neve Monoson, Israel) to Michael Lieb, 15 July 1985. Biblical references are cited according to the rendering Klein provides in his letter.

2   The Psychopathology of the Bizarre

1   The term "UFO" itself is worthy of consideration. In *Angels and Aliens: UFOs and the Mythic Imagination* (Reading, Mass.: Addison-Wesley, 1991), 12, Keith Thompson maintains that, because the adoption of the term UFO is so deliberately "antimythic" in its attempt to imbue the phenomenon with a "scientific" objectivity and resonance, it has assumed its own mythic status. According to Thompson, the term UFO as an acronym was invented in the 1940s by air force investigators who, finding the phrase *flying saucer* too "poetic" for its imaginal palate, decided to adopt UFO because the colorless objectivity of the term suited the "pervasive rationalism and scientism of the post-war era." Like the acronyms CIA (Central Intelligence Agency) and *laser* (light amplification by stimulated emission of radiation), UFO achieved popular associations far in excess of the bare denotation of three initials.

2   Lending credence to that enterprise is the J. Allen Hynek Center for UFO Studies. Based in Chicago, this organization carries forth the work of its founder, J. Allen Hynek (1910–86), one of the foremost authorities on the subject. The center publishes the *International UFO Reporter* and the *Journal of UFO Studies*.

3   For a discussion of this aspect, see in particular David M. Jacobs, "UFOs and the Search for Scientific Legitimacy," in *The Occult in America: New Historical Perspectives*, ed. Howard Kerr and Charles L. Crow (Urbana: University of Illinois Press, 1983), 217–30. See also J. Allen Hynek, *The UFO Experience: A Scientific Inquiry* (Chicago: Henry Regnery, 1972); and Timothy Good, *Above Top Secret: The Worldwide UFO Cover-Up* (New York: William Morrow, 1988).

4   Leon Festinger, Henry W. Riecken, and Stanley Schachter, *When Prophecy Fails* (Minneapolis: University of Minnesota Press, 1956). For corresponding studies of the various dimensions (sociological, psychological, and cultural) that UFOs involve, see, among others, Thompson, *Angels and Aliens;* John G. Fuller, *The Interrupted Journey* (New York: Dial, 1966); Edward U. Condon, *Scientific Study of Unidentified Flying Objects* (New York: Bantam, 1969); J. Allen

Hynek and Jacques Vallée, *The Edge of Reality: A Progress Report on Unidentified Flying Objects* (Chicago: Henry Regnery, 1975); Budd Hopkins, *Intruders: The Incredible Visitation at Copley Woods* (New York: Random House, 1987); David M. Jacobs, *The UFO Controversy in America* (Bloomington: Indiana University Press, 1975), and *Secret Life: Firsthand Accounts of UFO Abductions* (New York: Simon & Schuster, 1992); and Carl Raschke, "UFOs: Ultraterrestrial Agents of Cultural Deconstruction," *Archaeus: Cyberbiological Studies of the Imaginal Component in the UFO Contact Experience* 5 (1989): 21–32. Thomas Eddie Bullard presents a wide-ranging and illuminating study of the UFO phenomenon in "Mysteries in the Eye of the Beholder: UFOs and Their Correlates as a Folkloric Theme Past and Present" (Ph.D. diss., Indiana University, 1982). From the perspective of analytic psychology, see also Carl Gustav Jung, *Flying Saucers: A Modern Myth of Things Seen in the Sky* (Princeton, N.J.: Princeton University Press, 1978), which originally appeared as *Ein Moderner Mythus: Von Dingen, die am Himmel gesehen werden* (Zurich: Rascher, 1958). In this connection, see my "Ezekiel's Inaugural Vision as Jungian Archetype," *Thought* 54 (1989): 116–29.

5   Festinger, Riecken, and Schachter, *When Prophecy Fails*, 48.

6   Founded by the prophet Montanus, Montanism (also called the Cataphrygian Heresy) was a heretical movement that arose in the Christian church in Asia Minor in the second century. Shabbetaianism has its source in Shabbetai Tzevi, a false Messiah from Smyrna, Ottoman Turkey, who developed a large popular following in Europe and the Middle East in the seventeenth century.

7   One of the foremost authorities on UFOs, Jacques Vallée is by profession a computer scientist and astrophysicist whose works in the area of UFOs include *Anatomy of a Phenomenon: UFOs in Space* (New York: Random House, 1965), *Passport to Magonia: From Folklore to Flying Saucers* (Chicago: Henry Regnery, 1969), *Dimensions: A Casebook of Alien Contact* (New York: Random House, 1988), and *Confrontation: A Scientist's Search for Alien Contact* (New York: Ballantine, 1990). Vallée has also published widely in the area of computer science.

8   At the outset of *Dimensions*, e.g., Vallée records a trip to the Sainte Chapelle in the Palace of Justice in Paris. "One of the stained-glass panels of the Sainte Chapelle," he says, "shows the abduction of the prophet Ezekiel by an object that came as a whirlwind. He saw wheels within wheels and four strange creatures. He was carried away to a remote mountaintop, where he found himself in a state of wonder and confusion." This event is crucial for Vallée and the contactees or abductees he has interviewed in his many years of UFO investigations. "I have spoken to numerous witnesses," Vallée comments, "who told me they had been caught in a whirlwind, had seen strange creatures, and had been left wondering and confused." Although not destined to be memorialized in the stained glass windows of chapels, these people for Vallée are the children of the prophet Ezekiel (pp. 5–6). In the same vein, see Vallée's *Passport to Magonia*, 10–11.

9   Vallée cites from Agobard's *Liber de Grandine et Tonitruis* (chap. 11). See *Agobardi Lug-duinensis opera omnia*, ed. L. van Acker (Turnholti: Typograph Brepols, 1981).

10 Vallée, *Dimensions*, 13, and *Passport to Magonia*, 1–15. Developing the historical contexts further, Keith Thompson offers additional background. He notes, e.g., that fiery aerial phenomena bearing the shapes of military shields appeared during the battle between the Saxons and the Franks at Sigisburg in 776. Likewise, in 1118, the emperor Constantine witnessed a fiery cross suspended in the sky. In the nineteenth century, Thompson observes, Joseph Smith, founder of the Mormon Church, ventured the following report: "I saw a pillar of light exactly over my head, above the brightness of the sun, which descended gradually until it fell

upon me. . . . When the light rested on me I saw two personages whose brightness and glory defy all descriptions, standing above me in the air. One of them spoke unto me." Despite the absence of a "machine" in this report, Thompson is prompted to ask whether Smith's sighting belongs in the same category as modern UFO sightings. Like Vallée, Thompson concludes by offering for consideration a detailed rendering of the inaugural vision of Ezekiel. He does so, however, not to suggest a one-to-one relation between the biblical event and later visionary manifestations but to establish a kind of cultural tie in what becomes the technopoetic transformation of visionary material into the substratum of the UFO experience operative today (*Angels and Aliens*, 70–71). For additional historical perspectives, see also Jacques Bergier, *Extra-Terrestrial Visitations from Prehistoric Times to the Present* (Chicago: Henry Regnery, 1973); and David C. Knight, *UFOs: A Pictorial History from Antiquity to the Present* (New York: McGraw-Hill, 1979).

11   *The Best Little Whorehouse in Texas* (Universal City, Calif.: MCA, 1982), produced by Thomas L. Miller, Edward K. Milkis, and Robert L. Boyett, directed by Collin Higgins, screenplay by Larry L. King, Peter Masterson, and Colin Higgins, based on the musical stage play of the same name by Larry L. King and Peter Masterson. For an account of the plot and film production, see Roberta Le Feuvre, "*The Best Little Whorehouse in Texas*," in *Magill's Cinema Annual* (Englewood Cliffs, N.J.: Salem, 1983), 72–75. See also Larry L. King, *The Whorehouse Papers* (New York: Viking, 1982).

12   Frank Rich, review of *Close Encounters of the Third Kind, Time,* 7 November 1977, cited in "Steven Spielberg," in *Current Biography Yearbook, 1978,* ed. Charles Moritz et al. (New York: H. W. Wilson, 1978), 403. Produced in 1977 by Columbia Pictures, Culver City, Calif., *Close Encounters of the Third Kind* was written and directed by Steven Spielberg.

13   References are to Steven Spielberg's novel *Close Encounters of the Third Kind* (New York: Delacorte, 1977), 250–51. Observations are drawn both from the film and from the novel.

14   Ibid., 220–21.

15   Ibid., 236.

16   J. Allen Hynek, epilogue to *Close Encounters of the Third Kind,* 253–55. (In the epilogue, Hynek's name appears mistakenly as "T." Allen Hynek. I have silently corrected this error in my text.) For the distinctions among the various encounters, see Hynek, *The UFO Experience,* 31–34 and passim.

17   John E. Mack, *Abduction: Human Encounters with Aliens* (New York: Charles Scribner's Sons, 1994). For corresponding relations of such abductions, see Jacobs, *Secret Life.* The journal *Psychological Inquiry* devotes an entire issue (vol. 7, no. 2 [1996]) to the abduction experience. I am indebted to Michael E. Zimmerman of Tulane University for these references. Professor Zimmerman kindly shared with me his unpublished essay "Encountering Alien Otherness," which explores the philosophical and cultural ramifications of the abduction experience in depth. See also his "The 'Alien Abduction' Phenomenon: Forbidden Knowledge of Hidden Events," *Philosophy Today* 41, no. 2 (Summer 1997): 235–54.

18   Mack, *Abduction,* 20–21. Mack has in mind Thomas Kuhn's *The Structure of Scientific Revolutions,* 2d ed. (Chicago: University of Chicago Press, 1962).

19   Mack, *Abduction,* 420–21.

20   Kenneth Ring, "Toward an Imaginal Interpretation of 'UFO Abductions,'" *ReVision: The Journal of Consciousness and Change* 11, no. 4 (Spring 1989): 17–24. Ring's essay is part of an ambitious enterprise that collects the essays of an entire range of scholars (including Keith Thompson, Michael Grosso, Peter M. Rojcewicz, and Jacques Vallée) concerned with the UFO

phenomenon. Edited by Keith Thompson, the essays appear in *ReVision*, vol. 11, nos. 3 and 4 (1989). They are the product of a symposium ("Angels, Aliens, and Archetypes: Cosmic Intelligence and the Mythic Imagination") held in San Francisco, 21 and 22 November 1987.

21 Mack, *Abduction*, 29–50.

22 Citing the work of Mario Pazzaglini, a psychologist who has researched abductions for a number of years, Mack acknowledges that UFO-related experiences have been recorded over the centuries, extending back to Ezekiel's *visio Dei* (ibid., 7).

23 For a detailed account of the conference, see *Alien Discussions: Proceedings of the Abduction Study Conference*, ed. Andrea Pritchard et al. (Cambridge, Mass.: North Cambridge Press, 1994).

24 C. D. B. Bryan, *Close Encounters of the Fourth Kind: Alien Abduction, UFOs, and the Conference at M.I.T.* (New York: Alfred A. Knopf, 1995), 9, 424, 425. Although not directly concerned with Ezekiel's vision, Dennis Stillings begins his essay "Helicopters, UFOs, and the Psyche" (*ReVision* 11, no. 4 [Spring 1989]: 25–32) with an allusion to the spiritual "Ezekiel saw the wheel, / Way up in the middle of the air." Stillings's essay is important in its incorporation of the technological dimensions of the visionary and in its allusions to other films in the *Close Encounters* mode, among them *Apocalypse Now, Blue Thunder*, and *Firefox*.

25 Bryan, *Close Encounters of the Fourth Kind*, 416–49. In her poem "testimony" (*Response* [Los Angeles: Sun and Moon, 1996]), Juliana Spahr encapsulates the abduction experience to accord with the main themes suggested by Mack and Bryan, among others. Spahr's poetic rendering encompasses both the archetypal and the biblical dimensions of the abduction experience. At the same time, it associates that experience with the horrific and transcendental. The abductee is at once a victim of Auschwitz-like trauma and a voyager experiencing divine light. Archetypal in nature, the experience finds its source in the Bible, which contains reports encoded in scrolls secretly deposited in the CIA headquarters at Langley, Va.

26 The biographical summary is derived from Ronald D. Story, *The Space-Gods Revealed: A Close Look at the Theories of Eric von Däniken* (New York: Harper & Row, 1976), 1–6. See also Story's *Guardians of the Universe?* (New York: St. Martin's, 1980), 15–22; and Clifford Wilson, *Crash Go the Chariots* (New York: Lancer, 1972), esp. 59–60.

27 Erich von Däniken, *Errinerungen an die Zukunft: Ungeloste Ratsel der Vergangenheit* (Dusseldorf: Econ, 1968), translated by Michael Heron as *Chariots of the Gods? Unsolved Mysteries of the Past* (London: Souvenir, 1969; New York: G. P. Putnam's Sons, 1970). According to Story (*The Space-Gods Revealed*), *Errinerungen an die Zukunft* was written in 1966, the same year that the "God is dead" movement got under way. Perhaps responding to this movement, von Däniken sought "to provide the world with a new set of gods to worship" as a way of replacing the traditional deity, who was being murdered by the poison pens of contemporary theologians. This was likewise the period in which I. S. Shklovskii and Carl Sagan's *Intelligent Life in the Universe* (San Francisco: Holden-Day, 1966) was published. Their work contains many ideas that were later expressed (with distortions) by von Däniken. Shklovskii and Sagan "gave a certain legitimacy to the idea that extraterrestrial visits to earth may have taken place at various times before and after the advent of *Homo sapiens*." The most notable of von Däniken's predecessors, however, are the French authors Louis Pauwels and Jacques Bergier (*The Morning of the Magicians* [Paris: Gallimard, 1960]) and Robert Charroux (*One Hundred Thousand Years of Man's Unknown History* [Paris: Robert Laffont, 1963]), whose theories were proposed well before those of von Däniken but were not, according to Story, sufficiently acknowledged in von Däniken's writings.

28 Among von Däniken's other works, see *Gods from Outer Space,* trans. Michael Heron (New York: G. P. Putnam's Sons, 1970), *The Gold of the Gods,* trans. Michael Heron (New York: G. P. Putnam's Sons, 1972), *In Search of Ancient Gods: My Pictorial Evidence for the Impossible,* trans. Michael Heron (New York: G. P. Putnam's Sons, 1973), *Miracles of the Gods: A New Look at the Supernatural,* trans. Michael Heron (New York: Delacorte, 1974), and *Von Däniken's Proof,* trans. Michael Heron (New York: Bantam, 1978). According to *The Encyclopedia of UFOs* (ed. Ronald D. Story [Garden City, N.Y.: Doubleday, 1980], 383–84), "with sales approaching forty-two million copies," von Däniken's books make him "one of the most successful authors of all time."

29 Story, *The Space-Gods Revealed,* 5–6.

30 For a summary of the hypothesis, see *The UFO Encyclopedia,* ed. Margaret Sachs (New York: G. P. Putnam's Sons, 1980), 13–14. The theory of ancient astronauts has assumed institutional status in its espousal by the Ancient Astronaut Society, based in Highland Park, Ill. (ibid., 14). According to Story, the ancient astronaut hypothesis (or what Story calls the space-god theory) has early proponents in Helena Petrovia Blavatsky and Jean Sendy. Story traces von Däniken's ideas to a number of other writers who anticipated von Däniken as well (*Guardians,* 22–23).

31 The source of this idea is Euhemerus (ca. 316 B.C.), whose views gained wide currency in the histories of mythology produced during the Renaissance and beyond.

32 Von Däniken, *Chariots of the Gods?* (1970), 30–31, 37–40. See also his *Gods from Outer Space,* 137–38.

33 See *The Encyclopedia of UFOs,* 120–21; and *The UFO Encyclopedia,* 96–97.

34 See Morris K. Jessup, *UFO and the Bible* (1965; Clarksburg, W. Va.: Saucerian, 1970), which calls Ezekiel's vision "the most widely quoted and cited of those clearly direct statements" concerning UFOs. See also W. Raymond Drake, *Gods and Spacemen in the Ancient East* (New York: New American Library, 1968), 54, 180, 216–18, and *Gods and Spacemen of the Ancient Past* (New York: New American Library, 1974), 75–76; and Gerhard R. Steinhauser, *Jesus Christ: Heir to the Astronauts* (New York: Pocket, 1976), fifth illustration (unnumbered) between pp. 88 and 89. See additional references in *The UFO Encyclopedia,* 96–97, among them, R. L. Dione, *God Drives a Flying Saucer* (New York: Exposition, 1969), which argues that Ezekiel's visions were hallucinations caused by UFOs. For citations of additional works in the same category, see Lynn E. Catoe, *UFOs and Related Subjects: An Annotated Bibliography* (Detroit: Gale Research, 1978). For New Age converts to von Däniken's theories, see Shirley MacClaine, *Out on a Limb* (New York: Bantam, 1983), 240–44. The New Age movement is discussed below.

35 Adamski's account may be found in Desmond Leslie and George Adamski, *Flying Saucers Have Landed* (New York: British Book Centre, 1953), 171–224. See also the entry on Adamski in *The Encyclopedia of UFOs,* 2–4. One of Adamski's photographs of a Scout Ship "has been variously identified as: an old-model operating (surgical) theater lamp, a tobacco humidor with ping-pong balls, a chicken brooder (feeder), and the top of a canister-type vacuum cleaner made in 1937." See also Story, *Guardians,* 131–33. Timothy Good supports Adamski's contentions (*Above Top Secret,* 374).

36 See, e.g., George Adamski, *Flying Saucers Farewell* (London: Abelard-Schuman, 1961), 83–84.

37 See Drake, *Gods and Spacemen in the Ancient East,* 217–18, and *Gods and Spacemen of the Ancient Past,* 75–76.

38 Festinger, Riecken, and Schachter, *When Prophecy Fails,* 49.

39   Josef F. Blumrich, *Da tat sich der Himmel auf (Ezechiel Kapitel I Vers I: Die Raumschiffe des Propheten Ezechiel und ihre Bestätigung durch modernste Technik (The gates of heaven opened [Ezek. 1:1]: The spaceships of the prophet Ezekiel and their corroboration through modern technology)* (Düsseldorf: Econ, 1973), and *The Spaceships of Ezekiel* (New York: Bantam, 1974). Blumrich has a site on the Internet, http://www.earthportals.com/Earthportals/Portal_Ship/ezekiel.html, which lists other Web sites concerning UFOs in general and Ezekiel in particular. For a list of sites concerning "ancient UFOs," see also http://www.qtm.net/ğeibdan/oldufos/ancient.html. Blumrich's theories have been widely disseminated. See the references in Max H. Flindt and Otto O. Binder, *Mankind—Child of the Stars* (Greenwich, Conn.: Fawcett, 1974), 229; Alan Landsburg and Sally Landsburg, *In Search of Ancient Mysteries* (New York: Bantam, 1974), 134–43; and Story, *Guardians*, 39.

40   Blumrich is not the only NASA scientist with an interest in UFOs. See also Paul R. Hill, *Unconventional Flying Objects: A Scientific Analysis* (Charlottesville, Va.: Hampton Roads, 1995). Although Hill's book differs in both scope and outlook from Blumrich's, it applies similar "scientific" notions to the illumination of the phenomenon.

41   This information has been gleaned from *Contemporary Authors*, ed. Frances C. Locher (Detroit: Gale Research, 1980), 55–56; and *Who's Who in the West*, 24th ed. (New Providence, N.J.: Reed, 1993), s.v. See also, *The UFO Encyclopedia*, 38. In order to confirm this information, I called both the NASA Public Affairs Office, Washington, D.C., and the George C. Marshall Space Flight Center, Huntsville, Ala., on 15 February 1983. I was informed that Blumrich was in fact an employee of NASA, where he held the position of supervisory aerospace engineer at the Marshall Space Flight Center in 1962.

42   *Contemporary Authors*, 55–56.

43   The details of this account are derived from the statement titled "The Ezekiel Question: A Skeptic at Work" preceding the title page of Blumrich's *The Spaceships of Ezekiel.* See also the foreword to his book as well as his statements in Landsburg and Landsburg, *In Search of Ancient Mysteries*, 134. Likewise of interest is Flindt and Binder, *Mankind—Child of the Stars*, 229. Corresponding details are offered in *Contemporary Authors*, 55–56.

44   Josef F. Blumrich, "A Rising Tide of Structural Problems," *Astronautics and Aeronautics* 3 (June 1965): 54–57.

45   Blumrich, *The Spaceships of Ezekiel*, 1–2 (*Da tat sich der Himmel auf*, 10), 22, 4. See Roger A. Anderson, "Structures Technology—1964," *Astronautics and Aeronautics* 2 (December 1964): 14–20. According to the notice that accompanies the article, Anderson was at that time assistant chief to the Structures Research Division. An authority on structures for reentry vehicles, Anderson has contributed to the design and evaluation phase of major vehicle programs, such as the X-15, Mercury, and Apollo systems. Particularly as it addresses spacecraft and reentry vehicles, the language of his article is consistent with the approach and analysis of Blumrich's study of Ezekiel. Correspondingly, the illustrations that accompany Anderson's article find their counterpart in those found in Blumrich's book. I had occasion to meet and speak with Roger Anderson, who was in the audience when I gave an earlier version of this chapter at the College of William and Mary, Williamsburg, Va., 18 April 1986. When I met Anderson, he was chief of structures and dynamics at the Langley Research Center, NASA. In conversation, Anderson acknowledged his acquaintance with "Joe" Blumrich, who had earlier sent Anderson a copy of his book.

46   Blumrich, *The Spaceships of Ezekiel*, 3, 5–7.

47   In this regard, see Randall Fitzgerald, *The Complete Book of Extra-Terrestrial Encounters* (New

York: Macmillan, 1979), 31–32. In keeping with Blumrich's assertions concerning NASA's patents, Timothy Good (*Above Top Secret*) asserts that, along with the CIA and other government agencies, NASA has had a long-standing interest in UFOs. Good argues that the government has sought in various ways to keep the "reality" of UFOs a secret from the general public.

48 See Landsburg and Landsburg, *In Search of Ancient Mysteries,* 139. On his Web site (cited above), Blumrich, in fact, cites his patent number as 3,789,947, 5 February 1974. With his new invention, he would "facilitate considerably the mobility of wheelchairs for the physically handicapped."

49 Blumrich, *The Spaceships of Ezekiel,* 157–58 (the entire discussion runs over pp. 153–58).

50 Ibid., 18 (*Da tat sich der Himmel auf,* 24). Blumrich includes two Lutheran Bibles among his references: *Die Bibel oder die ganze Heilige Schrift des Alten und Neuen Testamentes nach der Ubersetzung von D. Martin Luther* (Stuttgart, n.d.) and *Biblia. Das ist: Die ganze Heilige Schrift,* trans. Martin Luther (Leipzig, 1842). He also cites a number of other Bibles, including the RSV, the New American Bible, and the Soncino Hebrew text/English translation version of Ezekiel (1957).

51 Blumrich, *The Spaceships of Ezekiel,* 54–55.

52 Ibid., 17–18 (*Da tat sich der Himmel auf,* 32).

53 Ibid., 72 (*Da tat sich der Himmel auf,* 110).

54 See in this regard *UFOs: A Scientific Debate,* ed. Carl Sagan and Thornton Page (Ithaca, N.Y.: Cornell University Press, 1972).

55 Donald H. Menzel, *Flying Saucers* (Cambridge, Mass.: Harvard University Press, 1953), 125, 130, and passim.

56 Ibid., 124–34, 301–10, 132.

57 Donald H. Menzel, "UFOs—The Modern Myth," in *UFOs: A Scientific Debate,* 177–81. For a further elaboration of the parhelion theory, see also Donald H. Menzel and Ernest H. Taves, *The UFO Enigma: The Definitive Explanation of the UFO Phenomenon* (Garden City, N.Y.: Doubleday, 1977), 22–26.

58 Used in the context of cyberspace, *avatar* becomes a term of art. According to the Internet definition provided by *The New Hacker's Dictionary,* "among people working on virtual reality and cyberspace interfaces, an *avatar* is an icon or representation of a user in a shared virtual reality." The term is also used in connection with UNIX machines (http://www.tpconsultants. com/tnhd/def_0050.htm). I adopt the term *avatar* both in its cyberspatial sense and in its conventional use as a concrete embodiment or manifestation of an abstract concept.

59 William Gibson, *Neuromancer* (New York: Ace, 1984), 51. In the same vein, see also Gibson's other books, including *Count Zero* (New York: Ace, 1986), *Burning Chrome* (New York: Ace, 1987), and *Mona Lisa Overdrive* (New York: Bantam, 1988).

60 See "Cyberpunk," in *The Cyberspace Lexicon: An Illustrated Dictionary of Terms from Multimedia to Virtual Reality,* by Bob Cotton and Richard Oliver (London: Phaidon, 1994), 52–53, 55: "'Cyberpunk' was originally the name given to the work of a group of science fiction writers who emerged in the 1980s. Specializing in 'low life/hi-tech' subject matter, they were also known as the Movement, the Mirrorshade Group, Radical Hard SF, the Eighties Wave, the Outlaw Technologists, and the Neuromantics" (p. 52).

61 Ibid., 52. For the suggestion of "neu/romancer" as "new romance," see Istvan Csicsery-Ronay Jr., "Cyberpunk and Neuromanticism," in *Storming the Reality Studio,* ed. Larry McCaffert (Durham, N.C.: Duke University Press, 1991), 182–93.

62 Michael Heim, "The Erotic Ontology of Cyberspace," in *Cyberspace: First Steps,* ed. Michael Benedikt (Cambridge, Mass.: MIT Press, 1991), 59–80.

63 David Tomas, "Old Rituals for New Space: *Rites de Passage* and William Gibson's Cultural Model of Cyberspace," in *Cyberspace: First Steps,* ed. Benedikt, 31–48.

64 Philo, "On Flight and Finding" and "Who Is the Heir of Divine Things," in *Philo,* trans F. H. Colson and G. H. Whitaker, 10 vols., Loeb Classical Library (Cambridge, Mass.: Harvard University Press, 1958), 4:437–39, 5:65. For the etymology of *cybernetics,* see *A Greek-English Lexicon of the New Testament and Other Early Christian Literature,* ed. William F. Arndt and F. Wilbur Gingrich (Chicago: University of Chicago Press, 1957), s.v. In its Latin form, *kybernētes* finds its counterpart in *gubernare* (control).

65 Norbert Wiener, *Cybernetics; or, Control and Communication in the Animal and the Machine* (New York: Wiley, 1948), 11–13.

66 Norbert Wiener, *The Human Use of Human Beings: Cybernetics and Society* (New York: Avon, 1954), 185–222.

67 Norbert Wiener, *God and Golem, Inc.: A Comment on Certain Points Where Cybernetics Impinges on Religion* (Cambridge, Mass.: MIT Press, 1964), 15–16.

68 Ibid., 15–17.

69 Ibid., 49–50.

70 See Gershom Scholem, *Major Trends in Jewish Mysticism* (New York: Schocken, 1954), 99, *Kabbalah* (New York: New American Library, 1974), 351–55, and *On the Kabbalah and Its Symbolism,* trans. Ralph Mannheim (New York: Schocken, 1966), 158–204. For a full and exhaustive study, see Moshe Idel, *Golem: Jewish Magical and Mystical Traditions on the Artificial Anthropoid* (Albany: State University of New York Press, 1990). The term *golem* (as that which is unformed, embryonic, folded, and wrapped together) appears in Ps. 139:16.

71 Donna Haraway, "A Manifesto for Cyborgs: Science, Technology, and Socialist Feminism in the 1980s," in *Feminism/Postmodernism,* ed. Linda Nicholson (New York: Routledge, 1990), 192.

72 William A. Covino, "Grammars of Transgression: Golems, Cyborgs, and Mutants," *Rhetoric Review* 14 (1996): 366.

73 Philip Hayward, "Situating Cyberspace: The Popularisation of Virtual Reality," in *Future Visions: New Technologies of the Screen,* ed. Philip Hayward and Tana Wollen (London: British Film Institute, 1993), 180–204. Hayward is particularly concerned with the powers of virtual reality. Virtual reality is a device that employs stereo-optical visors and position-sensing gloves to simulate "realistic" computer-generated environments of sight and sound (Cotton and Oliver, *Cyberspace Lexicon,* 209). Jaron Lanier, a computer scientist and the pioneer who coined the term *virtual reality,* enjoys a presence that distinguishes him as the veritable embodiment of the kind of bizarre visionary who has been the subject of our discussion throughout. He has been described as "amiable, round, and dreadlocked," "a Rastafarian hobbit," "the wizard of odd," "Bigfoot," "a new age Jabba the Hut," "an enigma," and "a Guru" (Anthony B. Perkens and Michael C. Perkens, "Jaron Lanier Gets Real," *Red Herring Magazine* [1993], http://www.herring.com/mag/issue01/guru.html).

74 Covino, "Grammars of Transgression," 364–65.

75 Cited in Michael D'Antonio, *Heaven on Earth* (New York: Crown, 1992), 294–96. William S. Burroughs called Timothy Leary (1920–96) "a true visionary of the potential of the human mind and spirit," and Allen Ginsburg has proclaimed him "a hero of American consciousness" (see http://www.roninpub.com/TimLea.html). There are many Web sites devoted to

Leary. In keeping with his own role as rider in the chariot, Leary, who talked of death as the ultimate trip and of outer space as the great new frontier, experienced his own chariot ascent after "deanimating," as he called it. On 21 April 1997, a rocket carrying a capsule with his ashes (along with the remains of Gene Roddenberry, the creator of "Star Trek," and other notables) was launched from Grand Canary Island off the Moroccan coast. Borne aloft in an American Pegasus rocket, the remains will orbit every ninety minutes for a period of two to ten years (*New York Times,* 22 April 1997, sec. 1, 1, 4).

76   Timothy Leary, "Cyberpunk: The Individual as Reality Pilot," in *Storming the Reality Studio,* ed. McCaffert, 246–53. Among the New Age devotees whom Leary invokes are the Jungians (see Timothy Leary, *Chaos and Cyber Culture,* ed. Michael Horowitz and Vicki Marshall [Berkeley, Calif.: Ronin, 1994]).

77   For enlightening discussions of the New Age, see, among other studies, J. Gordon Melton et al., *New Age Almanac* (New York: Visible Ink, 1991); *New Age Spirituality: An Assessment,* ed. Duncan S. Ferguson (Louisville, Ky.: Westminster/John Knox Press, 1993); and John White, *The Meeting of Science and Spirit: Guidelines for a New Age* (New York: Paragon, 1990). As might well be expected, there are a large number of New Age links on the Internet.

78   Information concerning Drunvalo Melchizedek is derived from Web sites the addresses of which are noted throughout. For published accounts, see, among other sources, Bob Frissell, *"Nothing in This Book Is True, but It's Exactly the Way Things Are": The Esoteric Meaning of the Monuments on Mars* (Berkeley, Calif.: North Atlantic, 1993; Berkeley, Calif.: Leap Frog, 1994).

79   "Who Is Drunvalo Melchizedek?" (from Density 4/Devin, 17 August 1995), Drunvalo Melchizedek and the Flower of Life (http://www.execpc.com/vjentpr/drunwho.html). Information on this site is excerpted from Frissel, "*Nothing in This Book Is True.*"

80   "Drunvalo's Spiritual Background," from "Conversations with Drunvalo," a videotape interview conducted by Auroria Genie Joseph (http://www2.cruzio.com/flower/1dr1.htm).

81   "Who Is Drunvalo Melchizedek?"

82   Drunvalo Melchizedek, "A Love Story" (lecture presented at the Archangel Michael Conclave, Banff Springs Hotel, Canada, March 1994), excerpted from the *Leading Edge Research Journal,* no. 71 (http://www. execpc.com/vjentpr/drunwho.html).

83   Interesting enough, for Drunvalo, the biblical antecedents of the *merkabah* are more nearly centered in the figure of Elijah, transported to the heavens in his chariot of fire (2 Kings 2:11), than in the figure of Ezekiel. On the other hand, Drunvalo is very much aware of the rabbinic traditions (including those of the *hekhalot* and the dangers of *ḥashmal*) surrounding Ezekiel's vision as well as the importance of the vision itself. See Jeff Wein, "The Merkabah in the Bible" (http://www2.cruzio.com/flower/1mm3.htm). Ezekiel's vision also underlies *The Book of Knowledge: The Keys of Enoch* (Los Gatos, Calif.: Academy of Future Sciences, 1978), an ascent document that refers to the Melchizedek Brotherhood (http://www.protree.com/npt-ufo/ufo-files/ufo-text-files/k/keys-of-Enoch).

84   "About the Flower of Life Workshop," prepared by Jeff Wein, Flower of Life facilitator (http://www2.cruzio.com/-flower/workshop.htm). Lasting six days, workshops have apparently taken place in Santa Cruz, Calif., Jerusalem and Tel Aviv, Israel, and Wellington and Christchurch, New Zealand, among other sites. See also "The Merkabah—a Vehicle of Ascension," from "Conversations with Drunvalo"; and "The Process of Planetary Ascension," an interview segment from the Flower of Life Workshop (http://www2.cruzio.com/-flower/1sh2.htm). Frissell also discusses the *merkabah* ("*Nothing in This Book Is True,*" 22–23 and passim).

85 Drunvalo Melchizedek, "The MER-KA-BA Meditation: The Teaching on Spiritual Breathing," transcribed by Gabriel Hettes, 21 December 1994 (http://www.spiritweb.org/Spirit/
merkabah-meditation.html).

86 "Drunvalo: Our History and the 1972 Sirian Experiment," excerpted from Frissell, *Nothing
in This Book Is True*" (see also http://earth.execpc.com/vjentpr//drunfris.html).

87 "The Grays," from "Conversations with Drunvalo."

88 *Leading Edge* interview with Drunvalo Melchizedek, 22 December 1995 (http://earth.execpc.
com/vjentpr/drunledg.html). See Frissell, *"Nothing in This Book Is True,"* 2–3.

89 When virtual reality ominously spills over into actual reality, the question of whether anyone
cares is no longer "rhetorical." On 26 March 1997, the mass suicide of a computer-related
movement known as Heaven's Gate was discovered in Rancho Santa Fe (near San Diego).
Distinguished by a belief in space aliens and UFOs, in conjunction with the appearance of the
Hale-Bopp comet, the Heaven's Gate movement maintained an elaborate Web site at http://
www.heavensgate.com. The movement received national and international news coverage.
Among other newspaper articles on Heaven's Gate, see *New York Times,* 28 March 1997, sec. 1,
1, 10–11.

3 Prophecy Belief and the Politics of End Time

1 Frank Kermode, *The Sense of an Ending: Studies in the Theory of Fiction* (New York: Oxford
University Press, 1967), 25.

2 Susan Harding, "Imagining the Last Days: The Politics of Apocalyptic Language," in *Account-
ing for Fundamentalisms: The Dynamic Character of Movements,* ed. Martin Marty and
R. Scott Appleby (Chicago: University of Chicago Press, 1994), 61.

3 Paul Boyer, *"When Time Shall Be No More": Prophecy Belief in Modern American Culture*
(Cambridge, Mass.: Harvard University Press, 1992), 1.

4 I shall not argue here distinctions between evangelicalism and fundamentalism. For a detailed and informative discussion of such matters, see Nancy T. Ammerman, "North American Protestant Fundamentalism," in *Fundamentalisms Observed,* ed. Martin E. Marty and
R. Scott Appleby (Chicago: University of Chicago Press, 1991), 1–65. According to Ammerman, during the first half of the twentieth century, *fundamentalist* and *evangelical* meant
roughly the same things. After World War II, distinctions began to arise (p. 4). As an
umbrella term, *evangelicalism* is itself something like a mosaic or kaleidoscope of denominations. George Marsden observes that all the denominations within the "evangelical mosaic"
reflect the Reformation attempt to return to the "pure Word of Scripture as the only ultimate
authority and to confine salvation to a faith in Christ, unencumbered by presumptuous human authority" ("The Evangelical Denomination," in *Evangelicalism and Modern America,*
ed. George Marsden [Grand Rapids, Mich.: William B. Eerdmans, 1984], x). Similar terms are
adopted by Erling Jorstad (*Popular Religion in America: The Evangelical Voice* [Westport,
Conn.: Greenwood, 1993]), who views *evangelicalism* as an "umbrella" encompassing evangelicalists and fundamentalists, among other "believers." Whereas evangelicalists espouse "an
experientially informed faith, focusing upon being born again," fundamentalists are more
nearly oriented toward "discovering and practicing strict loyalty to the teachings of the
inerrant Bible" (p. 9). For a discussion of the concept *biblical inerrancy* and its role in
fundamentalist hermeneutics, see James Barr, *Fundamentalism* (Philadelphia: Westminster,
1978). According to Martin Marty, fundamentalism "inevitably eludes precise definition"

(*Modern American Religion* [Chicago: University of Chicago Press, 1991], 159). The standard history of fundamentalism in its earliest phases is Ernest R. Sandeen's *The Roots of Fundamentalism: British and American Millenarianism, 1800–1930* (Chicago: University of Chicago Press, 1970). See also George Marsden, *Fundamentalism and American Culture: The Shaping of Twentieth-Century Evangelicalism, 1870–1925* (New York: Oxford University Press, 1980).

5   Timothy P. Weber, *Living in the Shadow of the Second Coming: American Premillennialism, 1875–1982* (Chicago: University of Chicago Press, 1987), 4. See also Weber's "Premillennialism and the Branches of Evangelicalism," in *The Variety of American Evangelicalism,* ed. Donald W. Dayton and Robert K. Johnston (Knoxville: University of Tennessee Press, 1991), 5–21.

6   In addition to Boyer (*"When Time Shall Be No More"*) and Weber (*Living in the Shadow of the Second Coming*), see, among others, Michael Barkun, *Disaster and the Millennium* (New Haven, Conn.: Yale University Press, 1974); Charles B. Strozier, *Apocalypse: On the Psychology of Fundamentalism in America* (Boston: Beacon, 1994); and Harding, "Imagining the Last Days," 57–78.

7   Strozier, *Apocalypse,* 159. When speaking of "other cultural forms," Strozier has in mind what he calls "New Age Apocalyptics," with a particular emphasis on the works of Carl Gustav Jung, magic, and UFOs (pp. 227–39).

8   Boyer, *"When Time Shall Be No More,"* 4.

9   As Martin Riesebrodt makes clear (*Pious Passion: The Emergence of Modern Fundamentalism in the United States and Iran,* trans. Don Reneau [Berkeley and Los Angeles: University of California Press, 1993]), "among the most subversive of the new religious movements and cults those that fundamentalism considered the most subversive of the Christian foundations of American society were Mary Baker Eddy's Christian Science; the Jehovah's Witnesses, or 'Russelites,' as they were called after their founder; theosophy; and spiritualism. They were regarded either as pure superstitions or dangerous falsifications of the Christian faith" (p. 51).

10 Herbert Hewitt Stroup, "Jehovah's Witnesses," in *The Encyclopedia of Religion,* ed. Mircea Eliade, 16 vols. (New York: Macmillan, 1987), 7:564–66. See also Stroup's pioneering study *The Jehovah's Witnesses* (New York: Columbia University Press, 1945). Of comparable interest are, among others, William J. Whalen, *Armageddon around the Corner: A Report on Jehovah's Witnesses* (New York: John Day, 1962); and M. James Penton, *Apocalypse Delayed: The Story of Jehovah's Witnesses* (Toronto: University of Toronto Press, 1985). Penton's bibliography of primary and secondary works is exemplary (pp. 361–82). Of interest is G. Hebert's informative entry in *The New Catholic Encyclopedia,* gen. ed. William J. McDonald, 17 vols. (Washington, D.C.: Catholic University of America, 1967), 7:864–65.

11 Stroup, "Jehovah's Witnesses," 564. For a concise but informative overview of apocalyptic thought in America from seventeenth-century Puritan New England on, see Charles H. Lippy, "Millennialism and Adventism," in *Encyclopedia of the American Religious Experience,* ed. Charles H. Lippy and Peter W. Williams, 3 vols. (New York: Charles Scribner's Sons, 1988), 7:831–44. Lippy provides an informative bibliography of some of the major works on the subject (p. 844).

12 Barbara Grizzuti Harrison, *Visions of Glory: A History and a Memory of Jehovah's Witnesses* (New York: Simon & Schuster, 1978), 42, citing Charles A. Beard and Mary R. Beard, *The Rise of American Civilization,* 4 vols. (New York: Macmillan, 1927). (*Visions of Glory* is an autobiographical, as well as a historical, account of the Jehovah's Witnesses by a former Witness.)

13 Stroup, "Jehovah's Witnesses," 564. See also Stroup, *The Jehovah's Witnesses,* 3–12.

14 Harrison, *Visions of Glory,* 43–44.

15 Harrison, *Visions of Glory,* 43–50.

16 Ibid., 54 (recorded in the court transcripts, Court of Common Pleas, Pittsburgh, Pa., the "jellyfish" remark was attributed to Charles Taze Russell by Mrs. Russell), 51–54.

17 Whalen, *Armageddon around the Corner,* 45–46.

18 Ibid., 48.

19 References are to Charles Taze Russell, *The Finished Mystery* (Brooklyn, N.Y.: Peoples Pulpit Association, 1917).

20 Penton, *Apocalypse Delayed,* 372.

21 For accounts of the controversy, see ibid., 50–55; Harrison, *Visions of Glory,* 170; and Whalen, *Armageddon around the Corner,* 52–53. Rutherford's release of *The Finished Mystery* to the assembled Bethel Watch Tower family "came as a 'bombshell' and, according to Watch Tower history, served to cause an open schism." Rutherford was censured in "a long, bitter, breakfast-table debate." Rutherford not only kept the printing of *The Finished Mystery* a secret from the organization's editorial board but printed it with the use of donated money that was never placed in the organization's accounts (Penton, *Apocalypse Delayed,* 51–53). In his own way, Rutherford was as colorful a figure as Russell. For discussions of Rutherford and his effect on the Witnesses, see Stroup, *The Jehovah's Witnesses;* Whalen, *Armageddon around the Corner;* Penton, *Apocalypse Delayed;* and Harrison, *Visions of Glory,* among others.

22 Both meanings are implicit in Ezekiel's full name *yehezk'el* (God strengthens, God will strengthen) *ben buzi* (son of contempt, son of my contempt).

23 Russell, *The Finished Mystery,* 367–68.

24 Ibid., 368.

25 Among other studies, see Marjorie Reeves, "The Development of Apocalyptic Thought: Medieval Attitudes," in *The Apocalypse in English Renaissance Thought and Literature: Patterns, Antecedents, and Repercussions,* ed. C. A. Patrides and Joseph Wittreich (Ithaca, N.Y.: Cornell University Press, 1984), 40–72.

26 Russell, *The Finished Mystery,* 373–74, 371.

27 Joseph Franklin Rutherford, *Vindication: The Name and Word of the Eternal God Proven and Justified by Ezekiel's Prophecy and Revealing What Must Speedily Come to Pass upon the Nations of the World* (Brooklyn, N.Y.: Watch Tower Bible and Tract Society, 1931). Book 1 of *Vindication* was published in 1931, bks. 2 and 3 in 1932. See also Rutherford's earlier *Prophecy: Many Mysteries of the Bible Made Plain and Understandable: The "Lightenings" of Jehovah and the Present Day Events Unlock the Secrets and Reveal to Man Eternal Truths* (Brooklyn, N.Y.: International Bible Students, 1929).

28 Rutherford, *Vindication,* 19–20. For Rutherford, the year 617 B.C., when Ezekiel was carried away captive, corresponds with the year A.D. 1914 as the beginning of the First World War. The year 612 B.C., when Ezekiel received the vision, corresponds with the year 1919, after the war had ended. The accuracy of Rutherford's dating of the events surrounding the prophet will not be debated here.

29 See Stroup, *The Jehovah's Witnesses,* 140.

30 Rutherford, *Vindication,* 5–20.

31 Ibid., 46–49.

32 Ibid., 22–24.

33 This observation was ventured over fifty years ago by Stanley High, "Armageddon, Inc.," *Saturday Evening Post,* 14 September 1940, 18–19, 50–58.

34 Stroup, *The Jehovah's Witnesses*, 147.

35 Ibid., 147–52. See Rutherford's *Enemies: The Proof That Definitely Identifies All Enemies, Exposes Their Methods of Operation, and Points Out the Way of Complete Protection for Those Who Love Righteousness* (Brooklyn, N.Y.: Watch Tower Bible and Tract Society, 1937). According to Stroup, "the Jews are also hated by the Witnesses. Although this feeling is common among them, it appears strange at first glance, inasmuch as the movement appeals to many Jews" (p. 154). For the complex relation between the Jehovah's Witnesses and the Jews, see Harrison, *Visions of Glory*, 159–61; and Penton, *Apocalypse Delayed*, 148–49, 187–88.

36 For a detailed account, along with diagrams, see Penton's illuminating chapter "Organizational Structure," in *Apocalypse Delayed*, 211–52. See also Stroup, *The Jehovah's Witnesses*, 138–41.

37 *"The Nations Shall Know That I Am Jehovah"—How?* (Brooklyn, N.Y.: Watch Tower Bible and Tract Society, 1971), 35–41. As a product of the organization, this work designates no single author on its title page. According to Penton, Chris Christenson of Vancouver, British Columbia, contributed to the research for the volume. Interestingly, Christenson "began to question the authority of the governing body of Witness disfellowshipping practices" (*Apocalypse Delayed*, 106).

38 *"The Nations Shall Know,"* 45–46.

39 Ibid., 49. The work notes its indebtedness for this idea to Rutherford's *Prophecy*. Titled "God's Organization," the fifth chapter of this work states: "The living creatures and the inanimate objects, or instruments, appearing in the vision [of Ezekiel], together give the appearance of an enormous living chariot-like organization extending high into the heavens, and over all of which God presides" (*Prophecy*, 121).

40 *"The Nations Shall Know,"* 53. The particular war emphasized in this work (as in earlier works) is once again the First World War, which is pivotal for the Jehovah's Witnesses. For Rutherford, as for his followers, the crucial year is 1919, when the momentum of the organization was at its height.

41 Ibid., 119–21.

42 In the traditions of evangelicalism, the Bible is customarily viewed as a harbinger of technological innovation. See, e.g., Arthur Skevington Wood, *Prophecy in the Space Age: Studies in Prophetic Themes* (Grand Rapids, Mich.: Zondervan, 1963). According to Wood, Christ is seen as the Lord of all space; an astronaut of sorts, he ascends into space and returns unharmed (pp. 9, 13–15).

43 Andrew Lang, "The Theology of Nuclear War: The Counterpoint," in *The Religious Right*, ed. Gary E. McCuen (Hudson, Wis.: GEM, 1989), 72.

44 Boyer, *"When Time Shall Be No More,"* 115. For an enlightening study of the effect of the atomic bomb on life in America, see Paul Boyer, *By the Bomb's Early Light: American Thought and Culture at the Dawn of the Atomic Age* (New York: Random House, 1985).

45 John W. Bradley, ed., *Hastening the Day of God: Prophetic Messages from the International Congress on Prophecy in Calvary Baptist Church, New York City, November 9–16, 1952* (Wheaton, Ill.: Van Kampen, 1953), 11–12 (manifesto), cited in Boyer, *"When Time Shall Be No More,"* 122.

46 Boyer, *"When Time Shall Be No More,"* 123–25. See 2 Pet. 3:7–12; Rev. 16:1–20:15; Joel 1–3; Zech. 14:12.

47 The fifth chapter of Boyer's *"When Time Shall Be No More"* is titled "Ezekiel as the First Cold Warrior." So prominent is this idea in evangelical discourse that, even in the post–cold war

period, Ezekiel is singled out as the prophet among prophets. See, e.g., the chapter titled "Ezekiel and You!" in Mark Hitchcock's *After the Empire* (n.p.: Hearthstone, 1992), 159.

48 It must be understood, of course, that the so-called chariot chapter (Ezekiel 1) and the so-called Gog and Magog chapter (Ezekiel 38) extend beyond their immediate "chapter headings" (an editorial device) to later chapters. In the case of Ezekiel 1, the inaugural vision is articulated and rearticulated in chaps. 2–10 and beyond, as it culminates in chaps. 40–48. In the case of Ezekiel 38, the Gog-Magog material is rearticulated in the oracles that constitute chap. 39. In fact, Ezekiel 38 and 39 must be viewed as a unit.

49 Commentary by John W. Wevers in *Ezekiel*, ed. John Wevers (London: Thomas Nelson, 1969), 284. See "Gog and Magog," in *The Anchor Bible Dictionary*, ed. David Noel Freedman, 6 vols. (New York: Doubleday, 1992), 2:1056. For references to Gog in Hebrew Scriptures, see Ezek. 38:2, 3, 14, 16, 18; 39:1, 11. For references to Magog in Hebrew Scriptures, see Gen. 10:2; 1 Chron. 1:5; and Ezek. 38:2, 39:6. Josephus identified Gog with the "Scythians," an equation, according to Jerome, widely accepted in rabbinic interpretation. From the very beginnings of Christian exegesis of the prophecies, Gog's precise identity puzzled the interpreters. For Augustine, Gog represented all unbelievers; Joachim of Fiore, among other medieval exegetes, looked on Gog and Magog as tribes beyond the Caucasus that would invade at the end of time. With the emergence of Islam and the Ottoman Turks in the thirteenth century, Gog was identified with these forces. The Turk-as-Gog theme was a prominent feature in American prophecy writing during the colonial period and beyond (Boyer, "*When Time Shall Be No More*," 153–54).

50 Meshech is probably the Assyrian "Mushku," south of Gomer, west of the Anti-Taurus mountains, and Tubal is probably the Assyrian "Tabal," south of Beth-togarmah, east of the Anti-Taurus mountains (Ezek. 27:13). The Hebrew renderings of "Persia, Ethiopia, and Libya" in the Authorized Version are *pares, cush,* and *put,* respectively. Put is perhaps Cyrene, east of Libya (Ezek. 27:10, 30:5). Gomer is probably the Assyrian "Gimirrai," east of the southernmost Halys River, southeast of Gomer (Ezek. 27:14). (See the notes to Ezekiel 38–39 in *The New Oxford Annotated Bible*, ed. May and Metzger. See also the annotations to Ezekiel 38–39 in Walther Zimmerli, *Ezekiel 2: A Commentary on the Book of the Prophet Ezekiel, Chapters 25–48,* trans. James D. Martin [Philadelphia: Fortress, 1983], passim; and Walther Eichrodt, *Ezekiel: A Commentary,* trans. Cosslett Quin [Philadelphia: Westminster, 1970], 519–29.)

51 For the mythic dimensions discussed here, see "Gôgh; Maghôgh," in *Theological Dictionary of the Old Testament*, ed. G. Johannes Botterweck and Helmer Ringgren, 7 vols. (Grand Rapids, Mich.: William B. Eerdmans, 1975), 2:419–25. These dimensions are distinguished by several motifs common to Hebrew Scriptures. Among them is the notion of Yahweh's "War with Chaos and War with the Nations." Hebrew Scriptures provide multiple instances of the overcoming of Chaos by Yahweh (e.g., Ps. 104:3–8, Isa. 50:2, Nah. 1:3–6, and Hab. 3:8). Such an act, in turn, finds its counterpart in Yahweh's triumph over his enemies in battle, a warfare in which the engagement between divine and human assumes cosmic proportions (e.g., Pss. 2, 110, 68:12–18; and Isa. 14:24–27, 17:12–14). In the Ezekiel oracle, Gog's attack on the land of Israel is accordingly conceived in cosmic terms. The attack provokes Yahweh's fury, which causes not only the earth and its inhabitants to shake but the mountains to fall to the ground (Ezek. 38:18–22). As the result of divine wrath, the entire world undergoes a terrible tumult. Prevailing over his enemies, Yahweh triumphs in battle, an event that results in the "Day of Yahweh." Although the expression itself does not occur in Ezekiel's oracle, the idea of

a day in which Yahweh will "manifest his wrath, sit in judgment, and carry out punishment through destruction" is clearly present. As a result of these actions, Yahweh will magnify himself and sanctify himself among the nations (Ezek. 38:23).

52 See *ṣaphon* in *The New Brown-Driver-Briggs-Gesenius Hebrew and English Lexicon*, 860b–61a. *Ṣaphon* is related to the verb *ṣaphan*, which means to "hide, treasure up." According to *The Anchor Bible Dictionary*, 4:1135, biblical Hebrew has two major words for *north: ṣaphon*, as that which is hidden or dark, and *shem'ol*, lit., "the left hand."

53 Walther Zimmerli, *Ezekiel 1: A Commentary on the Book of the Prophet Ezekiel*, trans. Ronald E. Clements (Philadelphia: Fortress, 1979), 119. "It has been pointed out," Zimmerli observes, "that the journey made by those taken into exile led first into Northern Syria, from there across to the Euphrates, and then along the Euphrates southwards (more precisely southeastwards). Thus Yahweh's coming followed this direction exactly." According to Rabbi S. Fish, the idea of God's coming from the north is "an indication of the forthcoming invasion of Judea by the Babylonians whose empire extended to the land of the Chaldeans to the north of the Holy Land" (*Ezekiel: With Hebrew Text*, ed. A. Cohen [London: Soncino, 1950], 2–3). Thus, Kieth Bernard Kuschel, *Ezekiel* (St. Louis: Concordia, 1994): "It was from the north that the Babylonians . . . had invaded the land of Israel. They had brought destruction on Jerusalem and would continue to do so for a time. Here, however, we have the statement by god 'I am behind it all' " (p. 9). Compare Ezek. 26:7: "For saith the Lord GOD; Behold, I will bring upon Tyrus Nebuchadrezzar king of Babylon, a king of kings, from the *north*, with horses, and with chariots, and with horsemen, and companies, and much people."

54 In Ugaritic mythology, *ṣaphon* "is the name of the holy mountain, used especially in Baal's abode, and was the place of the assembly of the gods." In fact, Baal calls himself the god of *ṣaphon* or "Baal *ṣaphon*." Closely related to the Ugaritic and Canaanite mythology is the biblical reference to God's abode in the far north (Isa. 14:13) (*The Anchor Bible Dictionary*, 4:1136).

55 *Ezekiel 1–20*, translated and with commentary by Moshe Greenberg (Garden City, N.Y.: Doubleday, 1983), 42. Greenberg demythologizes the reference by offering a naturalistic reading tied to windstorms (pp. 42–43). Correspondingly, see G. A. Cooke, *A Critical and Exegetical Commentary on the Book of Ezekiel* (Edinburgh: T. & T. Clarke, 1936), 9–10.

56 In a different context, the idea is also present in Job 37:22: "Fair weather cometh out of the north: with God *is* terrible majesty."

57 Halperin refers to "the dark side of the *merkabah*" in the context of his discussion of the so-called anti-*merkabah*, the idea of the *merkabah* as "evil" entity. In rabbinic circles, the dark side of the *merkabah* is particularly evident in gnostic modes of thought. See Halperin, *The Faces of the Chariot*, 238–47.

58 *Theological Dictionary of the Old Testament*, 2:423. The idea begins to assume a satanic dimension in Isaiah's oracle concerning the so-called shining star (Lucifer, "son of the morning") who is "brought down to hell, to the sides of the pit," because of his attempt to establish his power "in the sides of the north" (Isa. 14:12–15).

59 For an account of the publication history of the Gesenius *Lexicon*, see the preface to *A Hebrew and English Lexicon of the Old Testament . . . Based on the Lexicon of William Gesenius*, trans. Edward Robinson, ed. Francis Brown et al. (Oxford: Clarendon, 1907), v–vi.

60 The translation is from *Gesenius's Hebrew and Chaldee Lexicon to the Old Testament Scriptures*, trans. Samuel Prideaux Tregelles (London: Samuel Bagster, 1859), 752. Later editions of

Gesenius's Hebrew and Chaldee Lexicon eliminate this definition of *ro'sh*. In its rendering of *ne'shey ro'sh meshekh,* the Authorized Version translates *ro'sh* as an adjective, rather than as a noun, so that *ne'shey ro'sh* becomes "chief prince" (with *ro'sh* modifying *ne'shey*) as an attribute to Meshech (conceived as a possessive). (This translation obtains in the Revised Standard Version and the New Revised Standard Version.) If *ro'sh* is conceived as a proper noun (the name of a kingdom) in the possessive, then the phrase *ne'shey ro'sh* becomes "prince of *Ro'sh,*" with *meshekh* as a proper noun, no longer in the possessive, but part of a series of proper nouns that include *ro'sh, meshekh,* and *tubal.* Because the Hebrew text provides no punctuation, a reading such as "prince of *Ro'sh*" (as opposed to a series of proper nouns in apposition) becomes a distinct possibility. William Lowth, the eighteenth-century English churchman and exegete, argued for the "prince of *Ro'sh*" reading in his *Commentary on the Prophecies* (1714–25). So it appears in the Septuagint (Boyer, *"When Time Shall Be No More,"* 154).

61 According to Boyer, "the surge of Russian patriotism following the treaty of Tilsit (1807), by which Napoleon and Alexander I of Russia sharply reduced Prussia's territory and autonomy, created a congenial political environment in Germany for prophetic interpretations casting Russia in a sinister role" (*"When Time Shall Be No More,"* 154).

62 Ibid., 154. The early dispensationalist John Nelson Darby (1800–1882), e.g., identified Gog as Russia in his writings and lectures. Other writers, such as John Cummings, Samuel Baldwin, and Fountain Pitts, followed suit. In 1867, the Reverend Henry Cowles of Oberlin College maintained that the Russian policy of aggression in Europe and Asia foreshadowed the attack prophesied in the Bible. Along with the identification of Gog with Russia, such views also came under attack in the later nineteenth century.

63 Ibid., 97. For a brief account of Scofield's life and conversion, see Charles H. Keeney, "Rev. C. I. Scofield, D.D., Soldier, Lawyer, Pastor, Author," *World's Crisis,* 26 September 1936, 5.

64 Barr, *Fundamentalism,* 45, cited in Boyer, *"When Time Shall Be No More,"* 98. Boyer observes that, between 1909 and 1967, sales of the Scofield Reference Bible were between 5 and 10 million copies. A revised edition in 1967 sold an additional 2½ million copies by 1990.

65 Boyer, *"When Time Shall Be No More,"* 97–98. See also Kathleen C. Boone, *The Bible Tells Them So: The Discourse of Protestant Fundamentalism* (Albany: State University of New York Press, 1989), 79–80.

66 References are to *The Scofield Reference Bible: The Holy Bible: Containing the Old and New Testaments: Authorized Version, with a New System of Connected Topical References to All the Greater Themes of Scripture, with Annotations, Revised Marginal Renderings, Summaries, Definitions, and Index: To Which Are Added Helps at Hard Places, Explanations of Seeming Discrepancies, and a New System of Paragraphs,* ed. Rev. C. I. Scofield (New York: Oxford University Press, 1909), 883, n. 1. Bearing the same title and once again published by the Oxford University Press, "new and improved" editions appeared in 1917 and 1945. Essentially reprints of the 1909 edition, the later editions were published under the "consulting" editorship of Rev. Henry G. Weston and others. These editions of the Scofield Reference Bible do not alter Scofield's note concerning the Gog oracle. Under an editorial committee chaired by E. Schuyler English, *The New Scofield Reference Bible: Holy Bible, Authorized King James Version, with Introductions, Annotations, Subject Chain References, and Such Word Changes in the Text as Will Help the Reader* was published by the Oxford University Press in 1967. Although Scofield's original note concerning Gog has been "streamlined" a bit in this most recent edition, the sense of the note remains remarkably intact: "The reference is to the

powers in the north of Europe, headed by Russia. . . . Gog is probably the prince; Magog, his land. Russia and the northern powers have long been the persecutors of dispersed Israel, and it is congruous both with divine justice and with the covenants of God that destruction should fall in connection with the attempt to exterminate the remnant of Israel in Jerusalem. The entire prophecy belongs to the yet future day of the Lord" (pp. 881–82, n. 1). For a discussion of the Scofield Reference Bible that predates the new edition of 1967, see Frank Gaebelein, *The Story of the Scofield Reference Bible, 1909–1959* (New York: Oxford University Press, 1959).

67   Compare Josh. 12:21, 2 Kings 23:29, 2 Chron. 30:22–24, and Zech. 12:11.

68   Scofield no doubt looked on the czarist pogroms in his speculations that the attack would arise from Russian anti-Semitism. There was also speculation that, because so many Russian pilgrims visited Holy Land shrines, "there is no land under the sun [Russia] would [more] like to possess." After the Bolsheviks seized power in 1918, "history and prophecy seemed to converge; the identification of Russia and Gog took a powerful new lease on life" (Boyer, "*When Time Shall Be No More,*" 156). See also Dwight Wilson, *Armageddon Now! The Premillenarian Response to Russia and Israel since 1917* (Tyler, Tex.: Institute for Christian Economics, 1991), 16–17.

69   Among many examples of the end-time impulse to technologize the Gog oracle, see David L. Cooper, *When Gog's Armies Meet the Almighty: An Exposition on Ezekiel Thirty-Eight and Thirty-Nine* (Los Angeles: Biblical Research Society, 1970). In a section titled "The Weapons Used," Cooper responds to references to "the shields and bucklers, the bows and the arrows" (Ezek. 39:9, 10) as weapons by maintaining that language of this sort is Ezekiel's way of encoding the weapons of today (planes, machine guns, tanks, etc.): "Ezekiel had to speak of the future weapons of warfare in terms of those with which his auditors were familiar" (pp. 103–4).

70   Boyer, "*When Time Shall Be No More,*" 305.

71   Ibid., 311.

72   Grace Halsell, *Prophecy and Politics: Militant Evangelists on the Road to Nuclear War* (Westport, Conn.: Lawrence Hill, 1986), 21–27. Journalist, columnist, and White House staff writer under President Lyndon B. Johnson, Halsell darkened her skin and lived for six months as a black woman in Harlem. In response to that experience, she produced *Soul Sister* (Greenwich, Conn.: Fawcett, 1970). (*Contemporary Authors,* ed. Linda Metzger [Detroit: Gale Research, 1984], 13:241–42.)

73   Halsell, *Prophecy and Politics,* 33–35. For a more recent statement of these ideas, see Jerry Falwell, "The Twenty-First Century and the End of the World," *Fundamentalist Journal* (May 1988), 10–11.

74   A. G. Mojtabai, *Blessed Assurance: At Home with the Bomb in Amarillo, Texas* (Boston: Houghton Mifflin, 1986). For a discussion of Mojtabai's book, see Harvey Cox, "Fundamentalism as an Ideology," in *Piety and Politics: Evangelicals and Fundamentalists Confront the World,* ed. Richard John Neuhaus and Michael Cromartie (Lanham, Md.: University Press of America, 1987), 294–96.

75   In recent years, Pantex has become the center of both the assembly and the disassembly of nuclear weapons. For a history of Pantex as well as a full description of the facility and its present operations, see the Pantex site on the Web (http://www. pantex.com/ds/pxgeng.htm).

76   Mojtabai, *Blessed Assurance,* 48–50. According to Mojtabai, the locomotive is no longer white but now multicolored.

77 Ibid., 163–64, 168.

78 Hal Lindsey with C. C. Carlson, *The Late Great Planet Earth* (Grand Rapids, Mich.: Zondervan, 1970).

79 Weber, *Living in the Shadow of the Second Coming*, 211.

80 See Hal Lindsey, *There's a New World Coming: "A Prophetic Odyssey"* (Santa Ana, Calif.: Vision House, 1973), *The 1980's: Countdown to Armageddon* (New York: Bantam, 1981), *The Rapture: Truth or Consequences* (New York: Bantam, 1983), *Planet Earth—2000 A.D: Will Mankind Survive?* (Palos Verdes, Calif.: Western Front, 1994), and *The Final Battle* (Palos Verdes, Calif.: Western Front, 1995), among others.

81 Weber, *Living in the Shadow of the Second Coming*, 220–21. See Lindsey's account of his addresses to the American War College and the Pentagon in *The 1980's: Countdown to Armageddon*, 5–6. Learning of the global, military implications of Ezekiel's prophecy (among the other prophets that Lindsey discusses), various Pentagon officials informed Lindsey that they, too, had come to the same conclusions. Lindsey's "prophetic" nuclearization of the Bible confirmed them in their beliefs (p. 6).

82 This summary is derived from Weber, *Living in the Shadow of the Second Coming*, 213–14. In one form or another, the ideas espoused by Lindsey are present in other prophecy believers. See, among others, Charles E. Pont, *The World's Collision* (Boston: W. A. Wilde, 1956); Richard W. DeHaan, *Israel and the Nations in Prophecy* (Grand Rapids, Mich.: Zondervan, 1968); John F. Walvoord, *Armageddon, Oil, and the Middle East Crisis* (Grand Rapids, Mich.: Zondervan, 1974); Elwood McQuaid, *It Is No Dream: Bible Prophecy: Fact or Fiction?* (Bellmawr, N.J.: Friends of Israel Gospel Ministry, 1978); Richard W. DeHaan, *There's a New Day Coming: A Survey of Endtime Events* (Eugene, Oreg.: Harvest House, 1978); Edgar C. James, *Armageddon!* (Chicago: Moody, 1981); Louis Goldberg, *Turbulence over the Middle East* (Neptune, N.J.: Loizeaux Bros., 1982); Jack Van Impe, *11:59 . . . and Counting* (Royal Oak, Mich.: Jack Van Impe Ministries, 1983); and Grant Jeffrey, *Armageddon: Appointment with Destiny* (Toronto: Frontier Research, 1988). Van Impe Ministries has an Internet site ("Intelligence Briefing"). So all pervasive is end-time thinking in American life that there are multiple sites devoted to it on the Web. See, among others, the "Watcher's Website" (http://www.mt.net/ ⁻watcher/magog1.html) and corresponding links: "Angelic Conspiracy Homepage," "Antichrist & UFOS Linked!" "UFOS in Israel!" and "The Bible Prophecy Page" (along with its multiple links).

83 In more recent times, see *The 1980's: Countdown to Armageddon*, which provides a series of charts comparing American military power with Soviet military power, both keyed to Lindsey's reading of biblical prophecy (pp. 68–74). In *The Final Battle*, Lindsey has a section devoted to the arms race in modern times and the ties that this arms race has to Russian militancy and Islamic fundamentalists. This race, Lindsey observes, reflects "a timetable and a countdown to the fulfillment of Ezekiel 38 and 39, the two biblical prophecies that best describe Russia's last dash to the South and the beginning of the world's Final Battle" (pp. 177–80). Later in the book, Lindsey devotes a section to "Ezekiel's View of A.D. 2000." Addressing Ezek. 38:16, Lindsey says: "Picture this now: Ezekiel looking into the future with his ancient experiences sees this invasion force coming upon the Middle East like a cloud and sees a great shaking of the earth. Couldn't this well be a description of a nuclear strike?" Moving to Ezek. 38:20–22, Lindsey concludes: "I believe that Ezekiel is predicting accurately the phenomena [*sic*] of an escalating nuclear exchange" (pp. 242–43). Compare Ed Hindson, *End Times, the Middle East, and the New World Order* (Wheaton, Ill.: Victor, 1991); and

Grant R. Jeffrey, *Prince of Darkness: Antichrist and the New World Order* (Toronto: Frontier Research, 1994).

84 In *The Late Great Planet Earth*, Lindsey imagines Armageddon as "an all-out attack of ballistic missiles upon the great metropolitan areas of the world," a nuclear holocaust of greatest moment (p. 166).

85 Produced by Robert Amran Films and RCR Productions, *The Late Great Planet Earth* (1977) is directed by Robert Amran.

86 Reflecting the dispensationalist point of view, Lindsey looks on himself as the staunch supporter of Israel and the Jews. His *The Final Battle* provides a detailed account of anti-Semitism (rooted in an "ancient racial blood feud") from biblical times to the present (pp. 21–87). For an illuminating discussion of the "philosemitic/antisemitic" dimensions of prophecy belief, see Boyer (*"When Time Shall Be No More,"* 181–224).

87 Chaim Herzog died 17 April 1997. During a long and distinguished career, he was not only a soldier and diplomat but also an author of distinction. He founded Israel's military intelligence service in 1948. A member of the Labor Party, he was a member of Israel's parliament before serving as Israel's sixth president from 1983 to 1993 (*New York Times*, Friday, 18 April 1997, sec. 1, 15).

88 Boyer, *"When Time Shall Be No More,"* 168–69. So, in a chapter titled "The Yellow Peril" in *The Late Great Planet Earth*, Lindsey warns that "a vast Oriental army" is mobilizing to enter the Middle East in conjunction with its Red neighbors (pp. 81–87). For corresponding references in *The Late Great Planet Earth* to the other "perils," see also pp. 63, 68–69, 153. For racialism in Lindsey's other works, see, e.g., *There's a New World Coming*, 139–42.

89 Robert Glen Gromacki, *Are These the Last Days?* (Schaumburg, Ill.: Regular Baptist Church, 1970), 94–95.

90 At the very extreme of the impulse to emphasize racial categories in marking the enemy are the radical fundamentalist groups such as the Identity Baptists. Here, the Jewish/Israel ties are severed, and white Christian Americans become the spiritual and genealogical descendants of the biblical Israelites. Accompanying this new line of descent is an elaborate narrative moving from Genesis to Revelation. In this reinscription of biblical history, whites assume supremacy, and blacks and Jews (particularly in the allegiance to Zionism) are conceived as the enemy. Once again, Ezekiel's oracle against Gog assumes primary focus. Underscoring the beliefs of the Identity Baptists is a renewed militancy against those whose "racial identity" is held as suspect, if not threatening to the cause that the new "Israelites" uphold. (For enlightening discussions of this dimension of prophecy belief, see, among others, Phillip Finch, *God, Guts, and Guns* [New York: Seaview, 1983], 66–76; Lowell D. Streiker, *The Gospel Time Bomb: Ultrafundamentalism and the Future of America* [Buffalo: Prometheus, 1984], 95–109; and Michael Barkun, *Religion and the Racist Right: The Origins of the Christian Identity Movement* [Chapel Hill: University of North Carolina Press, 1994].)

91 Lindsey, *The 1980's: Countdown to Armageddon*, 32–33.

92 Lindsey, *Planet Earth—2000 A.D.*, 66–74. For an astute philosophical analysis of the anxieties and fears of those swept up in the impulse to demonize the "other," see Eric Hoffer, *The True Believer: Thoughts on the Nature of Mass Movements* (New York: Harper & Row, 1951), passim. Of particular interest is Hoffer's discussion of "hatred" as a prime motivation (pp. 89–99).

93 *Planet Earth—2000 A.D.*, 74–77.

94 Ibid., 73.

95 For biblical proof texts concerning the Rapture, see Lindsey, *The Rapture*, passim.

96 Lindsey's view of the Rapture subscribes to the premillennialist idea that there will be a seven-year period known as the Great Tribulation culminating in Armageddon, which, in turn, will be followed by the Second Coming and the millennial reign of Christ. Members of the true church will not suffer the woes of the Great Tribulation because they will have already been "raptured" before the onset of the millennium (R. Ludwigson, *A Survey of Bible Prophecy* [Grand Rapids, Mich.: Zondervan, 1975], 115–26). Founded on the belief that the true church will be removed or raptured from the earth to meet Christ in the air, the idea of the Rapture implies two phases of the Second Coming: first, before the Tribulation, Christ will come *for* his saints; then, after the Tribulation, he will come *with* his saints to subdue the Antichrist and establish his millennial kingdom. The Rapture itself may come at any moment (Weber, "Premillennialism and the Branches of Evangelicalism," 10). Lindsey calls the Rapture "the great snatch" (*The Rapture*, 29). Premillennialist dispensationalism represents a crucial dimension of fundamentalist belief. Although evident in the work of such early eschatologists as Joseph Mede in the seventeenth century, the main lines of dispensationalist thought among the fundamentalists are the heirs of Edward Irving, Henry Drummond, and especially John Nelson Darby in the nineteenth century (Boyer, "*When Time Shall Be No More*," 86–90).

97 For further development of these themes in the context of the Rapture, see Wood, *Prophecy in the Space Age*, 92–93. According to Wood, the Rapture encompasses the ascent not only of Enoch and Elijah but also of Saint Paul, who is transported to the third heaven (2 Cor. 12:2–4). So we shall be transported in "the wagons and chariots that Christ will send for us." The ultimate form of the Rapture is the Ascension of Christ himself (Acts 1:11). To these instances, Wood adds the coming forth of the throne-chariot of the Ancient of Days in Dan. 7:13, a counterpart of Ezekiel's throne-chariot.

98 Lindsey, *The Rapture*, 43, 45–46.

99 Mojtabai, *Blessed Assurance*, 180–81.

100 Weber, *Living in the Shadow of the Second Coming*, vii–viii.

101 Strozier, *Apocalypse*, 120–23.

102 Harding, "Imaging the Last Days," 63.

4 Arming the Heavens

1 Although it has often been argued that end-time thinking (particularly of the premillennialist sort) by its very nature eschews politics, precisely the opposite is true: the very idea of being "raptured" as an escape from the tribulations that will beset humankind implies its own kind of politics. Even if it is a politics of disengagement, such a view has an immense bearing on one's conduct in this world. But premillennialist end-time thinkers have generally endorsed a politics very much in accord with those who deem it essential to be prepared for the inevitable attack of the "other."

2 See *The God Pumpers: Religion in the Electronic Age*, ed. Marshall Fishwick and Ray B. Browne (Bowling Green, Ohio: Bowling Greene State University Popular Press, 1987), passim.

3 Harding, "Imagining the Last Days," 59. Her observation is drawn from Frances Fitzgerald's provocative "Reflections: The American Millennium," *New Yorker*, 11 November 1985, 88–113. Fitzgerald distinguishes between "gnostic" and "Manichaean" modes of thinking in American foreign policy. Reflecting a Manichaean outlook, the United States "is a land of prophets, not of philosophers." According to Fitzgerald, evangelicalism has been "not just the sub-

stance of American policy" but "the spirit of the nation itself" (pp. 110–12). For the significance of this idea in policy debates (particularly among conservatives), see Ed Dobson and Ed Hindson, "Apocalypse Now? What Fundamentalists Believe about the End of the World," *Policy Review* 38 (Fall 1986): 16–22. (*Policy Review* is published by the Heritage Foundation.)

4 Harding, "Imagining the Last Days," 58–59.

5 For autobiographical accounts of Colson's religious experiences (pre- and postconversion), see Charles Colson, *Born Again* (Lincoln, Va.: Chosen, 1976), and *Life Sentence* (Lincoln, Va.: Chosen, 1976).

6 Colson's political career commenced with his appointment as special counsel to President Nixon in 1969. Striving to protect Nixon from his opponents, Colson involved himself in such activities as the burglary and cover-up. Colson attempted to cover up the break-in at the Democratic Party headquarters at the Watergate office building in Washington, D.C. After serving his prison sentence, he began working with the Christian ministry organization Fellowship House. He founded Prison Fellowship Ministries in 1976. For additional information, see *Contemporary Authors*, vol. 54, ed. Jeff Chapman and John D. Jorgenson, New Revision Series (Detroit: Gale Research, 1997), 88–91.

7 Charles Colson with Ellen Santilli Vaughn, *Kingdoms in Conflict* (n.p.: William Morrow/ Zondervan, 1987), 9–13.

8 Ibid., 14–15.

9 Ibid., 18.

10 Ibid., 19–40.

11 Ibid., 40n.

12 Halsell, *Prophecy and Politics*, 16–17. For an enlightening discussion of Robertson in the context of the Religious Right and the Christian Coalition, see Harvey Cox, "The Warring Visions of the Religious Right," *Atlantic Monthly* 276, no. 5 (November 1995): 59–69. The author of several books concerning evangelicalism, Robertson has guided the growth of the Christian Broadcasting Network from its initial broadcast in 1961 to its worldwide syndication at the present time. He is founder and chair of the Christian Coalition and the American Center for Law and Justice, among other organizations, and he serves to foster many evangelical causes. He is also founder and chancellor of Regent University, Virginia Beach, Va. Among Robertson's books, see *The New World Order* (Dallas: Word, 1991) and *The Secret Kingdom* (Dallas: Word, 1992).

13 Robertson, *The End of the Age* (Dallas: Word, 1995). *The End of the Age* is only one among many such novels concerning end time. See Gary G. Cohen, *The Horsemen Are Coming* (Chattanooga: AMG, 1987); David P. Dolan, *The End of the Age* (Grand Rapids, Mich.: F. H. Revell, 1995); and Tim LaHaye and Jerry B. Jenkins, *Left Behind: A Novel of the Earth's Last Days* (Wheaton: Tyndale, 1995), and, the sequel, *Tribulation Force: The Continuing Drama of Those Left Behind* (Wheaton: Tyndale, 1996).

14 Robertson, *The End of the Age*, 168. One of the most complex deities of Hinduism, Shiva (which translates from the Sanskrit as "Auspicious One") is "both the destroyer and the restorer, the great ascetic and the symbol of sensuality, the benevolent herdsman of souls and the wrathful avenger." His female consort is known variously as Uma, Sati, Parvati, Durga, and Kali. Shiva has three eyes, "the third eye bestowing inward vision but capable of burning destruction when focused outward." Around his neck he wears a garland of skulls and a serpent (*Encyclopaedia Britannica* [http://www.eb.com]).

15 Robertson, *The End of the Age*, 169–70.

16   Pat Robertson, *The New Millennium: 10 Trends That Will Impact You and Your Family by the Year 2000* (Dallas: Word, 1990), 86–87.

17   Robertson, *The End of the Age*, 294–95, 365–74.

18   See, among others, Halsell, "Imagining the Last Days," 40–50; Boyer, "*When Time Shall Be No More*," 142–46, 162; Weber, *Living in the Shadow of the Second Coming*, viii–ix; and Bob Slosser, *Reagan Inside Out* (Waco, Tex.: Word, 1984), passim.

19   Halsell, "Imagining the Last Days," 42, 43. See also William Rose, "The Reagans and Their Pastor," *Christian Life* 30 (May 1968): 24–25, 44–48; Slosser, *Reagan Inside Out*, 48–53, 71.

20   Halsell, "Imagining the Last Days," 42–43; Slosser, *Reagan Inside Out*, 13–19. See also Lou Cannon, *President Reagan: The Role of a Lifetime* (New York: Simon & Schuster, 1991), 288.

21   Halsell, "Imagining the Last Days," 45–46.

22   Slosser, *Reagan Inside Out*, 90–99. As a result of these speeches, Reagan was assailed in the press as "The Right Rev. Ronald Reagan" (ibid., 99). For a transcript of his speeches, see ibid., 185–96. For Reagan's own account of his speeches, see Ronald Reagan, *An American Life* (New York: Simon & Schuster, 1990), 568–70. See also Paul D. Erickson, *Reagan Speaks: The Making of an American Myth* (New York: New York University Press, 1985), 76–86. In keeping with his commitment to prayer and the Bible, Reagan proclaimed both a National Day of Prayer and the Year of the Bible in 1983 (*United States Code, Congressional and Administrative News* [St. Paul: West, 1983], 3:A11–A13).

23   Halsell, "Imagining the Last Days," 47–49. See the discussion of Lindsey in chap. 3 above. For an account of Reagan's views regarding Armageddon as reflected in his speeches, see Kurt Ritter and David Henry, *Ronald Reagan: The Great Communicator* (New York: Greenwood, 1992), 14–31.

24   For Reagan's responsiveness to Lindsey's *Late Great Planet Earth*, see Boyer, "*When Time Shall Be No More*," 142; and Halsell, "Imagining the Last Days," 47. The phrase "hooked on Armageddon" is Cannon's (*President Reagan*, 289).

25   Cannon, *President Reagan*, 289–90. In 1989, Reagan told Cannon that signs of Armageddon can be seen everywhere, including changing weather patterns, natural disasters (such as earthquakes), and world events (p. 289).

26   See John Herbers, "Armageddon View Prompts a Debate," *New York Times*, 24 October 1984, sec. 1, 1, 25. See also the editorial "Reckoning with Armageddon," *New York Times*, 25 October 1984, sec. 1, 25. Likewise of interest is Kenneth L. Woodward, "Arguing Armageddon," *Newsweek*, 5 November 1984, 91. See also Yehezkel Landau, "The President and the Bible: What Do the Prophets Say to Our Time?" *Christianity and Crisis*, 12 December 1983, 474–75.

27   James Mills, "The Serious Implications of a 1971 Conversation with Ronald Reagan: A Footnote to Current History," *San Diego Magazine*, August 1985, 140–41, 258. For other accounts of Mills's article, see Halsell, "Imagining the Last Days," 44–46; Cannon, *President Reagan*, 289; and Boyer, "*When Time Shall Be No More*," 142. A Democrat, Mills was first elected to the Assembly in 1960 and served until elected to the Senate in 1966. He was elected president pro tempore of the Senate in 1971 (*Who's Who in American Politics*, ed. Paul A. Theis and Edward Henshaw, 3d ed. [New York: R. R. Bowker, 1971], 712).

28   The attack on Libya occurred on 14 April 1986, when U.S. Air Force and Navy bombers stationed in England attacked the Libyan cities of Tripoli and Benghazi to retaliate for the bombing of a West Berlin disco. For an account of the attack, see Cannon, *President Reagan*, 653–54. Calling Gadhafi "the crackpot in Tripoli," Reagan offered his own account of the attack in *An American Life*, 519–20. The text of Reagan's television address to the nation

concerning the attack is anthologized in Ronald Reagan, *Speaking My Mind: Selected Speeches* (New York: Simon & Schuster, 1989), 285–88. I discuss the attack on Libya in the context of the Nation of Islam later in this study.

29 Having been emperor of Ethiopia from 1930, Haile Selassie (1892–1975) was overthrown by a military coup in September 1974. The monarchy was abolished in 1975 and a socialist republic established in its place.

30 Mills, "A 1971 Conversation with Ronald Reagan," 141.

31 Merrill F. Unger, *Beyond the Crystal Ball* (Chicago: Moody, 1973), 96–104. Compare Hitchcock, *After the Empire*, 154–55. For end-timer Robert Glenn Gromacki, the event is reminiscent of the overwhelming of Satan's forces in *Paradise Lost*, as the devils fruitlessly conspire in their own demise by arming themselves against the Son of God at their overthrow (6.785–97). Consistent with this act of rearticulating end-time belief in literary terms, Gromacki draws on Milton's epic to portray what he calls "the satanically motivated nations" justifying their downfall in the words of Satan: "Better to reign in Hell than serve in Heaven" (1.263). At the core of Gromacki's allusion to *Paradise Lost* is the Chariot of Paternal Deitie as the divine instrument of God's power and glory. With the advent of this chariot, Gog and his forces are entirely overcome at the battle of Armageddon (*Are These the Last Days?* [Schaumburg, Ill.: Regular Baptist Church, 1970], 98).

32 See, e.g., Lindsey, *Planet Earth—2000 A.D.*, 71. Interestingly, Lindsey invokes the inaugural vision as chariot/UFO only to bypass it in his determination to demonize the very notion of UFOS as the products of an alien mentality, an idea discussed earlier. If one were to draw Lindsey's parallel to its obvious conclusion, then the "four angelic creatures" would be transformed into demons, and God's *merkabah* would assume the form of a threatening, alien vehicle. In Lindsey's thinking, the dark side of the *merkabah* resurfaces as a dimension of his eschatology.

33 According to Cannon, "Robert (Bud) McFarlane, who played a significant role in the formation of SDI, was convinced that Reagan's interest in antimissile defense was the product of his interest in Armageddon" (*President Reagan*, 290). The two (Armageddon and SDI) were linked in Reagan's mind.

34 One in a series of Brass Bancroft movies, *Murder in the Air* (Burbank, Calif.: Warner Bros., 1940) was written by Raymond Schrock and directed by Lewis Seiler. *Murder in the Air* received favorable reviews. See *The "New York Times" Film Reviews (1939–48)* (New York: New York Times, 1970), 3:1719.

35 For the effect of *Murder in the Air* on the later Reagan, see Michael Paul Rogin, *Ronald Reagan, the Movie, and Other Episodes in Political Demonology* (Berkeley and Los Angeles: University of California Press, 1987), 1–3; Philip M. Boffey et al., *Claiming the Heavens: The "New York Times" Complete Guide to the Star Wars Debate* (New York: Times Books, 1988), 4–5; Garry Wills, *Reagan's America: Innocents at Home* (Garden City, N.Y.: Doubleday, 1987), 76–77, 360–61; and Cannon, *President Reagan*, 292.

36 For a description of the technology of SDI, see Sanford Lakoff and Herbert F. York, *A Shield in Space? Technology, Politics, and the Strategic Defense Initiative* (Berkeley and Los Angeles: University of California Press, 1989), 93–94. Reagan offers his own account of SDI in *An American Life*, 14–15 and passim.

37 According to a recent article in the *New York Times* (18 May 1997, sec. 1, 15), a "Star Wars" defense system is still many decades (and billions of dollars) in the future. Most tests to date have proved largely unsuccessful.

38 Ronald Reagan, "A New Defense," in *Ronald Reagan, 1911–: Chronology—Documents—Biblio-graphical Aids,* ed. Irving J. Sloan, Presidential Chronology Series (Dobbs Ferry, N.Y.: Ocana, 1990), 155–65, cited in Boffey et al., *Claiming the Heavens,* 269–71.

39 Edward Tabaor Linenthal, *Symbolic Defense: The Cultural Significance of the Strategic Defense Initiative* (Urbana: University of Illinois Press, 1989), 10. According to Ben Bova, the "Star Wars" portion of the speech was apparently drafted by the president himself: "Most of the nation, most of the world, indeed most of Reagan's own advisers, were startled by this announcement. Many White House advisers had not been informed that the President was going to go public with the missile defense idea" (*Assured Survival: Putting the Star Wars Defense in Perspective* [Boston: Houghton Mifflin, 1984], 142–43).

40 Reagan, "A New Defense," 158–59, 163–65.

41 "The Strategic Defense Initiative, June 1985," Special Report no. 129 (Washington, D.C.: U.S. Department of State, 1985).

42 Linenthal, *Symbolic Defense,* 7–8.

43 Boffey et al., *Claiming the Heavens,* 11–12.

44 Edward Teller, "The Antiweapons," in *Better a Shield than a Sword: Perspectives on Defense and Technology* (New York: Free Press, 1987), 31–32.

45 Boffey et al., *Claiming the Heavens,* 11–28. "According to several individuals, Dr. Teller's secret briefings on the X-ray laser often glossed over the nuclear dimensions of the device. 'It's not a bomb, is it?' Defense Secretary Weinberger asked in reference to the X-ray laser shortly after Reagan's speech. An aide suggested he refer to it in upcoming congressional testimony as 'a nuclear event' " (p. 24).

46 The full title is *Star Wars, Episode IV: A New Hope* (San Rafael, Calif.: Lucasfilm, 1977), distributed by Twentieth Century Fox. The film was written and directed by George Lucas and produced by Gary Kurtz, with music by John Williams. The role of Princess Leia was acted by Carrie Fisher, Luke Skywalker by Mark Hamill, Obi-Wan Kenobi by Sir Alec Guinness, and Han Solo by Harrison Ford. Nominated for eleven Academy Awards (including Best Picture), *Star Wars* enjoyed great popularity when it first appeared. Its sequels are *The Empire Strikes Back* (Lucasfilm, 1980), written by Leigh Brackett and Lawrence Kasdan, directed by Irvin Kershner; and *Return of the Jedi* (Lucasfilm, 1983), written by Lawrence Kasdan and George Lucas, directed by Richard Marquand. The films were reissued in the mid-1990s.

47 According to Lakoff and York, the Strategic Defense Initiative was labeled "Star Wars" in the media because it was thought that SDI involved the orbiting of laser weapons (*A Shield in Space,* 29–30).

48 References are to the film version of *Star Wars* (1977), in accord with the complete script of the film published in George Lucas, *The Art of Star Wars, Episode IV: A New Hope, from the "Journal of the Whills,"* ed. Carol Titelman (New York: Ballantine, 1979). Citations from the complete script will be noted as *The Art of Star Wars.* Additional, supplementary references are to the novelized version of the film, ghostwritten by Alan Dean Foster from the screen-play, in George Lucas, *Star Wars: A New Hope, Episode IV, from the Adventures of Luke Skywalker,* anthologized in *The Star Wars Trilogy* (New York: Ballantine, 1976). Citations from the novelized version will be noted as *Star Wars: A New Hope.*

49 *The Art of Star Wars,* 49.

50 *Star Wars: A New Hope,* 66–67.

51 See Otto, *Das Heilige.*

52 *The Art of Star Wars,* 49.

53 *Star Wars: A New Hope*, 65.

54 *The Art of Star Wars*, 78, 52.

55 *The Art of Star Wars*, 49, 18, 134.

56 Edward Reiss, *The Strategic Defense Initiative* (Cambridge: Cambridge University Press, 1992), 153–64. See also Michael Vlahos, *Strategic Defense and the American Ethos: Can the Nuclear World Be Changed?* (Boulder, Colo.: Westview, 1986).

57 Reiss, *The Strategic Defense Initiative*, 153–55.

58 Citing H. G. Wells's *The War of the Worlds* (1898), Bova explores the science-fiction basis of the concept of the light-beam technology underlying Reagan's "Star Wars." In other contexts, "death-rays and disintegrator weapons were the standard sidearms for Buck Rogers, Flash Gordon, and a bevy of science fiction and comic strip heroes" (*Assured Survival*, 115–16).

59 Reiss, *The Strategic Defense Initiative*, 160–63.

60 Internet games, along with a host of other activities, are currently available at the "Star Wars Trilogy: Official Website" (http://www.starwars.com/home.html).

61 Cannon, *President Reagan*, 62–64, 292. "Such sharing of sophisticated technology seemed silly to politicians of every description, from Richard Perle to Walter Mondale" (p. 292).

62 Ibid., 292, 63–64.

63 Ibid., 61–64, 292–93.

64 From *The Day the Earth Stood Still* (Beverly Hills, Calif.: Twentieth Century Fox, 1951), directed by Robert Wise, screenplay by Edmund H. North, based on a story by Harry Bates. Michael Rennie appears in the role of Klaatu, Sam Jaffe is the scientist, and Patricia Neal is the young woman whom Klaatu befriends.

65 Linenthal, *Symbolic Defense*, 9.

66 Robert Serling, *Air Force One Is Haunted* (New York: St. Martin's, 1985). Reagan's acquaintance with this novel is alleged by Lou Cannon, who learned about it from "a mutual friend" (*President Reagan*, 293). Cannon does not identify the friend.

67 According to Cannon, "Reagan's enduring model for presidential performance in time of economic crisis was his first political hero, Franklin D. Roosevelt" (*President Reagan*, 108).

68 The code name for Air Force One is "Angel," an appropriately celestial appellation for such a vehicle. See J. F. terHorst and Colonel Ralph Albertazzie, *The Flying White House* (New York: Coward, McCann & Geoghegan, 1979).

69 Serling, *Air Force One Is Haunted*, 208–9.

70 For an account of Reagan's 1986 summit meeting with Gorbachev in Reykjavík, see Cannon, *President Reagan*, 763–72 and passim.

5 Heralding the Messenger

1 C. Eric Lincoln, "The Muslim Mission in the Context of American Social History," in *African American Religious Studies: An Interdisciplinary Anthology*, ed. Gayraud Wilmore (Durham, N.C.: Duke University Press, 1989), 346. See also Lincoln's pioneering study *The Black Muslims in America* (1961), 3d ed. (Grand Rapids, Mich.: Wm. B. Eerdmans; Trenton, N.J.: Africa World, 1994).

2 The distrust of the Nation of Islam has a long history extending back to the television documentary produced by Mike Wallace and Louis Lomax and aired 10 July 1959 on New York's WNTA-TV. Titled "The Hate That Hate Produced," this documentary introduced the movement as a religious group preaching a gospel of hatred toward whites and the su-

premacy of blacks as a reflection of the "rising tide of racism" among a segment of the black population in America.

3 The accounts of the Honorable Elijah Muhammad's life vary. My information is drawn from Lincoln, *The Black Muslims in America*, 179–80; E. U. Essien-Udom, *Black Nationalism: A Search for Identity in America* (Chicago: University of Chicago Press, 1962), 74–75; Mattias Gardell, *In the Name of Elijah Muhammad: Louis Farrakhan and the Nation of Islam* (Durham, N.C.: Duke University Press, 1996), 47–68; Clifton E. Marsh, *From Black Muslims to Muslims: The Resurrection, Transformation and Change of the Lost-Found Nation of Islam in America, 1930–1995*, 2d ed. (Metuchen, N.J.: Scarecrow, 1996); *The Negro Almanac: A Reference Work on the African American*, ed. Harry A. Plosk and James Williams, 5th ed. (New York: Gale, 1989), 1329–30; Arna Bontemps and Jack Conroy, *Anyplace but Here* (New York: Hill & Wang, 1966), 217–24; Louis E. Lomax, *When the Word Is Given* (New York: New American Library, 1963), 31–51; Malu Halasa, *Elijah Muhammad* (New York: Chelsea, 1990); Raymond L. Hall, *Black Separatism in the United States* (Hanover, N.H.: University Press of New England, 1978); Claude Andrew Clegg III, *An Original Man: The Life and Times of Elijah Muhammad* (New York: St. Martin's, 1997); and Arthur J. Magida, *Prophet of Rage: A Life of Louis Farrakhan and His Nation* (New York: Basic, 1996), 37–43. See also John Robert Howard, "Becoming a Black Muslim: A Study of the Commitment Process in a Deviant Political Organization" (Ph.D. diss., Stanford University, 1965); and Barbara Ann Norman, "The Black Muslims: A Rhetorical Analysis" (Ph.D. diss., University of Oklahoma, 1985). For background on the whole Muslim movement, see Monroe Berger, "The Black Muslims," *Horizon* 6 (1964): 49–64. Of corresponding interest is Gilles Kepel, *Allah in the West: Islamic Movements in America and Europe*, trans. Susan Milner (Stanford, Calif.: Stanford University Press, 1997). For the millenarian dimensions of Elijah Muhammad's thought, see Michael Parenti, "The Black Muslims: From Revolution to Institution," *Social Research* 31 (1964): 175–94.

4 Clegg, *An Original Man*, 10–11.

5 Halasa, *Elijah Muhammad*, 40–41. Black nationalism is profoundly indebted to the figure of Marcus (Mosiah) Garvey with his emphasis on pan-Africanism (see *Dictionary of American Negro Biography*, ed. Rayford W. Logan and Michael Winston [New York: W. W. Norton, 1982], 254–56). As Lincoln points out, the movement attributed to Master Fard has antecedents in the figure of Timothy Drew or the Noble Drew Ali, who established a Moorish Science Temple in Newark, N.J., in 1913 ("The Muslim Mission," 344–45). In fact, W. D. Fard claimed that he was Drew Ali reincarnated (Essien-Udom, *Black Nationalism*, 43). For an interesting discussion of Drew Ali and Marcus Garvey, see Berger, "The Black Muslims," 55–56. Also of interest is Theodore Draper, *The Rediscovery of Black Nationalism* (New York: Viking, 1970), 69–73.

6 Halasa, *Elijah Muhammad*, 43.

7 See Elijah Muhammad's own account of his condition in *Birth of a Saviour* (Chicago: Coalition for the Remembrance of Elijah, 1993): "I was poor. I was needy. I was hungry. . . . I had nothing. I had a wife and six children living in two little rooms and the third room was a little kitchen" (p. 41).

8 For a personal account of the transformation, see ibid., 41–42. The traditional date of Master Fard's appearance is set at 4 July 1930, which has its own obvious implications of independence for the Nation of Islam (Essien-Udom, *Black Nationalism*, 125). Elijah Muhammad joined the Nation of Islam in 1931, after his wife, Clara, visited the mosque where Master Fard was proclaiming his message.

9 According to Malcolm X, Master W. D. Fard's complexion was actually half black and half white. This allowed him not only to be accepted by blacks as one of their own but to move undiscovered among whites and thereby understand the ways of the enemy (Malcolm X with Alex Haley, *The Autobiography of Malcolm X* [New York: Ballantine, 1965], 167). See Elijah Muhammad's explanation in *The Theology of Time,* comp. Abass Rassoull (Hampton, Va.: UBUS Communications Systems, 1992): Master Fard's father was a black man and his mother a white woman. "He [Master Fard] said that His father knew that His Son could not be successful in coming into a solid white country being a solid Black Man"; therefore, he was sent among the white race to look like them (pp. 164–66). In an extant photograph of Master Fard that had been enshrined in the home of Elijah Muhammad, Master Fard is portrayed as a "fair-skinned Arab" (Essien-Udom, *Black Nationalism,* 126). The photograph is often re-produced in publications of the Nation of Islam and is installed at important occasions, such as Saviours' Day conventions. Additional photographs of an individual putatively identified as Master Fard may be found in Halasa, *Elijah Muhammad,* 51. The narratives surrounding Master Fard's genealogy are interesting. His mother is classified as "a Caucasian lady, a devil." Known as "Baby Gee," she is associated with the "woman clothed with the sun" in Rev. 12:1. Master Fard's father, known as Alphonso, is described as "a Jet Black Man of the Tribe of Shabazz." The date of Master Fard's birth is set at 26 February 1877. The city of his birth is Mecca (Essien-Udom, *Black Nationalism,* 125–27). Fard also maintained that he was educated in England and later at the University of Southern California (Halasa, *Elijah Muhammad,* 48–49). See also Martha F. Lee, *The Nation of Islam: An American Millenarian Movement* (Lewiston, N.Y.: Edwin Mellen, 1988), 30–31; Zafar Ishaq Ansari, "Aspects of Black Muslim Theology," *Studia islamica* 53 (1981): 137–76; Clegg, *An Original Man,* 20–22; Magida, *Prophet of Rage,* 44–57; and Gardell, *In the Name of Elijah Muhammad,* 50–53. According to Ansari, there is an entire range of theories about Master Fard's origins. Whereas some maintain that he was a Jamaican, others argue that he was a Palestinian, an Arab from Mecca, a Turk, an Indian, even a Jew ("Aspects of Black Muslim Theology," 139). In *The Black Jews of Harlem* (New York: Shocken, 1970), Howard Brotz conjectures that Master Fard was in fact one "Arnold Ford," who was part of the black Jewish movement that flourished in Harlem between 1919 and 1931. Of interest is Leon Forrest's novel *Divine Days* (New York: W. W. Norton, 1993), which focuses on Master Fard, conceived parodically as W. A. D. Ford, and "the vision of a Mothership, which arrived at the crack of dawn, swooping up Ford in a furor of jet propelling gas" (p. 79). I discuss the Mother Plane or Mother Ship throughout.

10 He was known variously as W. D. Fard, Farrad Muhammad, F. Muhammad Ali, Professor Ford, Master Wallace Fard Muhammad, Wallace Dod Ford, and Wali Farrad.

11 As such, *Shabazz* is the name bestowed on the lost tribe of the Diaspora. The name is believed to derive from the idea of sun worshiping or one who is a sun worshiper. From a biblical perspective, there might be some association with the narratives having to do with Sheba (e.g., 1 Kings 10:1–13; 2 Chron. 9:1–12). Elijah Muhammad's teachings concerning the tribe of Shabazz are available on audiotape. Gardell confirms the identification of Shabazz with Sheba, which in turn signifies *seven.* This number performs a generative function (*In the Name of Elijah Muhammad,* 155).

12 In *Birth of a Saviour,* Elijah Muhammad distinguishes between his teachings and the teach-ings of the Freemasons (pp. 45–46). According to Lincoln, Master Fard wrote two manuals for the movement. The first, *The Secret Ritual of the Nation of Islam,* was and continues to be transmitted orally. Although the second, *Teaching for the Lost-Found Nation of Islam in a*

Mathematical Way, was printed and given to registered Muslims, it was written in Fard's own "symbolic language" and required his interpretation (*The Black Muslims in America,* 14). Both manuals reflect the "mystical" dimensions of the teachings of the Nation of Islam, particularly the emphasis on the symbolic meanings of numbers. Corresponding influences are discernible, including the teachings of the Jehovah's Witnesses and Baptist fundamentalists (Gardell, *In the Name of Elijah Muhammad,* 54). According to Berger, the ideas underlying the religion and theology of the Nation of Islam under Elijah Muhammad closely resemble "the ideas and flavor" of Ismaili and Druze doctrines, two medieval heretical sects in Islam. These doctrines include secrecy, revelation of secrets to the initiated, belief in a succession of incarnations of God, cycles of history, and major emphasis on the coming of a savior. Nation of Islam members share with Ismailis and Druzes (who still survive in the Middle East and India) an interest in the occult sciences and the magic of numbers (p. 61). See, in this respect, Josef van Ess, "Drusen und Black Muslims," *Die Welt des Islams* 14 (1973): 203–13. Gardell also sees parallels between the doctrines of the Nation and those of the Ismailis and Druzes (*In the Name of Elijah Muhammad,* 182–84). Maintaining that the theology of the Nation of Islam has very little connection to traditional Islamic beliefs, Ansari sees antecedents in the theology of the Moorish Science Temple of Noble Drew Ali as well as other movements that emerged in the Northern urban centers of America during the second and third decades of the twentieth century ("Aspects of Black Muslim Theology," 167–70). See Arthur Huff Fauset, *Black Gods of the Metropolis: Negro Religious Cults in the Urban North,* 2d ed. (Philadelphia: University of Pennsylvania Press, 1975).

13 "In Islam, *Mahdi* means the one who is rightly guided (by God), and the doctrine holds that the *Mahdi* will come at the end of days to right all wrongs on earth" (Berger, "The Black Muslims," 56).

14 The account of W. D. Fard is derived from Bontemps and Conroy, *Anyplace but Here,* 216–23; Lincoln, *The Black Muslims in America,* 11–15; Erdmann Doane Beynon, "The Voodoo Cult among Negro Migrant in Detroit," *American Journal of Sociology* 43 (July 1937–May 1938): 894–907; and Malcolm X, *Autobiography,* esp. 206–7. In *The American Religion: The Emergence of the Post-Christian Nation* (New York: Simon & Schuster, 1992), Harold Bloom refers to Fard as "a major American visionary" (p. 248).

15 Malcolm X, *Autobiography,* 208. During the time that he was in prison, Malcolm X experienced his own visionary encounter with Master Fard. As he lay on his bed one night, he suddenly became aware of a man sitting beside him. He was attired in a dark suit. In complexion, "he wasn't black, and he wasn't white." Rather, he was "light brown-skinned," with "an Asiatic cast of countenance," and his hair was "oily black." Malcolm X would later come to realize that what he experienced was a vision of Master W. D. Fard, the Messiah (pp. 186, 189).

16 In commemoration of the birth of Master Fard, the Mahdi through whom this event is realized, the Nation of Islam celebrates Saviours' Day on 26 February. After the "departure" of Elijah Muhammad, the Nation likewise designated 7 October as Saviours' Day, to commemorate the birthday of Elijah Muhammad. (Consistent with the Nation's practices, I retain the spelling and punctuation of *Saviours'* in the phrase Saviours' Day.)

17 For Nation of Islam members, the name is everything. Elijah Muhammad himself had also been known as Gulam Bogans, Muhammad Rassouli, Elijah Muck Muhd, and various other names (Lincoln, *Black Muslims in America,* 179–80).

18 The explanations for Master Fard's disappearance differ. Among other accounts, see Lincoln, *The Black Muslims in America,* 15. According to Ayman Muhammad, the eldest child of Elijah

Muhammad, "Master Farad said that he was going up in the mountains, where the cavies (he referred to the Caucasian race of people as cavies)—where they would not be able to find him. Master Farad didn't specify no particular mountain. That was in 1934" (Steven Barboza, *American Jihad: Islam after Malcolm X* [New York: Doubleday, 1994], 269).

19 Charged with draft evasion and politically motivated claims of sedition, Elijah Muhammad was imprisoned as a means of subverting his cause. For an account of his imprisonment, see Elijah Muhammad, *Birth of a Saviour*, 40. See also Bontemps and Conroy, *Anyplace but Here*, 224; Gardell, *In the Name of Elijah Muhammad*, 70–71; and esp. Clegg, *An Original Man*, 84–87, 92–97.

20 Clegg, *An Original Man*, 93–94.

21 Sander L. Gilman, *Difference and Pathology: Stereotypes of Sexuality, Race, and Madness* (Ithaca, N.Y.: Cornell University Press, 1985), 132.

22 Clegg, *An Original Man*, 95–96.

23 Bontemps and Conroy, *Anyplace but Here*, 225–26.

24 Lincoln, *The Black Muslims in America*, 16–17.

25 Lincoln, "The Muslim Mission," 347–48. One is particularly struck by the personal testimony of H. Nasif Mahmoud, a partner in an international law firm based in Washington, D.C., and Pasadena and a professor at Howard University's law school: "Oh, he was so wise! Lee Iacoca, Joseph Kennedy, none of these men could shake a stick at Elijah Muhammad, and he supposedly had just a third-grade education. And he was impeded by a system that was against him! Joseph Kennedy had help. Lee Iacoca had help. Elijah Muhammad was fighting against a system full of Jim Crow discrimination, and he built an empire in the hells of North America that could be a symbol of what black Americans can do if they unite" (in Barboza, *American Jihad*, 91). For Elijah Muhammad, such unity was indeed tantamount to the building of an empire. Before Elijah Muhammad died, he presided over seventy-six temples nationwide and 50,000–100,000 people. His Nation included some 15,000 acres of farmland, a newspaper (*Muhammad Speaks*) with an estimated circulation of 500,000, several aircraft, a fish-importing concern, restaurants, bakeries, and supermarkets. Born to a family of former slaves in Georgia, Elijah Muhammad amassed a fortune estimated at $25 million (Barboza, *American Jihad*, 78). See also Gardell, *In the Name of Elijah Muhammad*, 60–61; and Curtis E. Alexander, *Elijah Muhammad on African-American Education* (New York: ECA, 1989), 55–71.

26 James Baldwin, *The Fire Next Time* (New York: Dial, 1963), 78. In a charming passage in *The Theology of Time*, Elijah Muhammad jests about the slightness of his physical stature in the context of the greatness of his spiritual stature implicit in his name: "Now comes Elijah and you read much of that. You start on Elijah in the Book of Kings in the Bible and you read about Him until you get into Revelations. This is a kind of important little fellow. . . . He is some tough Little Fellow" (p. 8).

27 The phrase *so-called Negro* is one that Elijah Muhammad uses throughout his writings as a designation underscoring the ignorance of those who do not know the true identity of those they refer to by the name *Negro*.

28 The details of this summary are derived from the following sources: Elijah Muhammad, *Message to the Blackman in America* (Philadelphia: House of Knowledge, 1965), *The Fall of America* (Newport News, Va.: National Newport News and Commentator, 1973), and *The Flag of Islam* (Chicago: FCN, 1983), passim; as well as Lincoln, *The Black Muslims in America*, 71–73; Essien-Udom, *Black Nationalism*, 122–42; Bontemps and Conroy, *Anyplace but Here*, 226–27; and Malcolm X, *Autobiography*, 164–67.

29 The phrase alludes to one of the foundation texts of the Nation of Islam, *The Supreme Wisdom: Solution to the So-Called Negroes' Problems*, 2 vols. (Chicago: Muhammad Mosque of Islam no. 2, 1957), which contains the basic tenets of the entire movement. These tenets are conveniently summarized by P. Chike Onwuachi, *Black Ideology in African Diaspora* (Chicago: Third World, 1973), 81–105.

30 Essien-Udom, *Black Nationalism*, 8.

31 Thus, Elijah Muhammad, *Message to the Blackman in America:* The explosion and consequent division into earth and moon was caused by the scientist, God, who desired his people to speak a single language (p. 31). Compare Bontemps and Conroy, *Anyplace but Here*, 226; Essien-Udom, *Black Nationalism*, 131–32. An element of Manichaean dualism seems to be at work in Elijah Muhammad's discussion of the nature of the black God of the moon in *The Flag of Islam*, 9, 26. According to Berger, "the Nation of Islam believes in a succession of Gods who followed the first Black God, the Supreme Being who created the universe sixty-six trillion years ago" ("The Black Muslims," 60). See Gardell's excellent discussion of these matters as well as what might be called the precosmogonic conceptions of the Nation (*In the Name of Elijah Muhammad*, 144–49). Gardell wisely observes the way in which the narrative of origins evolves and changes during the life of the Nation. In fact, he finds several narratives underlying the Nation's beliefs. All such narratives fall under the heading of what Gardell terms *blackosophy* (p. 172; see also pp. 144–86 and passim). Also of major importance is Clegg's account (*An Original Man*, 41–62 and passim).

32 Extending cosmogony into history, Elijah Muhammad provides a detailed analysis of the moon as an emblem of the plight of the black man in modern times. As such, the blasting away of the moon by the black scientist is comparable to black man's experience of being torn from his roots under slavery. The black man thereby loses all knowledge of self and kind. As a result, he is like the "dead moon" that needs to be replenished in order to regain life. "Here," declares Elijah Muhammad, "we the black man in America stand as a Moon devoid of wisdom and the knowledge of self and others." This deprivation is comparable to that of the moon deprived of water as a source of life. On the other hand, the crescent moon is also a positive symbol, just as are all the symbols on the flag of Islam: the sun (depicted in its fiery color as the background on the flag) represents freedom, the moon equality, and the star justice (*The Flag of Islam*, 1, 7–29). Elijah Muhammad's thinking is essentially mythopoetic and symbolic.

33 Lincoln, *The Black Muslims in America*, 71; Malcolm X, *Autobiography*, 164.

34 In suggesting the biblical underpinnings of Elijah Muhammad's thought, we must be aware of the fact that, for the Messenger, the Bible as revealed text assumed a position of ambivalence. Although Elijah Muhammad was profoundly influenced by its teachings, he approached it with what might be called a hermeneutics of suspicion, particularly as it was interpreted from what he considered the vantage point of a fallacious white supremacy employed to subjugate the black race. So he declared, "From the first day that the white race received the Divine Scripture they started tampering with its truth to make it suit themselves, and blind the black man. . . . The Bible is the graveyard of my poor people (the so-called Negroes) and I would like to dwell upon this book until I am sure that they understand that it is not quite as holy as they thought it was" (*Message to the Blackman in America*, 94–95). Viewing the Bible in this way, Elijah Muhammad did not hesitate to reinscribe it and refashion its meanings in his own terms.

35 Lincoln, *The Black Muslims in America*, 72.

36 The details concerning the story of Yakub are derived from the following sources: Elijah Muhammad, "The 15,000 Year History" (Chicago: Coalition for the Remembrance of Elijah, n.d.), *Message to the Blackman in America*, esp. 100–122, and *The Fall of America*, passim; as well as Bontemps and Conroy, *Anyplace but Here*, 227–28; Malcolm X, *Autobiography*, 164–67; Lincoln, *The Black Muslims in America*, 72–73; and Essien-Udom, *Black Nationalism*, 134–35. According to Elijah Muhammad, the history of Yakub and his people is to be found in the "Nation's Book, [composed] by the writers (23 Scientists)" some eighty-four hundred years before Yakub's birth (*Message to the Blackman in America*, 113–14).

37 "He had an unusual size head. When he had grown up, the others referred to him as the 'big head scientist' " (Muhammad, *Message to the Blackman in America*, 112).

38 Because of his association with Patmos as the site of Saint John the Divine's revelation, Yakub is likewise associated with John himself (Lee, *The Nation of Islam*, 40; see also Malcolm X, "Black Man's History," in *The End of White World Supremacy*, ed. Imam Benjamin Karim [New York: Seaver, 1971], 53).

39 According to Elijah Muhammad, these races were then and should now continue to be separated. In keeping with the creation of people in the civilization of today, each race has its own proper site of habitation. Thus, "the black man lives in Africa. The brown man is living in the islands of the Pacific. The yellow man is living on the continent known as Asia. The red man is here in North America," and, "before the coming of the white man, this was a red man's home" (*Muhammad Speaks*, 31 October 1962, 12, cited by Onwuachi, *Black Ideology in African Diaspora*, 99).

40 Muhammad, *Message to the Blackman in America*, 120–21.

41 Ibid., 121. The eschatological element in Elijah Muhammad's writings is one to which Bloom refers as "a violent form of Gnosticism" (*The American Religion*, 250).

42 John W. Roberts, *From Trickster to Badman: The Black Folk Hero in Slavery and Freedom* (Philadelphia: University of Pennsylvania Press, 1989), 103.

43 William J. Hynes, "Mapping the Characteristics of Mythic Tricksters: A Heuristic Guide," in *Mythical Trickster Figures: Contours, Contexts, and Criticisms*, ed. William J. Hynes and William G. Doty (Tuscaloosa: University of Alabama Press, 1993), 33–45. For the archetypal dimensions, see Carl Gustav Jung, "On the Psychology of the Trickster Figure," in *The Trickster: A Study in American Indian Mythology*, ed. Paul Radin (New York: Schocken, 1955), 195–211. (The volume in its entirety is of essential importance to the concept.)

44 Laura Makarius, "The Myth of the Trickster: The Necessary Breaker of Taboos," in *Mythical Trickster Figures*, ed. Hynes and Doty, 66–86.

45 Lawrence W. Levine, *Black Culture and Black Consciousness: Afro-American Folk Thought from Slavery to Freedom* (New York: Oxford University Press, 1977), 111–12, 113–14. In his probing examination of slave narratives, Levine provides fascinating accounts of both animal and human trickster tales. See in this respect the tale of Rabbit, who, in his relentless quest for power, sought enhanced potency by petitioning God himself (pp. 111–12).

46 Theophus H. Smith, *Conjuring Culture: Biblical Formations of Black America* (New York: Oxford University Press, 1994), 248 n. 49.

47 See, e.g., Elijah Muhammad, *The Fall of America*, 77, 203, 214. According to Malcolm X, the term *tricknology* goes back to Master Fard (*Autobiography*, 208).

48 Elijah Muhammad, *Message to the Blackman in America*, 101.

49 See Henry Louis Gates Jr., *The Signifying Monkey: A Theory of Afro-American Literary Crit-*

icism (New York: Oxford University Press, 1988). As Gates observes, "the Signifying Monkey is often called the Signifier, he who wreaks havoc upon the Signified" (p. 52).

50 Levine, *Black Culture and Black Consciousness*, 83–86.

51 Muhammad, *Message to the Blackman in America*, 122–24.

52 In *Message to the Blackman in America*, Elijah Muhammad specifically associates Yakub with Cain (p. 128).

53 The 1967 Saviours' Day message exists in the form of an audiotape. It has not been transcribed. As might be expected, Elijah Muhammad makes a connection between Jacob and Israel. "Israel," he says, "is not a good name. Israel is an evil name." The connection, however, is not made in the spirit of an anti-Jewish polemic. Rather, Israel as a nation represents for Elijah Muhammad the white race in general. Despite the specific association between Yakub and Jacob that Elijah Muhammad develops in his Saviours' Day message, it is interesting to note that this does not seem to be a connection that he explores elsewhere in much detail, at least in the works that I have examined.

54 Much of the virulence of the Nation of Islam toward Christianity focuses on the Catholic Church. As the spiritual foundation of the Catholic Church, the pope in particular is viewed as one of the beasts of Revelation. Glossing the pope in this case as the beast that rose from the earth (Rev. 13:11), Minister Clyde X of Muhammad's Mosque no. 28 in St. Louis performed (in 1962) a kabbalistic reading of the inscription *Vicarius Filii Dei* that the pope wears on his crown. This inscription, Minister Clyde X maintained, represents the number 666, the number of the beast in Rev. 13:16–18. To prove this interpretation, he assigned each of the letters of the inscription a numerical designation, all of which added up to 666 (Onwuachi, *Black Ideology in African Diaspora*, 167–68).

55 Elijah Muhammad, *Message to the Blackman in America*, 124–27.

56 Levine, *Black Culture and Black Consciousness*, 26. Although the specific text in question here is Ezekiel 37 (the valley of dry bones), Ezekiel 1 is entirely apt. Here, and elsewhere in my study, I make a point of reproducing the black English vernacular as recorded in the texts that I cite. I do so with full awareness that the reproduction of such vernacular is very problematic since the individuals who recorded the discourse in question were often not native to the group whose words they sought to transcribe. Acknowledging these limitations, I am convinced that, in order to be as faithful as possible to the "original," it is better to err on the side of the quest for "historical accuracy" than to impose on the transcriptions the criterion of "standard English." See Geneva Smitherman, *Talkin and Testifyin: The Language of Black America* (Detroit: Wayne State University Press, 1986).

57 In one of his speeches (partially broadcast in the documentary "The Hate That Hate Produced"), Malcolm X addresses members of the Nation of Islam with this idea in mind: "In the church we used to sing the song, 'Good news the chariot is coming.' But what we must bear in mind is that good news for one person is bad news for another."

58 The text of this spiritual is drawn from *The Negro Sings a New Heaven*, ed. Mary Allen Grissom (Chapel Hill: University of North Carolina Press, 1930), 32–33.

59 The text of this spiritual is drawn from *The Book of American Negro Spirituals*, ed. James Weldon Johnson and J. Rosamond Johnson (New York: Viking, 1925), 144–46.

60 In another version of the same spiritual, Ezekiel's vision represents the opportunity to engage not only in "theological speculation" but also in social commentary. The tone changes to embrace a new perspective, that of castigating the hypocrite for spending his time doing

nothing more than spreading false rumors about those who are truly religious ("He'll talk'about me an' he'll talk'about you"). See *Religious Folk-Songs of the Negro*, ed. R. Nathaniel Dett (Hampton, Va.: Hampton Institute Press, 1927), s.v.

61 Cited in Theo Lehmann, *Negro Spirituals: Geschichte und Theologie* (Witten and Berlin: Eckart, 1965), 364. For another version, see *Slave Songs of the Georgia Sea Islands*, ed. Lydia Parrish (Hatboro, Pa.: Folklore Associates, 1965), 154.

62 Lehmann, *Negro Spirituals*, 364–65. As a demonic symbol, such a train might also assume negative connotations. So it appears in one formulation: "This train is known as the Black Diamond Express Train to Hell; Sin is the engineer, pleasure is the head light, and the Devil is the conductor" (p. 365).

63 Smith, *Conjuring Culture*, 126. In this respect, the spiritual becomes conspiratorial as it assumes the form of a summons or an inducement to slaves to "steal" themselves away from bondage to freedom. Exploring this dimension, Smith asks us to recall the archaic meaning of the word *conjure* as that which means to "conspire," in the sense of swearing together. For a sociolinguistic perspective, see Smitherman, *Talkin and Testifyin*, 48–49. Professor Albert C. Labriola, Duquesne University, Pittsburgh, Pa., observes that coded language is "part of a larger phenomenon called the language of subjugated peoples." This phenomenon dates back more than a thousand years: "An interesting manifestation of this phenomenon includes the Saxon coded commentary on the conquering and oppressive Normans, at times rendered visually (e.g., in the marginal and, hence, marginalized visualizations of the Bayeux tapestry)" (letter to the author, 9 June 1995).

64 Levine, *Black Culture and Black Consciousness*, 32–33. See also Mechal Sobel, *Trabelin' On: The Slave Journey to an Afro-Baptist Faith* (Princeton, N.J.: Princeton University Press, 1979), 58–75.

65 Levine, *Black Culture and Black Consciousness*, 36–37. The accounts are drawn from "God Struck Me Dead: Religious Conversion Experiences and Autobiographies of Negro Ex-Slaves," ed. A. P. Watson, Paul Radin, and Charles S. Johnson (Nashville, 1945, typescript).

66 Levine, *Black Culture and Black Consciousness*, 43.

67 See Albert Raboteau, *Slave Religion: The "Invisible Institution" in the Antebellum South* (New York: Oxford University Press, 1977). In this connection, see also Gayraud S. Wilmore, *Last Things First* (Philadelphia: Westminster, 1982), and *Black Religion and Black Radicalism: An Interpretation of the Religious History of Afro-American People*, 2d ed., rev. (Maryknoll, N.Y.: Orbis, 1983).

68 Raboteau, *Slave Religion*, 312–13.

69 According to Riggins R. Earl Jr., such retribution was implicit in the way slaves viewed the Civil War, which they saw as the historical stage on which God enacted vengeance against the slavemaster: "Many spiritual songs of the slave community are replete with imagery that portrays God as the Ultimate Warrior who fights on behalf of the oppressed." In his conduct of that fight, God was both a destroyer and a redeemer (*Dark Symbols, Obscure Signs: God, Self, and Community in the Slave Mind* [Maryknoll, N.Y.: Orbis, 1993], 164).

70 Elijah Muhammad, *The Mother Plane* (Cleveland: Secretarius, n.d.).

71 According to Abass Rassoull, the transcriber and compiler of *The Theology of Time*, the collection in prepublication form was first presented to Elijah Muhammad for his approval in the fall of 1974, but apparently he did not feel that it was ready for publication at that time (p. xv). In *The Comer by Night* (Chicago: Honorable Elijah Muhammad Educational Foundation, 1986), Tynnetta Muhammad observes that the lectures were delivered in 1972. This is

confirmed by Abdul Wali Muhammad in the *Final Call* (newspaper), 31 January 1987, 2. Abdul Wali adds that the lectures were given in the Stony Island Avenue Mosque, Chicago. The time frame for the lectures is 24 June–29 October 1972. The lectures commemorate in part the opening of the mosque. In the collection as published, the lectures themselves are undated (except for the 1973 Saviours' Day address). The lectures are available on videotape.

72 Although the focus of my examination of Elijah Muhammad is on his writings, I must acknowledge at the outset the importance of the oral and visual dimensions of his message. During the period of his mission, Elijah Muhammad communicated his message principally through presentations at his mosques, through radio broadcasts, through audiotape, and through film (reel to reel) and videotape. The Nation of Islam is founded on this oral/visual culture as the most immediate and compelling mode of disseminating the message. As important as the writings are, then, print culture is finally a means of communication that is brought to the fore as a reinforcement of the oral/visual culture. This is correspondingly true in the Nation of Islam today. In this respect, see Smitherman: "Both in slavery times and now, the black community places high value on the spoken word." In fact, "the persistence of the African-based oral tradition is such that blacks tend to place only limited value on the written word, whereas verbal skills expressed orally rank in high esteem" (*Talkin and Testifyin*, 76).

73 Elijah Muhammad, *Message to the Blackman in America*, 17–18. In *The Theology of Time*, Elijah Muhammad comments on the name *Fard Muhammad* as follows: "Mahdi is His real Name, so He taught me. Mahdi means 'One who is Self-Powered' in the Arabic language. It means one who is self-controlled and controls others and who is a guide for others because he is self-guided." As a name, *Fard*, in turn, means " 'that which is the beginning; that which is opening' " (pp. 134–35). Correspondingly, in *Birth of a Saviour*, Elijah Muhammad observes that the morning prayer service is termed *fard* as that which is "absolutely binding." *Islam* itself, Elijah Muhammad comments, means "entire submission to God" (p. 36). In Arabic, the term *fard* (*feh, rah, dhuad*) carries multiple meanings. In Islamic law, it signifies a divine ordinance of God. It implies a command, an injunction, an order, a decree, or a religious duty as well as a statutory portion and lawful share. In worship, it carries the meaning of that which is obligatory, as a mandatory prayer, command, or injunction (Hans Wehr, *Dictionary of Modern Written Arabic*, ed. J. Milton Cowan [Ithaca, N.Y.: Cornell University Press, 1961], 705–6). Among its uses in the Qur'an, see Surah 24:1 (*An-Nur*): "We have revealed this Surah and made it obligatory [*fard*] as we have sent down clear injunctions in it that you may be warned" (*Al-Qur'an*, trans. Ahmed Ali [Princeton, N.J.: Princeton University Press, 1984]). According to Berger, *fard* is an Arabic word, the literal meaning of which is "the one who has no like," the unique one. By extension, "it is generally taken to mean Allah" ("The Black Muslims," 61).

74 Elijah Muhammad, *Message to the Blackman in America*, 17–18. The concept of targeting a specific people for destruction is already embedded in the biblical narrative having to do with God's smiting the firstborn of the Egyptians and sparing the Israelites (Exod. 12).

75 Elijah Muhammad, *Message to the Blackman in America*, 17–18. Elijah Muhammad says that he was instructed by Master Fard that "the world's time was out in 1914, but people could get an extension of time, depending upon their treatment of the righteous." Master Fard declared that "there can be no judgment until we (the so-called Negroes) hear Islam, whether we accept it or not" (p. 18).

76 *Muhammad Speaks*, 21 September 1973, 12.

77 Broome, "Ezekiel's Abnormal Personality."

78 Essien-Udom, *Black Nationalism,* 122.

79 See the discussion of the Mother Plane in Clegg, *An Original Man,* 65–66. According to Clegg, Master Fard Muhammad described the Mother Plane "as being constructed in Japan, where the world's largest dreadnoughts, the Yamoto class, were being designed in the 1930s" (p. 66). In the Mother Plane Clegg sees the dimension of "science fiction" associated with mythical spaceships of Martian movies popular at the time. He singles out such films as *A Trip to Mars* (1903), *England Invaded* (1909), *A Message from Mars* (1921), and *Mars Calling* (1923). The ultimate Martian thriller was the 1938 radio adaptation of H. G. Wells's *The War of the Worlds,* which frightened listeners throughout the country. The idea of inhabiting other planets, such as Mars, was consistent with the Nation's mythology (pp. 43–44, 66).

80 Essien-Udom, *Black Nationalism,* 141. It is, of course, dangerous to make large-scale generalizations about the "real presence" of the plane in the lives of all Nation of Islam members. Essien-Udom derives his information from the period of the 1950s. Some forty years later, the plane is still very much in evidence under Minister Farrakhan, as we shall see, but the move from the rhetoric of ideology and belief into the personal, day-to-day lives of individual members of the Nation of Islam must be undertaken with care.

81 Ibid., 141–42. For corresponding sentiments, see Peter Goldman, *The Death and Life of Malcolm X* (New York: Harper & Row, 1973), 41–44.

6 The Eschatology of the Mother Plane

1 Malcolm X, *Autobiography,* 185–86. Referring to *Paradise Lost,* Malcolm X comments as follows: "The devil, kicked out of Paradise, was trying to regain possession. He was using the forces of Europe, personified by the Popes, Charlemagne, Richard the Lionhearted, and other knights. I interpreted this to show that the Europeans were motivated and led by the devil or the personification of the devil. So Milton and Mr. Elijah Muhammad were actually saying the same thing" (p. 186). Recalling the history of Yakub, Malcolm X is applying that history to the figure of Satan in *Paradise Lost.* He is no defender of the satanic school: Elijah Muhammad's influence prevents him from misreading Milton's epic as a work that advocates Satan as hero.

2 The distinction between *maternal* and *paternal* is interesting in the context of a curious work that has appeared under the authorship of Bi'sana Ta'laha El'Shabazziz Sula Muhammadia and Master Tu'biz Jihadia Muhammadia, *In the Name of Almighty Ga'lah: The Holy Book of Life (Volume Two), Spiritual Government* (Detroit: Harlo, 1974). According to the author's note, this is an "introductory textbook of Spiritual Government, and Superior Knowledge," written to aid "the Blackman's rise to his proper place in life." Among the concerns of this book is "A History of Liyyum Lijab Muhammad," which is the name of the "scientist" who constructed "the Fathership and the largest portion of the Mothership." Addressing the distinction between the Fathership and the Mothership, the authors devote a good deal of attention to the form and constitution of both "machines" (pp. 123–30). I discovered the volume in the collections of the Regenstein Library of the University of Chicago. To my knowledge, there is no record of the publication of vol. 1 of this study, nor have I been able to find out anything about the authors of the book. Harlo Press itself has no record of the circumstances surrounding its publication.

3 The significance of Ezekiel is mentioned at various junctures in Lee, *The Nation of Islam,* esp. 47–48. See also Gardell, *In the Name of Elijah Muhammad,* 158–60.

4   Elijah Muhammad, *Message to the Blackman in America*, 265–73, 297–300.

5   Ibid., 290–91.

6   See the definition of *parousia* in *A Greek-English Lexicon of the New Testament and Other Early Christian Literature*, trans. and ed. William F. Arndt and F. Wilbur Gingrich (Chicago: University of Chicago Press, 1957), 635.

7   Elijah Muhammad, *Message to the Blackman in America*, 4–15. According to Essien-Udom, one of Elijah Muhammad's ministers gave the following description of God: "I learnt that God is not a spook and that He is in every respect a real man and that He is a Superior Being in the sense that He is all wise. He is a man in the sense that He is a superman" (*Black Nationalism*, 127). Laurence Levine observes, "The God the slave sang of was neither remote nor abstract, but as intimate, personal, and immediate as the gods of Africa had been" (*Black Culture and Black Consciousness*, 35). Perhaps something of this conception is at work in Elijah Muhammad's own theology.

8   The concept of the *scientist*, as we have seen throughout, is essential to Elijah Muhammad's thought. In *The Theology of Time*, Elijah Muhammad refers to himself as a "Messenger" who acts in accordance with the "Twelve Scientists of God." He declares that he is "the Last Scientist or Prophet" (p. 6).

9   Elijah Muhammad, *Message to the Blackman in America*, 290–94.

10 Like *Message to the Blackman in America*, *The Fall of America* contains essays ranging over several years and addressing a wide spectrum of topics. The second volume, however, is more nearly focused on the plight of the black person in America and the coming doom.

11 Elijah Muhammad, *The Fall of America*, 323.

12 Ibid., 236. In fact, Elijah Muhammad goes so far as to say that "the Mother Plane carries the same type of bomb on her that our black scientists dropped on the planet earth to bring up mountains out of the earth after the planet earth was created" (pp. 240–41). Once again, Elijah Muhammad appears to provide his own version of the big bang theory of the creation of the universe. This he transfers to the destruction of the planet as preparatory to its final regeneration. Similar ideas are reflected in *The Theology of Time*, 274–78.

13 Elijah Muhammad, *The Fall of America*, 236.

14 Ibid., 237–38.

15 Ibid., 238–39. Although one of the colors of the original black man, "the red," says Elijah Muhammad, "is not an equal power."

16 Ibid., 241. Elijah Muhammad cites the Qur'an (Sur. 50.16). Although not indicated in his reference, the text has the following: "We created man and surely know what misdoubts arise in their hearts; for We are closer to him than his jugular vein" (*Al-Qur'an*, trans. Ali). In Elijah Muhammad's application of this idea to the Mother Plane, there is a sense of the "influencing machine" that Broome examines in "Ezekiel's Abnormal Personality."

17 Elijah Muhammad, *The Fall of America*, 241–42.

18 Cited verbatim from the taped version of the Saviours' Day presentation, delivered in Chicago in 1967.

19 Elijah Muhammad, *The Theology of Time*, 479–80; see also 111–12.

20 Ibid., 540.

21 According to Lee, in fact, the time of apocalypse for the Nation was to have been 1965–66, during which a series of events occurred to suggest that the end was approaching. It is during this period that *Muhammad Speaks* is replete with visions of the end, coupled with an emphasis on Ezekiel's vision. In "The Truth as Revealed by Allah," by Elijah Muhammad, the

following statement occurs: "IT IS TRUE, according to religious scientists of both Christianity and Islam, as both agree that this year 1966 . . . is the fateful year of America and her people, and that the so-called Negroes should fly to Allah and follow Messenger Muhammad, for refuge from the dreadful judgments that Allah has said that He will bring, and which have already begun upon America" (*Muhammad Speaks*, 1 April 1966, 1–2, reprinted in Lee, *The Nation of Islam*, app. B, 158–59). When the apocalypse failed to occur, however, references to the event diminished (pp. 49–75).

22 *Muhammad Speaks*, 21 June 1973, 16–17.

23 *Muhammad Speaks*, 24 August 1973, 16–17. In a subsequent article, Elijah Muhammad applies similar ideas to the concept of the wheel, to which he refers as the "wheel of enslavement" of a black people subjugated by the white slavemasters (*Muhammad Speaks*, 14 September 1973, 12).

24 *Muhammad Speaks*, 24 August 1973, 12–13.

25 *Muhammad Speaks*, 24 August 1973, 12–13, and 21 September 1973, 12.

26 *Muhammad Speaks*, 24 August 1973, 12–13.

27 *Muhammad Speaks*, 7 September 1973, 12–13.

28 *Muhammad Speaks*, 14 September 1973, 12–13.

29 *Muhammad Speaks*, 21 September 1973, 12–14. Attesting to the forms of gnosis to which he is indebted, Elijah Muhammad discloses that at one time he was a Freemason and "swore . . . not to reveal the secrets" (*The Theology of Time*, 282).

30 *Muhammad Speaks*, 21 September 1973, 12–14.

31 Elijah Muhammad, *The Theology of Time*, 31.

32 In a series of reflections appended to *Birth of a Saviour*, Elijah Muhammad performs a gnostic reading of the concept *the beginning of time and the creation of God*. His text is the opening of Genesis: "In the beginning God created the heaven and the earth." Performing a kind of *gematria* on this text, he develops a "negative" or "mystical" theology through which blackness emerges from blackness. His concern is with the creation of self and selfhood for the black man (pp. 56–63). This dimension of his thought also underlies his treatment of Ezekiel's vision.

33 Elijah Muhammad, *The Theology of Time*, 509–18.

34 Ibid., 512.

35 Gardell, *In the Name of Elijah Muhammad*, 71. When interrogated by the FBI in 1942, Elijah Muhammad informed the government that "Allah has taught that blueprints of a plane which carries bombs, was [*sic*] given to the Japanese from the Holy City of Mecca, and that these blueprints had been there for thousands of years. These bombs would go into the earth for at least a mile and would throw up earth for a distance of at least a mile, so that it would make a mountain. I have reminded registered Muslims of this [*sic*] teachings" (file 105-63642-[?], 21 February 1957).

36 Elijah Muhammad, *The Theology of Time*, 512.

37 Elijah Muhammad, *The True History of Jesus as Taught by the Honorable Elijah Muhammad* (Chicago: Coalition for the Remembrance of Elijah, 1992). The collection is divided into two parts, "The True History of Jesus" (undated) and "The History of Jesus—Part II" (3 July 1966). There is also what appears to be a third part, "The True History of Jesus Revealed," but that is no doubt a section of the second part. The first part is narrative in form, the second discursive. One suspects that the whole dates back to the 1960s and perhaps earlier. Among Elijah Muhammad's other works on this subject, see *Our Saviour Has Arrived* (Newport

News: United Brothers Communications Systems, n.d.) and *Birth of a Saviour.* Compare Gardell's discussion of *The True History of Jesus* (*In the Name of Elijah Muhammad,* 234–37).

38 According to Gardell, the radio in the head signifies "the telepathic connection between a realized Self and the source of existence, as well as the wonderful abilities of a liberated mind" (ibid., 237).

39 Elijah Muhammad, *The True History of Jesus,* 1–14.

40 Ibid., 15, 16–37.

41 Ibid., 11.

42 Elijah Muhammad, *The Theology of Time,* 513.

7 Visionary Minister

1 The history of the Nation of Islam during the quarter of a century since the death of Elijah Muhammad is beginning to receive the treatment it deserves, but a great deal more needs to be done. For the purpose of the present undertaking, I am deeply indebted to the accounts provided by Lincoln, *The Black Muslims in America,* 254–76, and "The Muslim Mission"; Lawrence H. Mamiya, "From Black Muslim to Bilalian: The Evolution of a Movement," *Journal for the Scientific Study of Religion* 21 (1982): 138–52; Barboza, *American Jihad;* Lee, *The Nation of Islam,* 77–140; Marsh, *From Black Muslims to Muslims;* and Gardell, *In the Name of Elijah Muhammad.* See also Magida, *Prophet of Rage.* For the Nation of Islam's presence on the Web, see http://www.noi.org.

2 Warith Deen Mohammed is the seventh child of Elijah Muhammad. According to Barboza, Warith adopted the "legal spelling" of his father's surname, that is, Mohammed, as opposed to Muhammad (*American Jihad,* 95). It is customarily understood by the members of the Nation of Islam that Elijah Muhammad did not "die" in the literal sense. Rather, he "departed" for a season before his final return. For Louis Farrakhan's own statements on the subject, see ibid., 136.

3 Describing Warith Deen Mohammed at the present time, Barboza observes that, in contradistinction to his father, Warith lacks "the burning eyes of a zealot." Unpretentious in appearance, he dresses like a professor and emanates "about as much charisma as an accountant" (ibid., 95). Lincoln's estimate is far more magnanimous: "Wallace Deen Muhammad [Mohammed] is a dreamer, but he is a dreamer-cum-realist, and gentle, sensitive, and self-effacing. History may yet prove him to be one of the most astute religious leaders of his age." The scholar's scholar, he is a lifelong student of Islam, fluent in Arabic, and a master of qur'anic ideology ("The Muslim Mission," 349). See Gardell's treatment in *In the Name of Elijah Muhammad,* 99–118. For an enlightening treatment of the administration of Warith Deen Mohammed, see Marsh, *From Black Muslims to Muslims,* 67–78 and passim.

4 Lincoln, *The Black Muslims in America,* 263–64.

5 Barboza maintains that the financial aftermath of Imam Warith Deen Mohammed's seigniory of the Nation was deleterious to the business empire that his father had created. In fact, the Nation claimed bankruptcy and sold its Chicago mosque to Farrakhan's organization to pay off debts to its principal creditor (*American Jihad,* 96). On the other hand, Lincoln cites a series of business ventures that had been undertaken during Imam Mohammed's leadership ("The Muslim Mission," 352).

6 Lincoln, *The Black Muslims in America,* 264. In a 29 October 1990 interview, Warith Deen Mohammed phrases his estimate much more directly. He says that, although he has respect

for his father's "sincerity," Elijah Muhammad was in effect "misguided." "He came from the south with no high school education, and he had no way of knowing what the Islamic world believed in or what it didn't believe in." As Warith Deen Mohammed gained education, he began questioning the "myths" (including that of Master Fard) into which he had been indoctrinated as a child (Barboza, *American Jihad*, 99–100).

7   Lee, *The Nation of Islam*, 92.

8   See *Muhammad Speaks*, 9 May 1975, 1. I am grateful to Lee (*The Nation of Islam*, 83) for this reference.

9   For Malcolm X's own account of his conversion to the new order of all-inclusiveness, see the *Autobiography*, 343–82.

10 For an account of the experience of a white woman permitted to join the Nation of Islam under the leadership of Imam Warith Mohammed, see Dorothy Blake Fardan, *Message to the White Man and Woman in America: Yakub and the Origins of White Supremacy* (Hampton, Va.: United Brothers and Sisters Communications Systems, 1991). Fardan approaches the teachings of the Nation of Islam from a "Caucasian" perspective.

11 The inclusive dates of publication and the titles of the newspapers that issued from the Nation of Islam are as follows: (1) Under Elijah Muhammad, one issue of the *Islamic News* appeared in 1959, accompanied by the journal *Salaam*. Thereafter, *Muhammad Speaks* ran from 1960 (1961?) to 1975. (2) Under Imam Warith Mohammed, the newspaper became the *Bilalian News*, which ran from 1975 to 1981. In 1982, it was briefly called the *World Muslim News*, thereafter, the *American Muslim* [or *AM*] *Journal* (1982–85), then the *Muslim Journal* (1985–86/87[?]). Concurrently, there appeared *Progressions* (May, July, October, November, 1985). (3) Under Louis Farrakhan, the tone and spirit of the newspaper returned to that of *Muhammad Speaks*. In this form, it is now the *Final Call*, which has been running from 1979 to the present. According to Farrakhan, the title alludes to the *Final Call of Islam*, a newspaper that Elijah Muhammad himself produced in 1934 after Master Fard Muhammad had left his disciple with the mission. The phrase *final call*, however, is also derived from the Qur'an (Sur. 74.8), which suggests an eschatological urgency (Gardell, *In the Name of Elijah Muhammad*, 140). Information is likewise derived from Lee, *The Nation of Islam*, 106–7, 152; as well as the published "Serials Holding List" of the Vivian G. Harsh Research Collection of Afro-American History and Literature, Chicago Public Library. Phone consultations with staff at the *Final Call* headquarters and with Dartanyon at the Respect for Life (no. 1) bookstore have proved invaluable.

12 See Lincoln, "The Muslim Mission," 352.

13 Among such reforms, the Muslim temples were renamed masjids, and the Fruit of Islam (FOI) was abolished (see Lee, *The Nation of Islam*, 93; and Lincoln, *The Black Muslims in America*, 264–65).

14 Lincoln, *The Black Muslims in America*, 265.

15 Ibid., 265. Imam Mohammed's orthodox approach to Islamic universalism has won him many converts. According to Barboza, he has more followers than any other American Muslim leader ever had, estimated roughly at a million (*American Jihad*, 97).

16 Lee, *The Nation of Islam*, 88–89. See the *Chicago Tribune*, 1 September 1975, sec. 1, 1; and the *New York Times*, 1 September 1975, sec. 1, 1.

17 Although in a somewhat different context, this is precisely the idea that Louis Farrakhan applies to his own movement. According to Henry Louis Gates Jr., "Louis Farrakhan told me that the Nation of Islam might be understood as a kind of Reformation movement within the

black church—a church that had grown all too accommodating to American racism" ("The Charmer," *New Yorker,* 29 April, 6 May 1996, 119).

18 Lincoln, *The Black Muslims in America,* 268–69. For an enlightening account of the Lost-Found Nation of Islam as well as corresponding movements (including the Five Percent Nation of Islam and the Nubian Islamic Hebrews), see Gardell, *In the Name of Elijah Muhammad,* 215–31. Gardell is particularly helpful in delineating the theology and beliefs of the Lost-Found Nation of Islam. He also provides a list of the mosques under the tutelage of Silis Muhammad. According to Gardell, the Lost-Found Nation of Islam shares the Nation of Islam's belief in the Mother Ship but does not believe that Elijah Muhammad departed to it (pp. 219, 396 n. 71).

19 Frances Daniel, "Louis Farrakhan: America's Other Son," *N'Digo,* February 1966, 14.

20 Details are from Lincoln, *The Black Muslims in America,* 268–69; Daniel, "Louis Farrakhan," 14; William A. Henry III et al., "Pride and Prejudice," *Time,* 28 February 1994, 24–25; Gardell, *In the Name of Elijah Muhammad,* 119–31; Magida, *Prophet of Rage,* passim; and Clarence Page, *Showing My Color: Impolite Essays on Race and Identity* (New York: Harper Collins, 1996), 130–54. Titled " 'The Signifyin' Muslim,' " Page's analysis applies the concept of the signifying monkey or trickster to Farrakhan as a "charmer." At the same time, Page views Farrakhan as a "tragic figure" (p. 146). Further details of Farrakhan's life appear in a series of articles that *Newsweek* ran under the general heading "The Two Faces of Farrakhan," 30 October 1995, 28–47. Especially interesting in the series is Howard Fineman and Vern E. Smith's profile "An Angry 'Charmer,' " 34–35, 38. According to this profile, Farrakhan's mother was Sarah Mae Manning, an immigrant to New York City from St. Kitts. Her husband was a Jamaican, Percival Clark, who soon disappeared. Although he was Farrakhan's "actual" father, the son was given the surname of another West Indian, Louis Walcott, who was the true man in Mae's life. In later years, Farrakhan's mother confessed that she tried to abort him three times with a coat hanger. Louis Walcott eventually abandoned the family as well. According to Fineman and Smith, Elijah Muhammad became the surrogate for the father Farrakhan never knew (pp. 34–35). In an interview with Mike Wallace on "60 Minutes" (14 April 1996), Farrakhan said that his father was a "very light-skinned black man," possibly descended from white and black parentage.

21 With admission to the Nation, Louis Eugene Walcott became Louis X, until Elijah Muhammad bestowed the name *Farrakhan* on him in 1965. According to A. Marshall, Farrakhan "does not know what his Muslim name means." When he asked Elijah Muhammad for the meaning, Elijah Muhammad replied, "I have the meaning upstairs"; "one of these days I'll tell you its meaning. It's a good name" (*Louis Farrakhan: Made in America* [n.p.: BSB, 1996], 80).

22 Malcolm X was assassinated on 21 February 1965. Three Nation of Islam followers were convicted of the killing, but the question of ultimate responsibility remains unresolved. The whole issue flared up again in the spring of 1995, with the turmoil surrounding the figures of Betty Shabbaz (Malcolm X's wife) and Quibilah Shabbaz (Malcolm X's daughter), but the furor has since died down. Meanwhile, the Nation under Minister Farrakhan has reached out to reconcile differences.

23 Lincoln, *The Black Muslims in America,* 268–69; Daniel, "Louis Farrakhan," 14.

24 Lincoln, *The Black Muslims in America,* 268–69, 266–67. According to Steven Barboza, in 1993 the Nation had something over twenty thousand dues-paying members nationwide (*American Jihad,* 133).

25 Barboza, *American Jihad,* 125.

26 For an enlightening discussion of the biblical configurations of type and antitype underlying the notion of the "new" Nation, see Gardell, *In the Name of Elijah Muhammad*, 122–31.

27 "Backing Farrakhan Seen as Way of 'Hitting Back,' " *Washington Post*, 11 December 1989, 6–7.

28 See Henry, "Pride and Prejudice," 21–27, and the accompanying "Forum," 28–34. See Farrakhan's response to the *Time* article in the *Final Call*, 16 March 1994, 20–21. Other groups (such as homosexuals and lesbians) have on occasion likewise been subjected to castigation by Farrakhan.

29 See David Jackson and William Gaines, "The Power and the Money," *Chicago Tribune*, 12 March 1995, sec. 1, 1, 16–20; "The Business of Security," *Chicago Tribune*, 13 March 1995, sec. 1, 1, 14–18; "AIDS Hope or Hoax in a Bottle," *Chicago Tribune*, 14 March 1995, sec. 1, 1, 10–15; and "Ascent and Grandeur," *Chicago Tribune*, 15 March 1995, sec. 1, 1, 22–26. A concluding editorial ("Calling Farrakhan to Account") declares: "Thus demystified, Farrakhan turns out to be essentially no different from dozens of other hucksters, past and present, who have proven that in America you can turn a buck on religion as easily as on anything else. Quite literally the only thing that distinguishes Farrakhan is his rabidly racist theology" (*Chicago Tribune*, 21 March 1995, 18).

30 To date, the most comprehensive treatment of this vexed issue is that of Gardell, *In the Name of Elijah Muhammad*, 245–84. See also Magida, *Prophet of Rage*, 139–72.

31 In a 1994 interview that appeared between Farrakhan and a member of the *Final Call* staff, Farrakhan responds to the charge of anti-Semitism by distinguishing among the terms *Jew*, *Zionist, Semite,* and *Hebrew.* Although he has been outspoken in his criticisms of Jews and Zionists, he maintains that he is not anti-Semitic because he views himself as a Semite (*Final Call*, 2 March 1994, 34–36). More than that, he looks on himself as a Hebrew (see *Final Call*, 16 November 1994, 20–21). Whether conceived as *anti-Semitism* or as some other formulation, the climate of controversy, as reflected in the issues of the *Final Call*, has become increasingly confrontational over the years. According to Manning Marable, "Farrakhan has provoked the sharpest exchanges over black-Jewish relations in recent memory" ("In the Business of Prophet Making," *New Statesman*, 13 December 1985, 23–25). For earlier expressions of Farrakhan's views on the subject, see the tapes "The Jewish Stranglehold on Black Leadership," Chicago, 27 October 1985; and "The Black/Jewish Relationship," Northwestern University, Evanston, Ill., 30 May 1988.

32 However the Nation of Islam is to be viewed at the present time, it should be noted that, for all practical purposes, the virulence of the anti-Jewish rhetoric adopted on occasion by the present leadership is not characteristic of the outlook that informed the teachings of Elijah Muhammad, a fact that Farrakhan himself acknowledges in an article written in the form of an interview between himself and a member of the *Final Call* staff ("I Have Been Vindicated," *Final Call*, 4 May 1992, 20–21, 36–37). In this article, Farrakhan states that "the Honorable Elijah never mentioned Jews directly in his criticism of white people." Nonetheless, "he classified all whites as devils, regardless of their faith." Although Elijah Muhammad was fiercely opposed to the injustices against the black man perpetrated by the entire white race, including both Jews and Christians, Judaism in general and the Jews in particular are not singled out for special condemnation. In fact, Elijah Muhammad goes so far as to extol the Jews, who, in holding to the teachings of Moses, "are nearer to us than any other race in the way of worship" (Elijah Muhammad, *Our Saviour Has Arrived* [Newport News, Va.: United Brothers Communications Systems, n.d.], 33). What is true of worship is no less true of

dietary practice. Like the Muslims, the Jews (particularly the Orthodox Jews who keep kosher) are, according to Elijah Muhammad, "careful of the type of food that goes into their stomachs." Combining Muslim and Jewish practice, Elijah Muhammad says that "we give credit to the Jews for still trying to obey that which they were ordered to do by Allah, through the prophets that Allah sent to them." For this reason, Jews are the only people whose food is palatable to those who heed the dietary laws that Elijah Muhammad himself advocates (Elijah Muhammad, *How to Eat to Live*, 2 vols. [Newport News, Va.: National Newport News and Commentator, 1967], 1:1, 67–68). As indicated, this is not to suggest that Elijah Muhammad exempted the Jewish community from the righteous indignation he leveled at the white race as the oppressors of the black race. For him, both Christians and Jews are worthy of the destruction they have meted out to others (*Our Saviour Has Arrived*, 76). This is a strong condemnation, one made in the context of what is looked on as the sufferings that blacks have undergone as the result of persecution by whites. But its thrust is a generalized one against injustices both past and present. It should be noted that, before his break with the Nation of Islam, Malcolm X strongly criticized Jews as well. See "The *Playboy* Interview: Malcolm X Speaks with Alex Haley" (1963), in *Malcolm X as They Knew Him*, comp. David Gallen (New York: Carroll & Graff, 1992), 117.

33 Henry, "Pride and Prejudice," 25–26. On the other hand, Farrakhan was critical of Jewish leadership even in his earlier speeches and interviews. For example, in a 1973 interview with Joseph Walker, New York editor of *Muhammad Speaks*, Farrakhan criticizes Jews ("in control of every major Black organization") for seeing to it that black leaders spoke against Elijah Muhammad (*7 Speeches* [Newport News, Va.: Ramza Associates and United Brothers Communications Systems, 1974], 45, 63).

34 Louis Farrakhan, "A Crisis in Black Leadership," *Essence* 15 (June 1984): 87–88, 146, 148.

35 A similar outlook is present in Farrakhan's address ("What Is the Need for Black History") at Princeton University in 1984 (reprinted in *Back Where We Belong*, ed. Joseph D. Eure and Richard M. Jerome [Philadelphia: PC International, 1989], esp. 74–75). Ever since the split with the Jewish community, Farrakhan has attempted to engage in various forms of outreach. At his sixtieth birthday concert in Chicago (May 1993), Farrakhan performed Mendelssohn, an event he desired to be interpreted as a sign of outreach to the Jewish community. The concert was a repeat of a performance in April in Winston-Salem, N.C. Speaking of the performance, Farrakhan said, "Let us hope the music will open a way to dialogue and erase the bitterness and misunderstanding between the two communities [the Nation of Islam and the Jewish community]." See *Jet*, 7 June 1993, 56–57; and *Final Call*, 10 May 1993, 3, 21, and 2 June 1993, 3, 20–21, 28. Noting that Mendelssohn was in fact a Jew who converted to Christianity, Jewish leaders were unimpressed with this attempt at "outreach": "They wanted an offense that had been made with words to be corrected with words, in the form of a full, unequivocal apology" (Page, *Showing My Color*, 148).

36 On 25 January 1984, Reverend Jackson referred to Jews as "Hymies" and New York City as "Hymietown" in what he thought were remarks made off the record to Milton Coleman, the *Washington Post* reporter who disclosed the remarks. In response, Farrakhan charged Coleman (who is black) with being a "traitor" (see Gardell, *In the Name of Elijah Muhammad*, 251; Page, *Showing My Color*, 143–44; Adolph Reed, "The Rise of Louis Farrakhan," *Nation*, 21 January 1991, 37, 51–56; and Marshall, *Louis Farrakhan*, 136–46). In *Independent Black Leadership in America: Minister Louis Farrakhan, Dr. Lenora Fulani, Reverend Al Sharpton*, ed.

Gabrielle Kurlander and Jacqueline Salit (New York: Castillo International, 1990), Farrakhan recalls the extent to which "a vindictive spirit" prevailed "in certain members of the Jewish community" toward Reverend Jackson at that time (p. 37).

37 Henry, "Pride and Prejudice," 25. See Gardell, *In the Name of Elijah Muhammad,* 250.

38 Henry, "Pride and Prejudice," 25.

39 *New York Times,* 7 October 1985, sec. 1, 4, and 8 October 1985, sec. 1, 5. See also Marable, "In the Business of Prophet Making," 23.

40 *Final Call,* December 1985, 1 (no specific day of the month is indicated in this issue). From that time forward, Farrakhan claims that his path has been obstructed by Jews on numerous occasions. The most costly obstruction occurred in 1986 "when Jewish distributors, angry about his slurs, effectively torpedoed his plans for Nation of Islam cosmetics and toiletries sold under the Clean & Fresh label." (Such products were the efforts of Farrakhan's POWER [People Organized and Working for Economic Rebirth] initiative.) As the result of attempts to undermine his economic initiatives, Farrakhan said that he "recognized that the black man will never be free until we address the relationship between blacks and Jews" (Henry, "Pride and Prejudice," 26).

41 Henry, "Pride and Prejudice," 22–26. It is within this context that Khallid Abdul Muhammad, Nation of Islam national spokesman, issued his widely publicized remarks at Kean College, N.J., on 29 November 1993. Characterized as "hate speech," the remarks are very inflammatory: their mode is polemical, combative, and derisive. Excerpted by the Anti-Defamation League of B'nai B'rith in the *New York Times,* 16 January 1994, sec. 1, 27, they represent the extreme to which the rhetoric of confrontation is liable to extend to those perceived as the enemies of the cause that the Nation seeks to uphold. The public exposure that these remarks received created such a furor that they were denounced by Vice President Al Gore as "the vilest kind of racism" and were condemned by members of the U.S. Senate. Under pressure to respond to the substance of the remarks, Farrakhan publicly rebuked his aide for the "manner" in which they were made, but not for the "truths" that they embodied. One might well question the extent to which "manner" and "matter" can be distinguished. See, among other accounts, *Chicago Tribune,* 4 February 1994, sec. 1, 4–5. For analysis, see Magida, *Prophet of Rage,* 173–81; Gardell, *In the Name of Elijah Muhammad,* 263–67; and Marshall, *Louis Farrakhan,* 199–216.

42 *The Secret Relationship between Blacks and Jews* (Boston: Latimer Associates, 1991), vii–viii. Although the title page indicates that the book is "Prepared by the Historical Research Department of the Nation of Islam," there is no reference to a specific author. The title page also indicates that this is vol. 1, but no subsequent volume has appeared.

43 For detailed responses, see Michael C. Kotzin, "Louis Farrakhan's Anti-Semitism: A Look at the Record," *Christian Century,* 2 March 1994, 224–26; and Winthrop D. Jordan, review of *The Secret Relationship, Atlantic Monthly* 276, no. 3 (September 1995): 109–14. See also Gardell, *In the Name of Elijah Muhammad,* 260–61, 268–71; and Magida, *Prophet of Rage,* 181–86. The most comprehensive refutation is Harold Brackman, *Ministry of Lies: The Truth behind the Nation of Islam's "The Secret Relationship between Blacks and Jews"* (New York: Four Walls Eight Windows, 1994), published under the auspices of the Simon Wiesenthal Center. Brackman traces the genealogy of the book to Henry Ford's *The International Jew: The World's Foremost Problem* (1920–22). Compare *Jew-Hatred as History: An Analysis of the Nation of Islam's "The Secret Relationship between Blacks and Jews"* (New York: Anti-Defamation League, 1993).

44 In *Independent Black Leadership in America*, Farrakhan observes: "Now, most of those who call me anti-Semitic are not Semites themselves. These are Jews that adopted the faith of Judaism up in Europe; they're called *Ashkenazi* Jews. They have nothing to do with the Middle East—they're Europeans. . . . They are *not* Semitic people. Their origin is not in Palestine" (pp. 43–44). Recast in an inflammatory form, this assertion appears in Khallid Abdul Muhammad's remarks as the following: "I must say so-called Jew, because you're not the true Jew. You are Johnny-come-lately-Jew, who just crawled out of the caves and hills of Europe. . . . You're not from the original people. You are a European strain of people who crawled around on your all fours in the caves and hills of Europe, eatin' Juniper roots and eatin' each other" (*New York Times*, 16 January 1994, sec. 1, 27).

45 See, e.g., the following statement in *The Secret Relationship:* "Much like the Nazis at the concentration camps of Auschwitz, Treblinka, or Buchenwald, Jews served as constables, jailers and sheriffs, part of whose duties were to issue warrants against and track down Black freedom seekers" (p. 207).

46 In the "I Have Been Vindicated" interview between himself and a member of the *Final Call* staff, Farrakhan further elaborates on the idea of "chosenness" and concludes that it is really the black people who have been "chosen by God" (*Final Call*, 4 May 1992, 36–37). The entire interview is significant in its coverage of Farrakhan's response to the publication of *The Secret Relationship*.

47 See Henry Louis Gates Jr., "Black Demagogues and Pseudo-Scholars," *New York Times*, 20 July 1992, sec. 1, 15. More recently, see Gates, "The Charmer," 126–28. According to Michael C. Kotzin, the assumptions of the new anti-Semitism are discernible in the "long-discredited" book by Gary Allen, *None Dare Call It Conspiracy* (Rossmoor, Calif.: Concord, 1971). This book alleges that Jewish bankers (principally, the Warburgs as part of the Rothschild empire) "helped finance Adolph Hitler" (p. 40). Kotzin maintains that these allegations are "consistent with Farrakhan's longstanding attempts to deny victimization to the Jewish people." Ironically, Kotzin observes, *None Dare Call It Conspiracy* was originally distributed by "right-wing circles" that "merged anti-black racism with anti-Semitism" ("Farrakhan Has Forum to Renounce Hatred," *Chicago Sun-Times*, 16 October 1995, sec. 1, 29). Allen has published several other books in a similar vein.

48 My account of Minister Farrakhan's 1994 Saviours' Day address is drawn from the printed transcript, which appears in two parts under the title "Let Us Make Man," *Final Call*, 16 March 1994, 20–21, 29, and 30 March 1994, 20–21.

49 Farrakhan, "Let Us Make Man," 16 March 1994, 21.

50 See the headline of the *Final Call*, 22 September 1993 ("Minister Farrakhan to Jewish Leaders: LET MY PEOPLE GO"), and the accompanying article (pp. 3 and 20).

51 Farrakhan, "Let Us Make Man," 16 March 1994, 21.

52 In their pioneering study of the psychological and cultural forces that constitute "black rage," William H. Greer and Price M. Cobbs have in effect discovered what was known to Elijah Muhammad all along: in the emulation of the oppressor, the oppressed develops a contempt for self that reinforces his own oppression. Seduced by the "white man's magic" and the belief that "white is right," the black individual undergoes a "psychic division" that widens as his hatred of self grows (*Black Rage* [New York: Basic, 1968], 193–95). This is Yakub's secret, one that Elijah Muhammad sought to overcome by transforming the hatred of self into the love of self through an attempt to undermine those forces that sought to oppress it in the first place. In the new paradigm that emerges in Farrakhan's encounter with the "other" as it is repre-

sented by American Jewry and its allies, a transference occurs through which the Caucasian and the so-called Jew become interchangeable. To love oneself in this new paradigm involves the act of overcoming the white man masquerading as a Jew. One of the fascinating phenomena underlying the whole dynamic of repulsion and attraction is discernible within what might be called the psychic history of the Jews. The implications of this phenomenon have been ably explored by Sander L. Gilman in *Jewish Self-Hatred: Anti-Semitism and the Hidden Language of the Jews* (Baltimore: Johns Hopkins University Press, 1986). In his study of Jewish self-hatred, Gilman demonstrates the extent to which Jews themselves were stereotypically viewed as blacks and therefore "inferior" by those who sought to oppress them (see esp. pp. 1–15).

53 In a fascinating disclaimer made during a speech in Symphony Hall, Phoenix, on 21 August 1986, Farrakhan declared, "Now what I am saying to you when I speak to the Jews is not hatred. I am concerned that you do not end up being destroyed by God." As part of his disclaimer, he compares himself to a Jewish mother: "A good Jewish mother will spank you when you are wrong. A good Jewish mother does not care whether you desire a thing; if it is wrong she will speak against it. You may dislike her for a while but you will get over it." Playing on the ironies implicit in the comparison, he says: "I don't look like your Jewish mother, but I am better than she is by you" ("Self-Improvement: The Basis for Community Development," in *Back Where We Belong*, 192–93).

54 "The Charmer," 125, 126–28. For corresponding statements about Farrakhan's "Jewishness," see Magida, *Prophet of Rage*, 158.

55 Farrakhan, "Let Us Make Man," 16 March 1994, 21.

56 The National Park Service has conceded that its initial head count of 400,000 may be low. A Boston University study, commissioned by ABC News, has arrived at the larger figure. According to the National Park Service, the largest gathering in Washington was the 1.2 million who attended Lyndon B. Johnson's inauguration in 1965. Even if the 400,000 number for the march were accurate, it would still exceed the presidential inaugurations of George Bush in 1989 and Richard Nixon in 1973 as well as the 1963 March on Washington, which purportedly attracted 250,000 people (George E. Curry, "After the Million Man March," *Emerge*, February 1996, 37–48).

57 According to Curry, it was Chavis who had earlier reached out to Farrakhan during a time when many black leaders kept their distance. "When Chavis was fired by the NAACP, Farrakhan came to Chavis' rescue, making him one of only two paid staff members for the Million Man March, giving Chavis new life" (p. 47). The NAACP distanced itself from the march. The civil rights activist and NAACP board member Julian Bond called Farrakhan "notoriously and unapologetically antisemitic, anti-Catholic, anti-white, misogynist and anti-gay" (Kevin Merida, "In Farrakhan's Footsteps," *Washington Post National Weekly Edition*, 4–10 March 1996, 33).

58 See the list of Million Man March endorsers in the *Final Call*, 20 October 1995, 7. Mayor Edward Rendell of Philadelphia likewise was supportive of the march. Even Abraham H. Foxman, national director of the Anti-Defamation League, supported the march but called on blacks to distance themselves from Farrakhan (*New York Times*, 15 October 1995, sec. 1, 12).

59 Cornel West, "Historical Event," in *Million Man March/Day of Absence: A Commemorative Anthology*, ed. Haki R. Madhubuti and Maulana Karenga (Chicago: Third World, n.d.), 98–99.

60 "*Mission Statement*," the Million Man March, Day of Absence (Chicago: FCN, 1995), passim.

61 References are to Louis Farrakhan, "The Day of Atonement" (in *Million Man March/Day of Absence*, ed. Madhubuti and Karenga, 9–29), which is a transcript of the speech. Magida is critical of the speech, which he calls "a loose patchwork of themes that never quite cohere" (*Prophet of Rage*, 193). For Magida's account of the march and its aftermath, see pp. 190–202.

62 The reference to 1555 as the year that blacks first set foot on the shores of Jamestown as slaves is odd. The first recorded date for the arrival of slaves in a Dutch ship on the shores of Jamestown (which was founded in 1607) is August 1619. On this dating, see *Dictionary of Afro-American Slavery*, ed. Randall M. Miller and John David Smith (New York: Greenwood, 1988), 779–88, 819. Other sources attest to this date, ranging from such early studies as John R. Spears, *The American Slave-Trade: An Account of Its Origin, Growth, and Suppression* (Port Washington, N.Y.: Kennikat, 1900), 1–20; to Peter Kolchin, *American Slavery, 1619–1877* (New York: Hill & Wang, 1993), 3. On the other hand, the importation of slaves to the Americas existed well before that time. According to Colin A. Palmer (*The First Passage: Blacks in the Americas, 1502–1617* [New York: Oxford University Press, 1995]), the slaves who arrived in Jamestown in 1619 came 117 years after Africans were first enslaved in the Americas, in Hispaniola. For the Nation of Islam's own perspective on the dating, see Tynnetta Muhammad, "Founding of the Nation of Islam," *Million Man March: 1,000,000 (10,000 × 100)*, ed. Tynnetta Muhammad (Chicago: FCN, 1995), 16–26. This book contains other essays, letters, and testimonials of interest regarding the Nation of Islam and the Million Man March.

63 See Tynnetta Muhammad's ongoing analyses of "the Parable of 19" in the *Final Call* as well as her *The Comer by Night*. For an explanation of the importance of this number to the Nation, see Gardell, *In the Name of Elijah Muhammad*, 176–81.

64 Farrakhan, "Day of Atonement," 10.

65 Benjamin Banneker (1731–1806) was a mathematician, astronomer, scientist, almanac maker, and naturalist who in 1791 assisted Major Andrew Ellicott in the survey of the Federal Territory (now the District of Columbia), in which a new national capital was to be established (see *Dictionary of American Negro Biography*, 22–25). The Mall itself, of course, is based on the design of Pierre L'Enfant.

66 As we have seen, the founding principles of the Nation of Islam, as reflected in the writings of Elijah Muhammad, incorporate the symbolism of Freemasonry. For a discussion of Freemasonry and its relation to the Million Man March, see Tynnetta Muhammad, "Founding of the Nation of Islam." See also Gardell's discussion of Freemasonry and its ties to the Nation (*In the Name of Elijah Muhammad*, 148–49).

67 According to Barbara Franco ("Masonic Imagery," in *Aspects of American Printmaking, 1800–1950* [Syracuse, N.Y.: Syracuse University Press, 1988], 1–29), Masonic imagery permeated American culture throughout the late eighteenth and early nineteenth centuries. Intimately involved in the practice of architecture, Freemasonry drew on such visionary artifacts as Solomon's Temple, with its columns, pavement, and steps, to formulate Masonic symbolism. "For Americans, removed from the European centers of learning, Freemasonry served as a vehicle for the popularization and spread of new ideas," such as radical notions of equality and natural law. These notions underscored American arguments favoring revolution and political separation from Great Britain. Some nine signers of the Declaration of Independence were Freemasons. An interesting discussion of Freemasonry can be found under the heading "Freemasonry—History and Origin" in Lewis Spence, *An Encyclopedia of Occultism* (New York: University Books, 1960), 173–75.

68 Such a grounding finds its correspondence for Farrakhan in the original Seal of the United

States. According to the plans of those responsible for designing the Seal, the device was to have the eye of Providence in a radiant triangle on one side and a picture of Pharaoh sitting in an open chariot, passing through the divided waters in pursuit of the Israelites, on the other. Hovering over the sea was to be shown a pillar of fire in a cloud, representing God's divine presence. Raised from this pillar of fire, beaming down, was to be revealed Moses standing on the shore, extending his hand over the sea and causing it to overwhelm Pharaoh ("Day of Atonement," 9–11).

69   Ibid., 9–11.

70   So Farrakhan observes, "You're dead, Black man. But if you believe in the God who created this sun of truth and of light with 19 rays, meaning he's pregnant with God's spirit, God's life, God's wisdom," you will be reborn (ibid., 19).

71   Ibid., 11–21.

72   Louis Farrakhan, "The Vision of the Million Man March," delivered at the Vision Center, Los Angeles, 15 September 1995. The Vision Center is funded by the television actress Marla Gibbs. In a call to the Vision Center on 29 February 1996, I was told that the center was constructed as a meeting place of hope, one based on a "vision" of renewal for the community it seeks to serve. For additional comments on the indebtedness of the march to Yom Kippur, see *Final Call*, 27 September 1995, 21–23.

73   See Farrakhan, "Day of Atonement," 25. Whether this call for dialogue will ever be heeded remains to be seen. Events transpiring on the very eve of the march might certainly be viewed as counterproductive to the kind of outreach sought by Farrakhan's appeal for reconciliation. In a television interview taped 4 October and broadcast 13 October 1995, Farrakhan labeled *bloodsuckers* those Jews and other minority business owners (including Arabs, Koreans, and Vietnamese) who deliberately exploit black communities and give nothing back (Lynn Sweet, "Farrakhan Denounces 'Bloodsuckers,' Racial Views Draw Fire as Rally Nears," *Chicago Sun-Times*, 14 October 1995, sec. 1, 1, 20). During his speech at Mosque Maryam in Chicago in October, he maintained that he was "tricked" into making the "bloodsucker" remarks (*Chicago Sun-Times*, 25 October 1995, sec. 1, 3). In an interview with Larry King shortly after delivering his Million Man March speech, Farrakhan once again confirmed the "truths" of the speech delivered by Khallid Abdul Muhammad at Kean College (Michael C. Kotzin, "Confronting the 'Farrakhan Problem,'" *Chicago Tribune*, 23 October 1995, sec. 1, 15). After Farrakhan returned from Washington, D.C., he maintains that "there were many faxes from rabbis and some Jewish organizations that wished to begin dialogue" (*Newsweek*, 30 October 1995, 36).

74   I use the term *mainstream* advisedly here. According to Maulana Karenga, chair of the Department of Black Studies at California State University at Long Beach and a member of the executive council of the Million Man March, "we don't accept the mainstream. The mainstream is a polluted pool, and what we're trying to do is create a new ocean of possibilities" (David Jackson, "What Will Happen after the March?" *Chicago Tribune*, 14 October 1995, sec. 1, 2).

75   See Steven A. Holmes, "Farrakhan's Angry World Tour Brings Harsh Criticism at Home," *New York Times*, 22 February 1996, sec. 1, 1, 6; Mary A. Mitchell, "Farrakhan Dares U.S., Says It's Showdown Time with Congress," *Chicago Sun-Times*, 26 February 1996, sec. 1, 1–2; Byron P. White et al., "Farrakhan Defends World Tour, Nation of Islam Leader Welcomes 'Showdown,'" *Chicago Tribune*, 26 February 1996, sec. 1, 1, back page; and "Farrakhan Boasts of $1 Billion in Pledge from Libyan Leader," *New York Times*, 27 February 1996, sec. 1, 10. See

also Carl Rowan, "Enemy of His People," *Chicago Sun-Times*, 21 February 1996, 39. News-papers throughout the country featured articles on Farrakhan's trip. It should be remembered, of course, that Farrakhan has undertaken many such trips, so many, in fact, that he has been called the "traveler." In this regard, see Jabril Muhammad, *Farrakhan: The Traveler*, 2d ed. (Phoenix: Book Co., 1994). In one form or another, his trips have customarily become the focus of controversy.

76 As one who was present at the Saviours' Day rally on this occasion, I derive much of my information from my own notes. Interestingly enough, on this occasion Farrakhan also made the quite remarkable statement that, "when you become color conscious, you lose reality." Although not elaborating on this statement, he did acknowledge that it represented something of a departure for him. In his speech, he still maintained that the Jewish people are "beating up" on him, but he praised the "good Christian," the "good Muslim," and the "good Jew" for struggling against temptation, and he admonished them to follow the path of righteousness. In keeping with past statements, he also castigated the "homosexual" and the "lesbian."

77 As Manning Marable argued years ago, "the Farrakhan phenomenon is far more complex than the rubric of 'black anti-Semitism' allows." Marable maintains that "a careful analysis of Farrakhan's public addresses reveals a strong commitment to an anti-racist and anti-imperialist politics, which parallels the late social thought of Malcolm X." At the very least, he says, Farrakhan's writings and addresses merit a dispassionate and disinterested analysis in the context of black/Jewish relations during the course of the century ("In the Business of Prophet Making," 23). For an interesting movement in that direction, see Julius Lester, "Blacks, Jews, and Farrakhan," *Dissent* 41, no. 3 (Summer 1994): 365–69, which distinguishes between "black narrative strategies" and "Jewish narrative strategies." At their extreme, the purveyors of these narrative strategies embrace an outlook that "sees the world as an Armageddon with the Chosen (themselves and God) on the one side and the Rejected (everyone else) on the other." Such purveyors offer a "narrative of fear and exclusivity posturing as religious righteousness."

78 Henry, "Pride and Prejudice," 22.

8 Armageddon and the Final Call

1 One must distinguish between "public" and "private" as well as "domestic" and "foreign" acts of bearing witness. Farrakhan speaks of initially sharing the vision with only his most trusted confidantes, such as Kadijah Farrakhan (his wife) and Tynnetta Muhammad. Beyond that, his public acts of bearing witness have occurred in this country and abroad. In this country, these acts may be divided into those that have occurred before a select audience of his followers (assembled at the Final Call Administration and at Mosque Maryam in Chicago) and those that have occurred before the world at large, such as his Washington, D.C., press conference in 1989, to be discussed. Abroad, his act of bearing witness occurred in his address to the Libyans.

2 The Vision Complex is a meeting hall located at 4310 Degnan Blvd., Los Angeles.

3 The speeches cited are available on both videotape and audiocassette. They are produced and distributed by Final Call, Inc., Chicago. Although the information accompanying "A Special Men's Day Rally" indicates 16 September 1995, Sidney Muhammad advises me that the correct date is 17 September, exactly ten years after the vision. The only speech that has been

transcribed is "The Great Announcement: The Final Warning," which has been published as "A Final Warning" in *Final Call*, 30 November 1989, 16–17, 29–30; and *The Announcement: A Final Warning to the U.S. Government* (Chicago: Final Call, 1989, reissued 1991). Except for "The Great Announcement," all references to Minister Farrakhan's speeches are from my own transcriptions of audio- and videotapes.

4   The performative or theatrical dimension of Farrakhan's act of bearing witness is consistent with his own role as "performer." One thinks not only of his musical performances but of his role as actor in his own dramas. For example, he assumed the role of the prosecutor in a play he wrote called *The Trial* as an indictment of white supremacy. He also wrote and performed in a play titled *Orgena* (*a negro* spelled backward). He likewise recorded the song "A White Man's Heaven Is a Black Man's Hell."

5   "The Vision of the Million Man March," delivered at the Vision Complex, Los Angeles, 15 September 1995.

6   Fiona Dunlop, *Fodor's Exploring Mexico* (New York: Fodor's Travel Publications, 1995), 182. *The Columbia Lippincott Gazetteer of the World*, ed. Leon E. Seltzer (New York: Columbia University Press, 1962), 1895.

7   Tom Brosnahan et al., *Mexico: A Travel Survival Kit*, 4th ed. (Hawthorn, Australia: Lonely Planet, 1992), 259–63.

8   "Mysteries Revealed?" *Final Call*, 30 November 1989, 14–15.

9   *The Encyclopedia of Religion*, ed. Mircea Eliade, 16 vols. (New York: Macmillan, 1987), 15:152–53. The figure of Quetzalcoatl has elicited a range of fascinating studies. These include such works as D. H. Lawrence, *The Plumed Serpent* (London: Martin Secker, 1926); Rudolfo A. Anaya, *Lord of the Dawn: The Legend of Quetzalcoatl* (Albuquerque: University of New Mexico Press, 1987); Jose Lopez Portillo et al., comps., *Quetzalcoatl, in Myth, Archeology, and Art* (New York: Continuum, 1982); and the studies of David Carrasco, including *Quetzalcoatl and the Irony of Empire: Myths and Prophecies in the Aztec Tradition* (Chicago: University of Chicago Press, 1982).

10 Joseph Campbell, *The Masks of God*, 4 vols. (New York: Viking, 1969), 4:457–60; Frank Waters, *Mexico Mystique: The Coming Sixth World of Consciousness* (Chicago: Swallow, 1975), 122–26.

11 See Farrakhan, *The Announcement*, 5. See also his "The Wheel and the Last Days" (1986), "The Vision of the Million Man March" (1995), and "A Special Men's Day Rally" (1995); and Gardell, *In the Name of Elijah Muhammad*, 131–35.

12 See my discussion of this aspect in the introduction as well as my full treatment in *The Visionary Mode.*

13 Among many other works, see Evelyn Underhill, *Mysticism: A Study in the Nature and Development of Man's Spiritual Consciousness* (London: Methuen, 1967), passim. In a fascinating examination of this journey in terms of the Nation's theology, Gardell views the ascent as a confirmation of the deification not only of Master Fard Muhammad but, through him, the Honorable Elijah Muhammad and, through him, Minister Louis Farrakhan, in a train of succession. According to this succession, Farrakhan "is on the black path of deification" (*In the Name of Elijah Muhammad*, 134–35).

14 So we recall the Father's commissioning of the Son in *Paradise Lost:* "Go then thou Mightiest in thy Fathers might, / Ascend my Chariot, guide the rapid Wheels . . . / Pursue these sons of Darkness, drive them out / From all Heav'ns bounds into the utter Deep" (6.710–16). In the same vein, Farrakhan's account is (to use the title of Milton's poem to his father) an *ad patrem.*

15 As indicated, the details of the vision are drawn from the various accounts cited above. Probably the most explicit of these is the description contained in "The Vision of the Million Man March," but the other accounts are important as well. For an account of the earthquake, see *Chicago Tribune*, 20 September 1985, 1–2.

16 According to Farrakhan ("The Vision of the Million Man March"), this press conference was delivered at the request of Kwame Turré.

17 Details are drawn from the tapes, especially the account in "The Vision of the Million Man March" and "The Wheel and the Last Days."

18 Ezekiel is addressed as the "Son of man [*ben'adam*]" by the enthroned figure (Ezek. 2:1) and repeatedly thereafter. The designation comes to assume apocalyptic overtones.

19 As transcribed from the tape "The Wheel and the Last Days."

20 See Lieb, *The Visionary Mode*, 42–84.

21 Signifying sections or units, the term *rak'a* is from a root meaning "to bow."

22 As transcribed from the tape "The Wheel and the Last Days." Farrakhan's experience at Big Mountain has been placed in a cross-cultural context by Wauneta Lone Wolf (Sioux), whose background and credentials are chronicled in the *Final Call*, April 1985, 7. In one of her columns in the *Final Call*, she makes a point of suggesting that, by means of the experience of the Wheel, both the red and the black nations have been able to achieve unity. She alludes to several gatherings when Farrakhan and his companions arrived at Flagstaff, Ariz., to travel to Big Mountain, as a site of the return of the Wheel. This experience she associates with what she calls the Hopi myths of the Great Star and the Great Purification. Explaining these myths in some detail, she suggests that their universality underlies her own experience of the Mother Ship as a highly moving and personal event. Her visions lead her to bear witness to the fact that she has been "acknowledged by the presence of God through the 'Wheel'" ("The Mothership on Big Mountain: Presence of Sacred Wheel Indicates 'the Eagle Wants to Land,'" *Final Call*, 30 September 1986, 30).

23 *The Announcement*, 7. As far as I can determine, the 24 October 1989 press conference is Farrakhan's first public "announcement" of the vision and its message to the world at large. He had long shared the vision with his own following and of course with Colonel Mu'ammar Gadhafi. Although the vision returned to his consciousness at his earlier press conference (mentioned above), he does not appear to have referred to it at that time. According to what he says in "The Vision of the Million Man March" speech, the significance of the vision was only beginning to dawn on him then. During the press conference, he maintains that he wondered to himself, "Was this what I was supposed to do?"

24 *The Announcement*, 5.

25 The specific details of this trip and his meeting with Colonel Gadhafi are recorded, among other places, in "The Vision of the Million Man March."

26 *The Announcement*, 7–8.

27 See *Final Call*, 27 April 1986, 15, 18. According to an editor's note that accompanies the text of the speech, limitations of time did not permit Farrakhan to deliver the speech in its entirety in Tripoli. By traveling to Tripoli, Farrakhan defied a State Department ban on such trips. The speech defends Farrakhan's defiance of this ban. This is not the first trip that Farrakhan made to Tripoli. He embarked on an earlier trip in 1984. Colonel Gadhafi's responsiveness to Farrakhan and the Nation of Islam has resulted in a history of financial, as well as moral, support. The close association between the Nation of Islam and the government of Libya extends back to the time of Elijah Muhammad (see Lee, *The Nation of Islam*, 71–72).

28 For an account of the bombing of Libya, see "U.S. Jets Bomb Libyan Targets" (18 April 1986), in *Facts on File: World News Digest with Index* (New York: Facts on File, 1986), 48, no. 2369: 257–61. See also Cannon, *President Reagan,* 653–54.

29 *The Announcement,* 8–9. Details of these occurrences are present in the other speeches as well.

30 *The Announcement,* 8–9. See also "The Vision of the Million Man March," among the other speeches recorded here. According to Farrakhan, the *Atlanta Gazette,* the *New York Times,* and the *Washington Post* later contained articles that repeated the words that Elijah Muhammad issued on the Wheel. The dates of the issues of these respective articles are not indicated, but the issues themselves appeared in 1987.

31 *The Announcement,* 10. Here, as elsewhere, the conspiratorial dimensions of Farrakhan's outlook underlie the thrust of his denunciations of a society that has been guilty of centuries of "chattel slavery," followed by a century of "free slavery." While acknowledging in his oxymoron *free slavery* that the visible chains of chattel slavery have been removed in the past century, Farrakhan nonetheless argues with the righteous indignation of the prophet infused with "bitterness" and "the heat of [his] spirit" (Ezek. 3:14) that the oppression is as real and as heartfelt as it ever was. In fact, he would no doubt maintain that the oppression is even more pronounced now that the chains are not apparent to those who are most responsible for having imposed them. Unremittingly adversarial, such a stance is founded on an indignation fueled by centuries of suffering and exploitation. As much as society might balk at the conspiratorial assumptions that inform Farrakhan's point of view, he is determined to stand his ground. For him, the attack on Libya is nothing less than a harbinger of the attack on the black people in the world he knows.

32 Ibid., 11.

33 Ibid., 15.

34 Ibid., 16.

35 *Al-Qur'an,* Ali.

36 As indicated in the previous chapter, the argument concerning Jewish control of the Federal Reserve finds expression in Gary Allen, *None Dare Call It Conspiracy* (Rossmoor, Calif.: Concord, 1971). According to Allen, the financier Paul Warburg "masterminded the establishment of the Federal Reserve" in order to place control over the U.S. economy in the hands of international bankers (p. 65).

37 Interestingly, almost as an afterthought to his sermon, Farrakhan harkens back to the press conference he delivered in Washington, D.C., in 1989. Within that context, he mentions visiting the editorial offices of the *Washington Post* to discuss his views some two years after the press conference. Although reporters from the *Post* attended the press conference, Farrakhan notes that they never printed anything about it. During his visit to the editorial offices, one of the editors asked in scorn, "What about this 'Wheel' business?" Farrakhan, in turn, responded not only by listing all the U.S. presidents that had seen it but also by providing additional "evidence" of its existence. In response, Farrakhan maintains that he heard the editor of the *Post* whispering excitedly to a colleague, "He know, he knows, he knows!"

38 The idea of associating Ezekiel's vision with the vision of God's "cloud and pillar of fire" in Exodus is repeatedly voiced by Farrakhan in his depiction of the Wheel. Such an association is evident not only in the speeches but also in the writings. See, e.g., Louis Farrakhan, *A Torchlight for America* (Chicago: FCN, 1993), 151–57.

39 The phrase *above top secret* is drawn from a letter that Senator Barry Goldwater wrote on 28 March 1975 requesting information on UFOs stored at the Wright Patterson Air Force Base. As the letter indicates, Senator Goldwater is aware that this information is classified as "above top secret." Deriving the title of his book *Above Top Secret* from Senator Goldwater's letter, Timothy Good explores the UFO phenomenon in detail. Good, of course, was present at the Frankfurt conference, where he delivered a paper. His book is sold in the Nation of Islam's Respect for Life bookstores.

40 Repeatedly throughout the issues of the *Final Call*, similar themes are sounded. See, e.g., the issues of 30 May, 30 June, and 30 September 1986; 31 January, 14 February, 30 June, 16 September, and 7 October 1987; 30 November 1989; 19 August 1991; and 8 and 22 September, 26 October, and 16 November 1992.

41 Tynnetta Muhammad enjoys a place of great prominence in the present Nation of Islam hierarchy. According to Gardell, she is understood to be the "wife" of the Messenger. Because Elijah Muhammad never officially remarried after the death of his wife, Clara Muhammad, the idea of Tynnetta Muhammad (formerly Denear) as the Messenger's wife might imply that she "was one of the 'Islamic' wives, or the so-called secretaries of the Messenger" (Gardell, *In the Name of Elijah Muhammad*, 125).

42 Jabril Muhammad, *Farrakhan: The Traveler*, 2d ed. (Phoenix: Book Co., 1994).

43 Jabril Muhammad, "Farrakhan, Ezekiel's Wheel, UFOs, Libya: Do They Relate?" *Final Call*, 30 May 1986, 5, 10.

44 For a list of Tynnetta Muhammad's publications and tapes, see the end page of *The Comer by Night* (p. 128).

45 Jabril Muhammad, "Farrakhan, Ezekiel's Wheel," 5, 10.

46 Jabril Muhammad, "A Message in UFO Sightings," *Final Call*, 30 June 1986, 19, 33–34.

47 Jabril Muhammad, "What Do UFO Sightings Mean?" *Final Call*, 16 September 1987, 27.

48 Jabril Muhammad, "God's Solution to the World's Problems," *Final Call*, 7 October 1987, 27.

49 *The Comer by Night*, 108. In *The Comer by Night*, Tynnetta Muhammad says that Master Fard Muhammad "introduced Himself through a mathematical code of letters, numbers and words to reveal His identity." In this spirit, she has been guided to focus her research "on the phenomenal teachings of Elijah Muhammad, which base Islam's Root upon Mathematics as the Universal Language of Truth" (pp. 16–17). See, e.g., her discussion of the Parable of the 19 in "World Prophecies Link to Current Events," *Final Call*, 30 May 1986, 10. Such discussions form a central part of *The Comer by Night*, in which she credits the Muslim scholar Dr. Rashad Khalifah with "the discovery of the underlying Mathematical Code of the Holy Qur'an's Revelation based on the number 19" (p. 16). See Gardell, *In the Name of Elijah Muhammad*, 176–81.

50 Tynnetta Muhammad, *The Comer by Night*, 78.

51 Ibid., 109, 111, 45.

52 Framing the head of Elijah Muhammad is an astral body that is no doubt a representation of Halley's comet, referred to often in *The Comer by Night*. For a detailed discussion of the symbolism of the crescent moon and star (as well as the sun as background symbol), see Elijah Muhammad, *The Flag of Islam*.

53 Tynnetta Muhammad, *The Comer by Night*, 59; see also vi–vii.

54 Ibid., 86, 86–90. The Parable of 19 is operative in the "holy names" of Master Wallace Fard Muhammad, the Honorable Elijah Muhammad, and Minister Louis Farrakhan. Engaging in her own act of *gematria*, Tynnetta Muhammad explores the numerological continuity that

exists in the transmission of divine wisdom and power from the beginnings of the movement to its present incarnation (pp. 108–12). Such concerns assume a cultic dimension in her numerological analysis of the Muslim prayer service, which likewise centers on the meanings of the Wheel. Replicating the very motions of the Wheel in her gestures and movements as she enacts the prayer service, Tynnetta Muhammad speaks of moving her body "in a complete circle" as a way of making her Lord "the exclusive object of worship." By means of the "Miraculous Mathematical Code" that unlocks the mysteries of such worship, she seeks to provide insight into the mystical process through which she ascends like a bird in flight to the divine creator, whose very throne is the Mother Plane. The ascent to this throne is the "Majestic Night Journey" that Tynnetta Muhammad undertakes as a way of being initiated "into the signs of God's Secret Knowledge and Wisdom" (pp. 113–14).

55 It must be noted here that, in keeping with the confrontational posture explored above, Tynnetta Muhammad's reflections embody a strong anti-Jewish polemic. In her book, name-calling and accusation abound. The Jews are to be faulted for their "dirty religion," for selling blacks into slavery, for leveling death threats and exhibiting other hostilities against Farrakhan, and for labeling him "a Hitler, an anti-Semite and a hater of Jews, when it was some of their own Jewish colleagues conspiring with Hitler that helped sell their own people out to the gas chambers" (p. 61). Coupled with this is a strong anti-Zionist thrust: "Today the Jews have perpetrated untold crimes of aggression, murder and fraud under the political movement called Zionism" (p. 48). Giving such anti-Jewish sentiments an "interpretive" twist, Tynnetta Muhammad provides her own commentary on the Yakub narrative that associates it with the Jacob/Israel narrative in Genesis (pp. 46–47).

56 Tynnetta Muhammad, *The Comer by Night*, 42.

Conclusion

1 For *rings* as *rims*, see the *New Oxford Annotated Bible with Apocrypha*, ed. Metzger and Murphy, s.v. For the meanings of *gav* as well as the other terms noted, see *The New Brown-Driver-Briggs-Gesenius Hebrew and English Lexicon*, s.v.

2 See the full discussion of this notion in Halperin's *The Faces of the Chariot*, 238–470.

3 Broome, "Ezekiel's Abnormal Personality."

Index

Abduction: Human Encounters with Aliens,
48–49

Abductions, 16, 48, 49, 50, 51, 72, 73, 208–9, 213,
239 n.35, 247 n.8, 248 n.17, 248 n.20, 249
nn. 22, 25

Abduction Study Conference, 50

Abolitionists, 76

Adamski, George, 53, 250 n.35

'Aggadot, 12, 233

Agobard, archbishop of Lyons, 44

Air Force One Is Haunted, 124–25

Alètheia, 13–14, 16, 34–35, 158, 233; and science,
55, 63–64

Allen, Gary, 296 n.36

Amber, 237 n.16

Ames, William, 242 n.11

Ancient astronaut hypothesis. *See* Astronauts,
ancient

Ancient astronaut society, 55

Anderson, Roger A., 251 n.45

Angelou, Maya, 190

Announcement, The, 212, 214–15, 222

Ansari, Zafar Ishaq, 273 n.12

Antichrist, 6, 33, 75, 94–97, 105–7, 161, 263 n.82,
265 n.96

Anti-Defamation League of B'nai B'rith, 288
n.41

Apocalypse. *See* End time

Armageddon, 5, 75, 80, 89–93, 95, 105; and Nation of Islam, 157, 213, 217, 225–26; Ronald
Reagan on, 108–9, 111, 114, 115, 124–25. *See
also* End time

Armstrong, Thomas, 43, 53

Ascension, 207

Ashton, T. S., 35

Astronautics and Aeronautics, 56

Astronauts, ancient, 52–54, 163, 249 n.27, 250 n.30
Atonement, 193–96
Autobiography of Malcolm X, 133, 155, 272 n.9, 273 n.15
Avatars of *technē*, 17, 32, 65, 70, 72, 74, 77, 82, 92, 93, 101, 110, 112, 125, 232, 239 n.37, 252 n.58, 280 n.1

Babylonian Captivity, 80
Bakker, Jim, 109
Baldwin, James, 136, 168
Banneker, Benjamin, 192, 291 n.65
Barboza, Steven, 283 n.2, 285 n.24
Barr, James, 255 n.4
Barry, Marion, 190, 203
Bates, Leon, 91
Bauer, Melchior, 31–35, 41
ben'adam, 295 n.18
Ben-Sira, 11
Ben Zoma, Rabbi, 12
Best Little Whorehouse in Texas, The, 45, 248 n.11
Bible: demythologization of, 92; and end time, 74; and foreign policy, 101, 108–11; and Nation of Islam, 133, 275 n.34; nuclearization of, 5, 84–99, 101, 112, 263 n.81; and politics, 96; and Protestant Reformation, 23; and race, 5–6; and technology, 91, 95–96; 258 n.42. Books: 1 Chronicles, 11; 1 Corinthians, 203; 2 Corinthians, 207; Daniel, 77; Ecclesiasticus, 11; Ephesians, 80; Exodus, 133, 220, 296 n.38; Ezekiel, 8, 14, 79, 85–89, 93, 146, 170–71, 207, 209, 214, 259 n.48, 260 n.53; Genesis, 98, 140, 187, 189; Habbakuk, 40, 79; Isaiah, 40, 75, 89, 260 n.58; Jeremiah, 87; Job, 68; 260 n.56; Joel, 85; 2 Kings, 98, 147; Luke, 97, 161, 225; Matthew, 28–30, 90, 225; 2 Peter, 85, 173; Psalms, 28; Revelation, 23, 85, 89–90, 98, 99, 113, 144–45, 148, 171, 192, 210, 241 n.5, 277 n.54; Samuel, 40; 1 Thessalonians, 98; 2 Thessalonians, 97; Zechariah, 85
Bilal Iban Rabah, 179
Black Culture and Black Consciousness, 143
Black Muslims in America, 135
Black Nationalism, 138, 271 n.3

Black Rage, 289 n.52
Blavatsky, Helena Petrovia, 78, 250 n.30
Blumrich, Josef F., 54–64, 159, 251 n.39
Bond, Julian, 290 n.57
Boone, Pat, 108
Boyer, Paul, 75, 84–85, 91, 259 n.49, 261 nn. 61, 64
Brackman, Harold, 288 n.43
Bradley, Omar, 190
Bredesen, Harald, 108
Broome, Edwin C., 15–16, 41, 124, 151
Brotz, Howard, 272 n.9
Bryan, C. D. B., 49–51
Bullinger, Henry, 23, 241 n.6
Burroughs, William S., 253 n.75
Bush, George, 101, 104, 215

Campus Crusade for Christ, 94
Cannon, Lou, 121, 270 n.66
Carey, John, 242 n.13
Carlyle, Thomas, 245 n.33
Catholics: attitudes toward, 33, 52, 70, 81, 184, 256 n.10, 277 n.54, 290 n.57
Challenger (space shuttle), 211
Chariot, 83–85, 147, 149, 237 nn. 12, 15. See also *Merkabah*
Chariot of Paternal Deitie, 21–30, 38, 68, 155–56, 209, 240 n.3, 268 n.31
Chariot of Zeale, 23–28
Chariots of the Gods?, 52–53
Chavis, Benjamin, Jr., 190, 203, 290 n.57
Children of Frankenstein, The, 3, 236 n.4, 246 n.45
China, 94, 96
Christendom, 81, 83, 84
Christian Identity, 264 n.90
Church of Jesus Christ of Latter Day Saints, 76, 247 n.10
CIA, 223, 252 n.47
Close Encounters of the Fourth Kind, 49–50
Close Encounters of the Third Kind (film), 45–48, 50
Cobbs, Price M., 289 n.52
Coleman, Milton, 287 n.36
Colson, Charles W., 101–4, 107, 266 n.6
Comer by Night, The, 224, 227

Communism, 107
Cooper, David L., 262 n.69
Covino, William A., 69
Cowles, Henry, 261 n.62
CROE (Coalition for the Remembrance of Elijah), 153–54, 173
Cuomo, Mario, 184
Curry, George E., 290 n.57
Cybernetics, 67, 253 n.64
Cyberpunk, 66, 69, 252 n.60
Cyberspace, 65–70, 73, 252 n.58
Cyborg, 68, 73

Darby, John Nelson, 261 n.62
Day the Earth Stood Still, The, 122–23
DeHaan, M. R., 85
Dimensions, 44, 247 nn. 7, 8
Disciples of Christ, 108
Divine Days, 272
Dole, Robert, 104
Drake, W. Raymond, 53
Drew, Timothy, 271 n.5
Druze doctrines, 273 n.12

Earl, Riggins R., Jr., 278 n.69
Eilmer (the Flying Monk), 244 n.19
Electrum, 12, 233
Elēktron, 12–13
Elijah, 98
Ellul, Jacques, 14–15, 34, 238 n.28
End of the Age, The, 105–8
End time, 5, 74; and alien abductions, 49; in America, 75–76; as b'aharit hayamim, 88; and the Bible, 74; and foreign policy, 101, 105, 111–12; and Israel, 94–98; and Nation of Islam, 153; and nuclear weapons, 84–99; and politics, 107, 109, 114, 265 n.1; and race, 96–97, 105–6, 148–49, 232; as Rapture, 98–99; and Revelation, 89–90; and UFOs, 113; and Web sites, 263 n.82. See also Armageddon
Enemies, 82, 258 n.35
Enoch, 98, 207
Epcot Center, 121
Erasmus, 35
Eschatology, 74, 84–85, 148–49
Essien-Udom, E. U., 152–53, 280 n.80

Ethiopia, 85, 92, 93, 105, 107, 111, 259 n.50, 268 n.29
Euhemerism, 53
Euhemerus, 250 n.31
Evangelicalism, 5, 75, 76, 84, 85, 87, 88, 90, 91, 93, 100, 103, 104, 108, 109, 255 n.3, 265 n.3, 266 nn.12, 14. See also Fundamentalism
Excalibur, 116, 120
Ezekiel (prophet), 14–15, 80, 85, 89, 236 n.3, 257 n.22
Ezekiel, vision of, 1–3, 9, 10, 8–13, 80–81, 230–32, 258 n.39; as alien abduction, 50–51, 239 n.35, 247 n.8; and ancient astronaut hypothesis, 52–54, 250 n.34; as anxiety dream, 15; in black folk culture, 146–49, 152; bureaucratization of, 81–84; and cosmology, 245 n.29; and end time, 85–99; and Exodus, 296 n.38; eyes of the, 230–32, 236 n.9; and First World War, 80; and Gog oracle, 85–91, 112, 259 n.48; and Jehovah's Witnesses, 79–84; as locomotive, 37–38, 148; and Louis Farrakhan, 199–201, 206–10; and Mother Plane, 156–71, 211–12, 228; and Nation of Islam, 150–53, 159–71, 198–99, 225–29; and nuclear weapons, 84–99; as parhelion, 64–65; as plane, 150, 152; psychological foundations of, 14–16, 151–52, 210; and Ronald Reagan, 110–14; and science, 63–64, 158; as UFO, 42–45, 50, 56–64, 153, 248 n.10, 251 n.39, 268 n.32; as warning, 213; as Wheel, 216; wheels of, 146–47, 230, 236 n.9
Ezekiel's Vision Explained, 245 n.29

Faid, Robert, 91
Falwell, Jerry, 5, 91–93, 108–9
Fardan, Dorothy Blake, 284 n.10
Farrakhan, Kadijah, 201, 210
Farrakhan, Louis, 6, 130, 202, 204; on Armageddon, 213, 217; on atonement, 193–96; bears witness, 199–201; on blackness, 189; conspiracy theories of, 215–18; conversion of, 181; and Elijah Muhammad, 181; as hierophant, 191–93; on Israel, 217; and Jewish community, 183–89, 195–96, 205, 215–18, 221, 286 n.31, 287 n.35, 293 n.77; in Libya, 214–15, 295 n.23; life of, 181–82, 285 n.20; and Mal-

Farrakhan, Louis (*cont.*)
    colm X, 181–82; media coverage of, 183–84; and Million Man March, 190–97, 200–201, 203, 220–21; and Mother Plane, 206–12, 218–19; naming of, 285 n.20; on numerology, 191–94; performative role of, 201–4, 294 n.4; as Quetzalcoatl, 205–6; in Tepoztlan, 205–6, 224; on tricknology, 187–89; as trickster, 285 n.20; and vision of Ezekiel, 199–201, 206–10; and vision of Wheel, 211–12, 220–21. Works: "Day of Atonement," 191–96; "Great Announcement: The Final Warning," 200–201; "Shock of the Hour, The," 217–19; *7 Speeches*, 286 n.33; "Special Men's Day Rally, A," 201, 220; "Vision of the Million Man March, The," 201, 220; "Warning, Mr. Reagan: God's Power is Real," 224; "Wheel and the Last Days, The," 200, 211–12
FBI, 173, 215, 223
Festinger, Leon, 43–44
Final Call, 182, 184, 201, 205, 212, 214, 221–26, 284 n.11
Fineman, Howard, 285 n.20
Finished Mystery, The, 79–80
Fire and brimstone, 112
Fire Next Time, The, 168
First World War, 80–81, 257 n.28, 258 n.40
Firth, Katherine, 23
Fish, Rabbi S., 260 n.53
Fishbane, Michael, 8, 236 n.6
Fisher, George, 79
Flying machines, 33, 244 n.19
Flying Saucers, 64
Forrest, Leon, 272 n.9
Fowler, Alistair, 242 n.13
Franco, Barbara, 291 n.67
Frederick the Great, 31, 34
Freemasonry, 133, 192, 272 n.12, 291 nn. 66, 67
Freud, Sigmund, 15, 239 n.31
Fruit of Islam, 133, 184, 202
Fundamentalism, 76, 100, 255 n.4. *See also* Evangelicalism

Gadhafi, Mu'ammar, 110, 196, 214, 267 n.28, 295 n.23
Gaines, William, 286 n.29

Gardell, Mattias, 173, 275 n.31, 283 n.38, 285 n.18, 294 n.13, 297 n.41
Garvey, Marcus, 131, 181, 271 n.5
Gates, Henry Louis, Jr., 186, 188–89, 284 n.17
Gematria, 297 n.54
Gesenius, Heinrich Wilhelm Friedrich, 88
Gesenius's Hebrew and Chaldee Lexicon, 260–61 n.60
Gibson, William, 65–66
Gilman, Sander L., 135, 290 n.52
Ginsburg, Allen, 253 n.75
Gnosis, 163, 171, 282 nn. 29, 32
God and Golem, 67
Gog oracle (Ezek. 38), 5, 6, 75, 85–91, 93, 95, 102, 103, 105, 110, 112–15, 120, 124, 259 nn. 48, 49, 51, 261 nn. 62, 66, 262 n.69, 264 n.90
Goldwater, Barry, 297 n.39
Golem, 68, 253 n.70
Good, Timothy, 252 n.47, 297 n.39
Gore, Al, 288 n.41
Goren (Israeli Defense Force rabbi), 39
Gort, 122–24
Graham, Billy, 101, 108
Graham, Daniel O., 116
Great Awakening, 76
Great Dark Brotherhood, 70–72
Great Deception, 98
Great Depression, 131, 199
Great Migration, 131
Great White Brotherhood, 70–72
Greenberg, Moshe, 87, 260 n.55
Greer, William H., 289 n.52
Gromacki, Robert Glenn, 96–97, 268 n.31
Guillory, John, 240–41 n.4
Gunpowder, 242–43 nn. 14, 15
Gunpowder Plot, 242 n.14. *See also* Milton, John

Hagigah, 12–13, 233
Haig, Alexander M., Jr., 104
Halley's comet, 228
Halperin, David J., 15, 87, 233, 260 n.57
Halsell, Grace, 91–92, 105, 262 n.72
Har-meggido, 89–90
Haraway, Donna, 68–69

Harding, Susan, 101, 107
Harrison, Barbara Grizzuti, 76–78
Hart, Clive, 33–34
Hart, Helen, 33–34
Ḥashmal, 12–13, 61, 231, 233, 254 n.83; and
 Mother Plane, 170; and UFOs, 46, 51, 124
Hassain, Margary, 151, 166, 169
Hayward, Philip, 69, 253 n.73
Ḥayyot, 51, 231
Heaven's Gate, 255 n.89
Heidegger, Martin, 13–14, 34–35, 238 nn. 25,
 27
Heidegger's Confrontation with Modernity, 238
 nn. 25, 27
Heifetz, Jascha, 188
Heim, Michael, 66
Hekhalot, 68, 207, 254 n.83
Hertlein, Axel, 225
Herzog, Chaim, 95–96, 264 n.87
High Frontier, 116
Hill, Paul R., 251 n.40
Hinduism, 106–7
Hiroshima, 84
History and Warfare in the Renaissance Epic,
 243 n.15
Horowitz, Vladimir, 188
HUAC (House Committee on Un-American
 Activities), 114
Hussein, Saddam, 196
Hynek, J. Allen, 48, 50, 246 n.2
Hynes, William J., 142

Identity Baptists, 264 n.90
IDF (Israeli Defense Force), 39
Idolatry, 25
*Illustration of Ezekiel's Vision of the "Chariot,"
 An*, 37–38
Industrial Revolution, 4, 35–36, 67
Inertia Projector, 114–15
Internet, 69
In the Name of Mighty Ga'lah, 280 n.2
Islam, 107
Ismaili, 273 n.12
Israel, 6, 37, 39, 40, 84–99, 102, 112, 165, 171, 185,
 187, 196, 214, 217, 220, 254 n.84, 277 n.53, 298
 n.55

Jackson, David, 286 n.29
Jackson, Jesse, 183–85, 190, 287 n.36
Jefferson Memorial, 192
Jehovah's Witnesses, 5, 76–84, 256 n.9, 258 n.35
Jessup, Morris K., 53
Jewish Defense League, 184
Jewish Self-Hatred, 289–90 n.52
Jews: attitudes toward, 7, 33, 90, 94, 95, 96, 107,
 182, 183–90, 195–97, 204, 205, 215, 217, 220,
 221, 222, 258 n.35, 264 nn. 86, 90, 272 n.9, 277
 n.53, 286 nn. 31, 32, 296 n.36, 298 n.55
Joachim of Fiore, 80
John the Divine, Saint, 23, 89–90, 210
Johnny Lee X, 182
Jorstad, Erling, 255 n.4
Jung, Carl Gustav, 256 n.7

Kalb, Marvin, 109
Karenga, Maulana, 292 n.74
Keech, Marian, 43, 53
Kermode, Frank, 74
Khalifah, Rashad, 297 n.49
King, Martin Luther, Jr., 184
Kingdoms in Conflict, 102
Kissinger, Henry, 104
Klein, Ralph W., 236 n.9
Klein, Ze'ev, 39–40
Knight, G. Wilson, 38–39, 41
Knoppers, Laura Lunger, 240 n.3
Koch, Edward, 184
Kotzin, Michael C., 289 n.47
Kuhn, Thomas, 48
Kuschel, Kieth Bernard, 260 n.53
Kybernētes, 67, 253 n.64

Labriola, Albert C., 278 n.63
Lactantius, 242 n.11
Lang, Andrew, 84
Lanier, Jaron, 253 n.73
Lasers, 1, 39, 91, 105, 116–17, 118, 120, 121, 124,
 246 n.1, 269 nn. 45, 47
Late Great Planet Earth, The, 94–96, 109
Lawrence Livermore National Laboratory, 117,
 121
Leary, Timothy, 69, 253 n.75
Lee, Martha F., 179

Lehmann, Theo, 147
Leonardo da Vinci, 244 n.20
Lester, Julius, 293 n.77
Levi, Edward, 104
Levine, Laurence W., 143, 146, 281 n.7
Libya, 85, 92, 93, 107, 110–11, 196, 214–15, 216,
    225, 226, 267 n.28, 293 n.1, 295 nn. 23, 27, 296
    nn. 28, 31
Lieb, Michael, 233–34
Lincoln, C. Eric, 130, 135–36, 140, 179, 182, 271
    n.5, 283 n.3
Lincoln Memorial, 192
Lindsey, Hal, 5, 94–99, 109, 263 nn. 81, 83, 268
    n.32
Livermore, Mary, 148
Living in the Shadow of the Second Coming, 263
    n.82
Locomotive, 35–38, 246 n.35
Lone Wolf, Wauneta, 295 n.22
Löwe (rabbi of Prague), 68
Lowth, William, 261 n.60
Lucas, George, 117–21
Luther, Martin, 23–24

Maʿaseh merkabah, 8, 12, 16, 68, 233. *See also*
    Chariot; *Merkabah*
Machina ex Deo, 239–40 n.38
Mack, John E., 48–50
MAD (Mutual Assured Destruction), 114, 116.
    See also MAS
Magida, Arthur J., 291 n.61
Magog. *See* Gog oracle
Mahdi, 134, 150, 152, 157, 161–62, 273 n.13
Mahmoud, H. Nasif, 274 n.25
Makarius, Laura, 142
Malcolm X, 133, 150, 155–56, 179, 272 n.9, 273
    n.15, 277 n.57, 285 n.22
Marable, Manning, 286 n.31, 293 n.77
Marʾot, 232
Marrs, Texe, 91
Marsden, George, 255 n.4
Marty, Martin, 255 n.4
Marx, Leo, 35–36, 245 n.33
MAS (Mutual Assured Survival), 117. *See also*
    MAD
McGee, J. Vernon, 85

Mecca, 133, 139
Meditation, 176
Melchizedek, Drunvalo, 70–73, 254 n.83
Memoirs of My Nervous Illness, 238 n.31
Mendelssohn, Felix, 287 n.35
Menzel, Donald H., 64–65
Merkabah, 11, 85, 232, 240 n.2, 268 n.32; dark
    side of, 233, 260 n.57; Internet as, 69; and
    New Age mysticism, 69–73, 254 n.83; as tank,
    39–41. *See also* Chariot; *Maʿaseh merkabah*
Mesmerism, 76
Millennium, 8, 77, 102, 106, 265 n.3
Miller, William, 77
Million Man March, 190–97, 201, 203, 220–21
Mills, James, 109–12
Milton, John, 4, 21–30, 241 n.9, 244 n.20.
    Works: *Apology for Smectymnuus*, 22–24; *De
    doctrina christiana*, 26, 243 n.17; *History of
    Britain*, 244 n.19; *Paradise Lost*, 21–22, 24–
    30, 37–38, 68, 72, 124, 155–56, 240 n.2, 294
    n.14; *Readie and Easie Way*, 241 n.9; *Reason
    of Church-Government*, 25, 241 n.9. *See also*
    Chariot of Paternal Deitie
Ministry of Lies, 288 n.43
Mohammed, Warith Deen (Imam), 178–82,
    204–5, 283 nn. 2, 3, 284 n.15
Mojtabai, A. G., 93–94
Mondale, Walter, 109
Montanism, 247 n.6
Moody, Dwight L., 89
Moses, 141–42, 150, 187, 207
Mother Plane, 7, 153; and eschatology, 172–76;
    and the FBI, 282 n.35; internalization of, 213,
    224; as Jesus Christ, 173–75; as maternal
    chariot, 209; and Million Man March, 203;
    and science fiction, 280 n.79; as UFO, 162–63,
    208–9, 221, 280 n.79; and vision of Ezekiel,
    162–71, 211–12, 228; as war machine, 158–59,
    218–19; as wheel, 169–71; as Wheel of
    Wheels, 207–9, 211–12
Mother Plane, The, 149
Muhammad (prophet), 179
Muhammad Speaks, 149, 151, 164–71, 179, 182,
    284 n.11. *See also* Final Call
Muhammad, A. Akbar, 202
Muhammad, A. Alim, 202

Muhammad, Abdul Wali, 222–24, 279 n.71

Muhammad, Ayman, 273 n.18

Muhammad, Elijah, 6, 137; and ancient astronaut hypothesis, 163; as biblical interpreter, 167–71, 275 n.34; on black history, 138–45; conversion of, 133–34, 150–51, 271 n.8; on cosmogony and history, 275 nn. 31, 32; death of, 178; early life of, 130–32; on eschatology, 171–77; as Ezekiel, 151, 199, 208; and Fard Muhammad, 279 n.73; and FBI, 173, 282 n.35; imprisonment of, 134–35, 274 n.19; on Israel, 277 n.53; on Jesus Christ, 172–76; and Jewish community, 183, 286 n.32; and Louis Farrakhan, 181; and *ma'aseh merkabah,* 164; and Malcolm X, 133; as Messenger, 171, 176; on Mother Plane, 158–64, 199, 213, 282 n.35; naming of, 134; as prophet, 136, 151; and "Radio in the Head," 174; and science, 281 n.8; and sense of humor, 224 n.26; as spiritual father, 209; theology of, 159–60; on tricknology, 143, 160; on vision of Ezekiel, 149–52, 158–71; on wheel, 282 n.23. Works: *Birth of a Saviour,* 282 n.32; *Fall of America,* 149, 156, 161–64, 168; *Flag of Islam,* 275 n.32; *How to Eat to Live,* 286 n.32; *Message to the Blackman in America,* 144, 149–50, 156–61, 222; *Our Saviour Has Arrived,* 286 n.32; *Theology of Time, The,* 149, 165, 171–76; *True History of Jesus, The,* 173–75, 282 n.37

Muhammad, Emmanual Abdullah, 180

Muhammad, Fard, 130–34, 150–52, 161–62, 176, 199, 226, 271 n.5, 272 n.9, 297 n.49

Muhammad, Jabril, 224–26

Muhammad, John, 180

Muhammad, Khallid Abdul, 288 n.41, 289 n.44, 292 n.73

Muhammad, Munir, 153

Muhammad, Sidney, 293 n.3

Muhammad, Silis, 180, 285 n.18

Muhammad, Tynnetta, 202, 210, 224–29, 278 n.71, 297 n.41

Muhammadia, Bi'sana Ta'laha El'Shabazziz Sula, 280 n.2

Muhammadia, Tu'biz Jihadia, 280 n.2

Muller, Herbert J., 3, 236 n.4, 246 n.45

Murder in the Air, 114

Murrin, Michael, 243 n.15

Myers, Lew, 202

Mysterium, 236 n.10

Mysticism, 16, 68–73

NAACP, 190, 290 n.57

Nagasaki, 84

NASA, 56, 59, 251 nn. 41, 47

Nation of Islam, 6–7; in Chicago, 134; and Christianity, 152, 277 n.54; cosmogony of, 138–39, 160, 275 nn. 31, 32; dissent within, 180–81; and end time, 153, 281 n.21; establishment of, 132–33, 151; and Freemasonry, 282 n.29, 289 n.66; growth of, 135–36; and Islamic world, 179; and Jewish community, 185–86, 195–96, 215–18, 286 n.32, 288 n.41, 289 n.44; and Libya, 295 n.27; newspapers of, 284 n.11; and numerology, 226–28, 297 nn. 49, 54; oral/visual culture of, 279 n.72; reform of, 178–81; and UFOs, 222–23, 225; and vision of Ezekiel, 225–29; and Yakub figure, 140–45. Works: *Secret Relationship between Blacks and Jews, The,* 185–86, 217. *See also* Farrakhan, Louis; Muhammad, Elijah; Muhammad, Fard

"Nations Shall Know That I Am Jehovah"—How?, 82–84

Neuromancer, 65–66

New Age, 4, 17, 69–73, 95, 223, 256 n.7

"New anti-Semitism," 186. *See also* Jews: attitudes toward

New Millennium, The, 106

Nietzsche, Friedrich, 235 n.1

1980's: Countdown to Armageddon, The, 97, 263 n.83

Nixon, Richard M., 102, 266 n.6

None Dare Call It Conspiracy, 296 n.36

Nuclear weapons, 5, 84–99, 115–17

Numerology, 191–92, 226–28, 273 n.12, 297 n.49

Odium Dei, 26–28, 33, 152, 156–57, 232, 242 n.11

Oldfield, Barney, 110

'Ophannim, 230

Original Man, 139–40

Otis, George, 108

Page, Clarence, 285 n.20

Palmer, Colin A., 291 n.62

Pantex, 93–94, 262 n.75

Paradise Lost (Milton), 21–22, 24–30, 37–38, 68, 72, 124, 155–56, 240 n.2, 268 n.31, 280 n.1, 294 n.14

Parhelion, 64–65

Parks, Rosa, 190

Passport to Magonia, 44

Patmos, 145

Penton, M. James, 256 n.10, 258 n.37

Persian Gulf War, 101

Philo, 67

Phrenology, 76

Planet Earth—2000 A.D., 97

Poetic Authority, 240–41 n.4

Poiēsis, 14, 17, 64, 30–31

Powell, Colin L., 122, 215

Premillennialism, 98, 99, 265 n.96

Pritchard, David E., 50

Prophecy belief, 5, 74, 75, 84, 85, 91, 93–97, 100, 101, 102, 104, 108, 110, 112, 113–14, 236 n.5, 264 n.90

Psyche, 15

Quetzalcoatl, 205–6

Qur'an, 133, 157, 189, 192, 212, 219, 222, 279 n.73, 281 n.16, 284 n.11

Raboteau, Albert, 148

Race: and the Bible, 5–6; and end time, 96–97, 105–6; and the enemy, 96–97, 264 nn. 88, 90; as ethnocentrism, 143–44; scientific discourse on, 139

Railroad. *See* Locomotive

Rainbow Coalition, 183–84

Rapture, 98–99, 104, 108, 109, 265 nn. 96, 97. *See also* Tribulation

Rassoull, Abass, 278 n.71

Reagan, Ronald, 6, 108–17, 121–25, 214–15; as actor, 114; and Armageddon, 108–9, 111, 113–15, 268 n.33; as avatar, 110, 112; as born-again Christian, 109; and domestic and foreign policies, 113–17, 121–22; and end time, 108–18; and evangelicalism, 108–9; and Ezekiel, 110, 111–12; and Gog, 112–15; as governor,

108–11; and Israel, 112; and Libya, 110; as president, 112–25; and SDI, 115–17, 269 n.39

Reformation, 22–30, 180, 241 n.9

Reiss, Edward, 121

Reynolds, Edward, 242 n.11

Riders in the chariot, 16–17, 34, 49, 68, 69, 77, 98, 146, 152, 207, 212, 232, 239 n.33

Riecken, Henry W., 43–44

Riesebrodt, Martin, 256 n.9

Ring, Kenneth, 248 n.20

Robertson, Pat, 5, 104–8, 266 n.12

Roman Catholic Church, 81

Rōs, 88–90

Ro'sh, 88–90, 261 n.60

Russell, Charles Taze, 77–80

Russia, 88–90, 92–95, 96, 97, 98, 99, 103, 104, 105, 109, 112, 115, 121, 261 nn. 61, 62, 262 nn. 66, 68

Rutherford, Joseph Franklin, 79–82, 257 nn. 21, 28

Sagan, Carl, 249 n.27

Saphon, 86–88, 260 nn. 52, 54

Sartor Resartus, 245 n.33

Satan, 24–26, 68

Saviours' Day, 144, 164, 181, 184, 187, 189, 272 n.9, 273 n.16, 277 n.3, 278 n.71, 281 n.18, 289 n.48, 293 n.76

Schachter, Stanley, 43–44

Schreber, Daniel Paul, 15, 41, 151, 238–39 n.31

Schultz, George, 214

Scofield, Cyrus, 88–90, 262 n.68

Scofield Reference Bible, 88–90, 261–62 n.66

SDI (Strategic Defense Initiative), 6, 115–17, 120–25, 223–24, 269 n.47

Second World War, 114

Secret Relationship between Blacks and Jews, The, 185–86, 217

Secret Ritual of the Nation of Islam, The, 272 n.12

Selassie, Haile, 111, 268 n.29

Sense of an Ending, The, 74

Serling, Robert, 124

Seventh-Day Adventists, 76

Shabbaz, Betty, 285 n.22

Shabbaz, Quibilah, 285 n.22

Shabazz (tribe), 132, 139, 272 n.11
Shabbetaianism, 247 n.6
Sharpton, Al, 201
Sharrief, Raymond, 137
Shawcross, John T., 240 n.2
Shiva, 106, 266 n.14
Shklovsii, I. S., 249 n.27
Skywalker, Luke, 117–20
Slave narratives, 142–43
Slavery, 291 n.62, 296 n.31
Smith, Chuck, 91
Smith, Joseph, 247 n.10
Smith, Theophus H., 143
Smith, Vern E., 285 n.20
Southern Christian Leadership Conference, 190, 203
Southside Mosque (Chicago), 134
Soviet Union, 112, 121
Spaceships of Ezekiel, The, 54
Spahr, Juliana, 249 n.25
Spielberg, Steven, 45–48
Spirituals, 146–47, 278 n.63
Stanton, Gerald, 91
"Star Wars." *See* SDI (Strategic Defense Initiative)
Star Wars (film trilogy), 117–21, 233
Stein, Arnold, 28
Steinhauser, Gerhard R., 53
Stillings, Dennis, 249 n.24
Story, Ronald D., 249 n.27
Strangely, Mona, 45, 51, 52, 252
Stroup, Herbert Hewitt, 81, 258 n.35
Strozier, Charles B., 75–76, 256 n.7
Studies in Scriptures, 77, 79
Sturgeon, Menta, 78–79
"Swing Low, Sweet Chariot," 147, 219

Tal, Israel, 39–40
Tanks, 39–41, 246 n.39
Teaching for the Lost-Found Nation of Islam, 272–73 n.12
Technē, 13–14, 17, 55, 74, 77, 82, 92–93, 96, 101, 112, 125, 147, 232; and cyberspace, 70, 73; and gnosis, 96; as invention, 31; and the Jehovah's Witnesses, 77, 82; and the locomotive, 37–38; and *merkabah* tanks, 40–41; and

nuclearization of Bible, 92–93; as psyche, 15; and Ronald Reagan, 112, 125. *See also* Avatars of *technē*
Technological Society, The, 238 n.28
Technology, 13–14, 77–78, 91, 99, 238 nn. 27, 28
Technopoetics, 239–40 n.38
Telepathy, 176, 283 n.38
Teller, Edward, 116–17, 120
Temple of Islam, 132–33
Tepoztlan (Mexico), 205–6
Terauchi, Kenju, 223
Theological Dictionary of the Old Testament, 259–60 n.51, 260 n.58
Thompson, Keith, 246 n.1, 247 n.10
Tomas, David, 66–67
Tribulation, 98, 104, 109, 265 n.96 (ch. 3), 265 n.1 (ch. 4). *See also* Rapture
Tricknology, 143, 160, 187–89
Trickster, 285 n.20

UFOS (unidentified flying objects), 42–43, 122–23, 246 n.1; and alien abductions, 48–51; and ancient astronaut hypothesis, 52, 53; as archetype, 249 n.25; and cybernetics, 67; and cyberspace, 65–67, 70; and cyborgs, 68; and demonization of the "other," 97; and Drunvalo Melchizedek, 70–73; and elite culture, 48; and end time, 97–98, 113, 123–24; and Ezekiel, 9, 42, 46, 52, 53, 247 n.10; and the golem, 68; and the *hekhalot*, 68; history of, 44, 247–48 n.10; and the ineffable, 9; kinds of encounters with, 48; and the *merkabah*, 68; and the Mother Plane, 162–63, 208–9, 221, 280 n.79; and mountebanks, 42–43; and the Nation of Islam, 153, 208–9, 222–26; and the New Age, 69–73; as parhelion, 64–65; and popular culture, 44–45; psychology of, 48–50; and the SDI, 124; technology of, 50, 54–65, 217 n.43; as "tola," 43
Unger, Merrill F., 133
United World Federalists, 121–22
University of Islam, 132, 178

Vader, Darth, 117–21
Vallée, Jacques, 44, 46, 50, 247 n.7
Valvasone, Erasmo di, 242 n.13

Van Impe, Jack, 263 n.82
Vindication, 80
Virtual reality, 253 n.73, 255 n.89
Visionary Mode, The, 236 n.10
Visions of Glory, 76, 256 n.12
Von Däniken, Erich, 52–53, 249 n.27

Walker, Joseph, 287 n.33
Wallace, Mike, 285 n.20
Walmesley, Charles, 245 n.29
Washington, George, 192
Washington Monument, 192–93
Watchtower, The, 77
Watergate, 102, 266 n.6
Watt, James, 101
Weber, Timothy P., 75
Welles, Orson, 95
Wells, H. G., 270 n.58, 280 n.79
West, Cornel, 190

Wheels, 80, 146–47
When Prophecy Fails, 43–44
"*When Time Shall Be No More,*" 75, 255 n.3
Whisenant, Edgar, 91
White, Lynn Townsend, Jr., 239–40 n.38
Wiener, Norbert, 67–68
Will to power, 3, 235 n.1
Wilson, Willie F., 203
Wood, Arthur Skevington, 265 n.97
Woodworth, Clayton, 79

Yakub, 140–45, 160, 175, 187–89, 221, 276 nn. 37, 38, 289 n.52, 298 n.55
Yom Kippur, 195–96, 220

Zeal, 28–31, 81, 243 n.17
Zimmerli, Walter, 86–87
Zimmerman, Michael E., 13–14, 238 nn. 25, 27, 248 n.17

Michael Lieb is Research Professor of Humanities and
Professor of English at the University of Illinois at Chicago.
His previous books include *Milton and the Culture of
Violence* and *The Visionary Mode: Biblical Prophecy,
Hermeneutics, and Cultural Change.*

Library of Congress Cataloging-in-Publication Data
Lieb, Michael.
Children of Ezekiel : aliens, UFOs, the crisis of race, and the
advent of end time / by Michael Lieb.
Includes index.
ISBN 0-8223-2137-8 (hardcover : alk. paper).
ISBN 0-8223-2268-4 (pbk. : alk. paper)
1. Bible. O.T. Ezekiel I, 4–28—Influence—History.
2. Technology—Religious aspects—History of doctrines.
3. Nation of Islam (Chicago, Ill.)—History.   4. End of the
world—History of doctrines.   I. Title.
BS1545.2.L54   1999   001.9—dc21   98-27097 CIP